Parametric Optimization Techniques of Integrated Circuits

Vazgen Melikyan • Kang Li

Parametric Optimization Techniques of Integrated Circuits

Springer

Vazgen Melikyan
Synopsys Armenia CJSC
Yerevan, Armenia

Kang Li
Xidian University
Xi'an, Shaanxi, China

ISBN 978-3-032-20399-1 ISBN 978-3-032-20400-4 (eBook)
https://doi.org/10.1007/978-3-032-20400-4

This Springer imprint is published by the registered company Springer Nature Switzerland AG
The registered company address is: Gewerbestrasse 11, 6330 Cham, Switzerland

Preface

The book systematically expounds the main results obtained by the authors in the field of parametric optimization techniques of integrated circuits (ICs). The proposed methods are based on machine learning (ML) and by applying new search approaches and performing a small number of simulations, discover an IC candidate satisfying the proposed behavioral constraints.

The main goals of most proposed methods and solutions of parametric optimization of ICs are to improve important parameters and characteristics (performance, power consumption, occupied area on the die) of circuits by transforming the given circuit and constraints into a single-objective or multi-objective optimization problem, apply swarm intelligence optimization algorithms and ML strategies with the goal of finding the global minimum with the least number of simulations.

The book is anticipated for scientists and engineers specializing in the field of IC design and optimization as well as for students and postgraduate students studying disciplines related to IC design.

Yerevan, Armenia — Vazgen Melikyan
Xi'an, Shaanxi, China — Kang Li

Competing Interests The authors have no competing interests to declare that are relevant to the content of this manuscript.

Introduction

Integrated circuits (IC) are widely used in several areas crucial to human life. Modern ICs are an integral part of such fields as artificial intelligence-controlled autonomous vehicles to space satellites and medical devices implanted in the human body. To ensure the high performance and functional diversity of these systems, the number of electronic components in ICs continues to increase. Parallel to the increase in the number of components, their physical dimensions and supply voltages are being scaled to reduce die area and power consumption. To meet the demands of high speed, small area and low power consumption, semiconductor technologies have been continuously scaled, resulting in decrease of transistor sizes down to 3 nm and supply voltages down to 600 mV. The scaling of supply voltages however is much slower compared to channel length which results in increasing electrical fields between the device terminals in advanced node technologies. As a result of this scaling, new phenomena emerge whose effects on circuits are reflected in significant parametric changes, while the effects of secondary phenomena previously neglected in larger process nodes become more pronounced, leading to deterioration of circuit reliability. As an example, one of the most important factors affecting the reliability of ICs is the phenomenon of aging, in the result of which the parameters of the circuits worsen over time and may even lead to the failure of the IC. There are also other examples which have decisive impact on ICs functionality and reliability.

Due to the mentioned reasons, currently one of the most important and time-consuming phases of IC design is parametric optimization. It boils down to a selection of physical dimensions of individual components, so that the necessary measurements of output signals of the circuit satisfy the imposed constraints. The complexity of the process is conditioned by the increase of the influence of secondary factors on the normal operation of circuit components, among which are parasitic elements, leakage currents, etc. That circumstance, coupled with the need for similar ICs to be embedded into different environments and to satisfy behavioral constraints, creates situation where similar circuits need to be optimized for different processes and operating conditions. Therefore, the development of means of automating the process of IC parameter selection becomes a topical issue.

Existing precise methods of IC parametric optimization are based on transformation of the given design constraints into an optimization problem and numerous IC simulations. These kinds of optimization problems, when considered in the context of IC optimization, give rise to a number of challenges:

- Fluctuations in the values of behavioral parameters happening with the change of the technological process, supply voltage, and temperature.
- In some cases, conflicting constraints given in the technical specifications.
- The time-consuming nature of IC simulations, which greatly complicates the ability to evaluate the effectiveness of possible solutions, limiting the number of calls to the modeling tool.
- The nonlinear nature of the dependence of input/output parameters due to the presence of secondary factors, which cannot be expressed using polynomials and can change in different parts of the parameter search domain.

Given these constraints, finding the optimum begins to depend on several factors, including the use of effective search strategies and the proper selection of the values of the initial solution options that serve as the starting point of the search process. However, the existing approaches still are not sufficient from the standpoint of practical applications. A need arises to develop new methods for automated parametric optimization of ICs, which, by applying new search approaches and performing a small number of simulations, will discover an IC candidate satisfying the proposed behavioral constraints.

In the book approaches, methods, models, solutions, and software tools of automated parametric optimization of ICs have been proposed, which transform the given circuit and constraints into a single-objective or multi-objective optimization problem, apply swarm intelligence (SI) optimization algorithms (OA) and ML strategies with the goal of finding the global minimum with the least number of simulations. They meet modern requirements and significantly increase the convergence rate to the optimum by introducing additional, computationally non-time-consuming components into the system, while making fewer calls to the simulation tool. A hybrid algorithm based on SI approaches has been proposed, which, in order to ensure the diversity of individuals in the population, applies genetic algorithm strategies in the global search phase, in particular, the combination of wheel selection, uniform crossover, and multipoint mutation approaches, and uses the four techniques of "Harris hawks" OA during the local search, the choice among which is made through internal parameters. As a result, the proposed approach significantly outperformed the other existing algorithms. A primary population initialization algorithm, contributing to the increase of convergence rate of OAs was proposed, which, by carrying out proportional sampling along the existing search domain and not depending on random factors and additional calculation processes, ensuring reproducibility of the results, in the problems of minimization of the observed benchmark functions and optimization of many ICs, compared to other methods provided better results. A system for replacing the IC simulation software tool with an artificial neural network during optimization has been proposed, in which a part of the training data is generated using a calculation process based on the generative-adversarial network and Spearman's correlation coefficients, which allows to reduce the number of necessary simulations even further without the loss of performance.

Approaches to reduce the effects of aging phenomena in ICs have been proposed, which provide the necessary operating conditions with an increase in power consumption, occupying area and duration of simulations within the permissible limits. A method of compensating the effects of aging phenomena in voltage-controlled generators was proposed, because of which, due to the circuit controlled by a digital unitary code, it was possible to significantly increase the frequency of the output signal. A method for monitoring and detecting aging phenomena in ICs has been created, which makes it possible to detect aging phenomena by obtaining a digital code at the output of the circuit and use it in compensating circuits. The method was implemented only with metal-oxide-semiconductor transistors, and by not requiring additional input/output pins and testing on the chip, it leads to a reduction in area, testing time, cost of IC, and increased reliability. A method to reduce the effect of aging phenomena in three types of operational amplifiers: two-stage, folded-cascode, and rail-to-rail, was proposed, using only thin gate oxide transistors. In this case, by adding cascode transistors and digital switches, as well as by changing the amplitude of the power down signal, protecting the transistors from stress, it was possible to neutralize amplification factors decrease. A method of increasing design accuracy of the aging-aware standard cell libraries was proposed, in which case, by considering the possible phase shifts between the input signals of the cells, it was possible to obtain an increase in design accuracy at the expense of increasing the simulation time.

Methods, models, solutions, and software tool have been proposed to improve the accuracy of power supply and management systems embedded in integrated memory circuits and reduce the effects of aging phenomena, which ensure the necessary operating conditions at the expense of a permissible increase in power consumption, occupied die area, and simulation duration. Four-phase and six-phase clock signal-controlled charge pump circuits consisting only of pMOS transistors have been proposed, in which the effect of aging phenomena has been reduced at the expense of die area increase. The proposed circuits provide an aging protected charge pump solution that is also more efficient due to a more precise clocking scheme which allows for a better control over the charging and boosting cycles. A method of dynamic current limiting in voltage regulators has been proposed, which reduces the effects accelerating the aging of the output stage pMOS transistor, ensuring less output power deviation in high temperature conditions at the expense of die area increase. The proposed system builds upon existing foldback current limiting techniques adding enhanced protection against hot carrier injection and negative bias temperature instability effects. A supply voltage detection circuit has been proposed to control voltage level converters in control systems embedded in integrated memory devices, which eliminates their unpredictable failure state under low voltage conditions. The proposed circuit matches the behavior of the level shifters that are used in the control systems to ensure minimally necessary margin loss of the operational voltages.

Methods, models, and software tools have been proposed for reducing aging effects in IC receiver circuits while improving operational conditions at the cost of increased area and simulation time. A method was proposed to reduce aging effects in comparators based on cascode operational amplifiers. By incorporating voltage-regulation transistors, the proposed design improved gain compared to conventional and baseline circuits. A technique to counteract aging in charge pump circuits was introduced.

Methods, models, solutions, and software tools have been proposed to improve the efficiency of information transmission systems between ICs, which meet modern requirements and provide the necessary operating conditions by increasing the occupied area and simulation time. A multi-output level shifter circuit has been developed, in which efficiency is enhanced by increasing the circuit's functionality and reducing the impact of aging. This has made it possible to decrease the degradation of power consumption and delay, with an increase in the physical design area. A bandgap reference circuit based on transistors with a thin gate oxide layer has been proposed, in which parametric degradation caused by aging effects is reduced. An embedded approach to low-power mode management has been created, featuring continuous monitoring of differential data.

Methods, models, and software tools have been proposed for detecting and mitigating interactions between IC components have been proposed, which ensure time savings during detection and mitigate the consequences of interactions and overall timing parameters with minimal deterioration of their characteristic parameters, and with minimal increase in die area and power consumption. A method based on ML algorithms has been developed for predicting interactions between elements in ICs. This method ensures time savings on signal integrity analysis with minimal data collection and circuit analysis. A method for modifying the corresponding wires, considering the architecture of the clock tree and the design features of the circuit, has been proposed. A method for evaluating and modifying data-carrying wires has been created, considering the peculiarities of circuit design, which, through optimized combinations of shielding and repeaters, ensures a reduction in the impact of interactions, a reduction in the total setup and hold times at the cost of more power consumption and die area.

The authors express deep gratitude to PhD students, Arman Vardumyan, Sevak Ghukasyan, Hrayr Sahakyan, Sergo Harutyunyan, Garnik Voskanyan, and Gor Abgaryan, who have participated in the development of described means.

The authors would also like to thank our colleague Ruzanna Goroyan who assisted in the preparation of this monograph.

Contents

Abbreviations

AdaGrad	Adaptive gradient descent
AI	Artificial intelligence
ANN	Artificial neural network
B	Bulk
BANN	Branched Artificial Neural Network
BI	Bus inversion
BS	Bus shift
BSINV	Bus shift-inverse
BTI	Bias temperature instability
CDR loop	Clock and data recovery loop
CHE	Channel hot electron
CI	Chaotic initialization
clk	Clock
CMOS	Complementary metal-oxide-semiconductor
CP	Charge pump
CSINV	Circular shift-inverse
CT	Clock tree
D	Drain
DAHC	Drain avalanche hot carrier
DCC loop	Duty cycle correction loop
DRG-GA	Design rule-guided genetic algorithm
DS	Drain-source
E	Equalizer
EC	Equalizer circuit
ED	Eye diagram
EEC	Element encoding compliance
EM	Electromigration
FEXT	Far end crosstalk
G	Gate
GA	Genetic algorithm
GAN	Generative adversarial network

GI	Group intelligence
HCI	Hot carrier injection
HD	Hamming distance
HHOA	"Harris Hawks" optimization algorithm
HRP	High resistance point
HSD	High-speed decoder
HT	Hold time
IBE	Interactions between elements
IMD	Integrated memory device
in/out	Input-output
LHS	Latin hypercube sampling
LS	Level shifter
MAD	Maximum amplitude deviation
ML	Machine learning
MOM	Metal-oxide-metal
M_{PD}	Power-down switch
MSE	Mean squared error
MTTF	Mean time to failure
NBTI	Negative-bias temperature instability
Neg-C	Negative capacitance
NEXT	Near end crosstalk
NV	Non-volatile
NVM	Non-volatile memory
OA	Optimization algorithm
OpAmp	Operational amplifier
ox	Oxide
par	Parasitic
PBTI	Positive bias temperature instability
PLL	Phase-locked loop
PM	Phase margin
PSA	Partial search algorithms
PVT	Process-voltage-temperature
RBER	Raw bit error rate
RCS	Resistance coordination system
RMN	Resistor matching node
RMS	Root mean square
RO	Ring oscillator
RVS	Reference voltage source
SC	Standard cell
SGD	Stochastic gradient descent
SGHE	Secondary generated hot electron
SHE	Substrate hot electron
SI	Signal integrity
SIA	Signal integrity analysis
SL	Serialization loop

ST	Setup time
sup	Supply voltage
sup.nom	Nominal supply
SW	Signal wire
$T_{clk\text{-}q}$	Clock to Q delay
TDDB	Time-dependent dielectric breakdown
TH	Threshold voltage
th.be	Threshold voltage due to substrate effect
th.nom	Nominal threshold voltage
UGB	Unit gain bandwidth
UGF	Unity gain frequency
USB	Universal serial bus
VCO	Voltage-controlled oscillator
VG	Voltage generator
VR	Voltage regulator
WOA	Whale OA

Chapter 1
Development of Design Means for Automated Parametric Optimization of Digital Integrated Circuits

1.1 General Issues of Developing Design Means for Automated Parametric Optimization of Digital Integrated Circuits

1.1.1 Necessity to Develop Design Means for Automated Parametric Optimization of Digital Integrated Circuits

It is known that in recent decades, the development of semiconductor technologies has proceeded in accordance with Moore's Law [1], due to which there are currently ICs consisting of metal-oxide-semiconductor (MOS) transistors with a channel length of up to 3 nm on the market [2]. Along with the decrease in size, the productivity of ICs increases due to the possibility of placing a larger number of functional blocks per unit area [3]. At the same time, the power consumption of ICs increases due to thinning of the oxide layer of transistor gates, as well as the loss of its insulator properties. Due to the mentioned shortcomings, leakage currents increase [4, 5]. Such a rapid development of the IC industry, in turn, has created conditions for the transformation of a number of other technological areas, including AI [6], virtual intelligence devices [7], the Internet of Things [8], self-driving cars [9], etc. The introduction of ICs in crucial areas of human life has led to a sharp increase in the demand for the design of nodes that meet technical specifications of various degrees of complexity.

One of the most important and time-consuming stages of IC design is parametric optimization. The latter comes down to choosing the physical dimensions of individual components (transistors, capacitances, resistors, etc.) in such a way that the required measurements of the circuit output signals will satisfy the specified one-sided or two-sided constraints. It is mainly performed manually by designers, based on previous experience and information about the operation of the circuit. The complexity of the process is due to the increased influence of secondary factors on the

V. Melikyan, K. Li, *Parametric Optimization Techniques of Integrated Circuits*,
https://doi.org/10.1007/978-3-032-20400-4_1

normal operation of transistors, including parasitic elements [10, 11] and leakage currents [12, 13]. They can lead to significant distortions in the output signal (Fig. 1.1) [14–16].

It is known that each foundry provides separate models for components used in the design of ICs with the technological nodes belonging to it. This circumstance, combined with the need to embed similar ICs in different environments and satisfy different behavioral constraints, creates a situation in which similar circuits need to be optimized repeatedly for different processes and operating conditions. Considering the large number of ICs being designed simultaneously, there is a need to develop means to automate the process of optimizing them.

The problem of selecting the IC parameters can be formulated mathematically as a single-objective or multi-objective optimization problem. In this case, having the input and corresponding output values, it will be necessary to find those combinations of data entries whose output will coincide with the optimum [17, 18]. The first group of input data is the constraints imposed on the design process of the

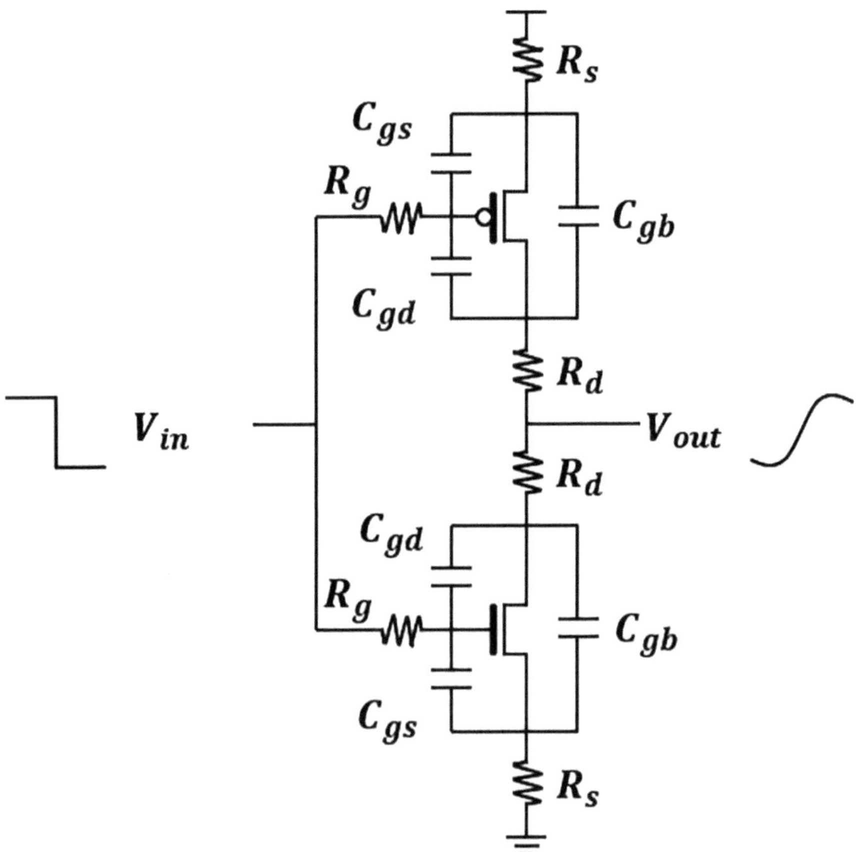

Fig. 1.1 Simplified view of parasitic cells in the inverter

considered ICs, in the form of technical specifications, which represent the limits of the permissible values of the output parameters characterizing the circuit's performance. It is represented as m-dimensional vector $=\{z_1, z_2, \ldots, z_m\} \in Z$, $Z \subset (Z_l, Z_u)$, where z_i $(i = 1, 2, \ldots, m)$ is the value of the i-th output parameter, Z is the set of permissible values, with upper $X_u$and lower X_l limits. The next set of input data is the set of parameters characterizing the dimensions of the components included in the circuit (length, width, number of transistors). It can be represented as n-dimensional vector $x = \{x_1, x_2, \ldots, x_n\} \in X$, $X \subset (X_l, X_u)$, where x_i $(i = 1, 2, \ldots, n)$ is the value of the dimensions of the i-th component, X is the complete set of values, with the upper X_u and lower X_l limits of the possible permissible values of the parameters. And the output values are such values of the dimensions of the components, by setting which the constraints on the values of the output parameters presented in the technical specifications, set for the circuit will be satisfied: $y = \{y_1, y_2, \ldots, y_n\}$. The process of selecting values is carried out on the basis of simulation performed using any modeling software tool and processing the set of values of measurements performed for the necessary output parameters. If the optimization of each measured value is considered as a separate problem, then the multi-objective optimization scenario is applied and the goal is to find the best Pareto-optimal solution [19–21]. Otherwise, if the single-objective problem of optimizing a single value obtained as a result of combining the measured values using some function is considered, then it is necessary to find the global minimum point on the available plane, avoiding local minima (Fig. 1.2) [22, 23].

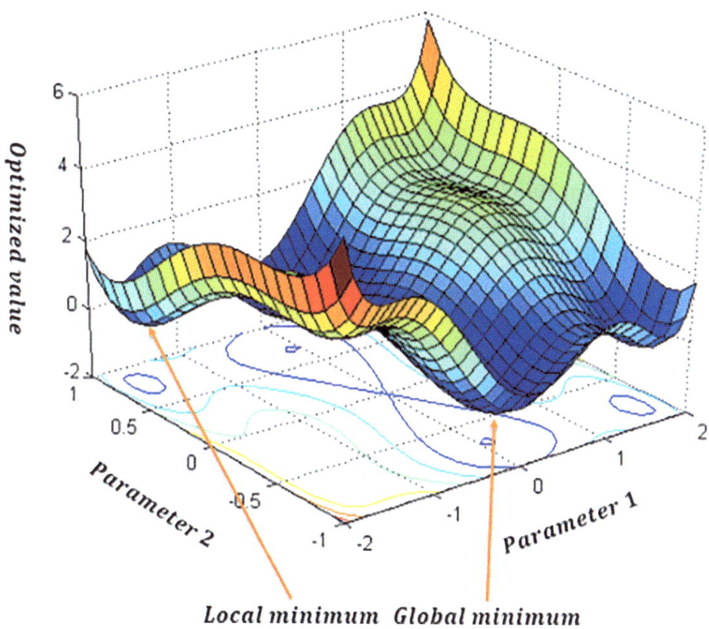

Fig. 1.2 View of plane minima of the single-objective optimization problem

Therefore, in the generalized case, given the technical specifications, the IC parametric optimization problem can be formulated as:

$$\min.\text{Cost}\left(f\left(x\right)\right) = \min.\text{Cost}\left[f_1\left(x\right), f_2\left(x\right), \ldots, f_m\left(x\right)\right]$$

$$\text{satisfying } g_i\left(x\right) \le 0, i = 1, 2, \ldots, p \text{ constraints,} \tag{1.1}$$

where $f_i(x)$ $i = (1,2,\ldots,m)$ is a nonlinear function that establishes relationship between x and y vectors to perform the i-th measurement, Cost($f(x)$) is the weight value obtained as a result of combining the measurements taken, and $g_i(x)$ $i = (1,2,\ldots,p)$ is the set of constraints imposed on the optimization problem. The measurement combination function can be linear, exponential, Gaussian, etc. [24].

Similar optimization problems, when considered in the context of IC optimization, give rise to a number of challenges [25–29].

- Fluctuations in the values of behavioral parameters happening with the change of the technological process, voltage, and temperature.
- In some cases, conflicting constraints given in the technical specifications,
- The time-consuming nature of IC simulations, which greatly complicates the ability to evaluate the effectiveness of possible solutions, limits the number of calls to the modeling tool.
- The nonlinear nature of the dependence of input/output parameters due to the presence of secondary factors, which cannot be expressed using polynomials and can change in different parts of the parameter search domain.

Given these constraints, finding the optimum begins to depend on a number of factors, including the use of effective search strategies and the proper selection of the values of the initial solution options that serve as the starting point of the search process (Fig. 1.3) [30, 31].

From the above, it follows that the development of the semiconductor industry and the increase in the influence of secondary factors have created a number of complications in the process of effective selection of the dimensions of components

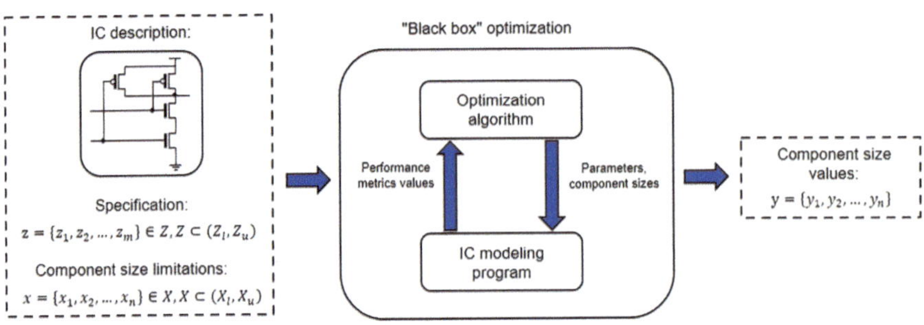

Fig. 1.3 View of the IC optimization process

included in ICs. Taking into account the irreplaceable role of ICs in various spheres of human life and the number of circuits corresponding to various technological processes being designed at a time, the development of means for automating the process of selecting IC parameters becomes a topical issue. There is a need to develop new methods for automated parametric optimization of ICs, which, using new approaches to initializing the primary search points and looking up around them, will more effectively carry out the search for an IC candidate that meets the proposed behavioral constraints. At the same time, they will save operating time and computational resources by making fewer calls to the modeling tool.

1.1.2 Existing Design Means for Automated Parametric Optimization of Digital Integrated Circuits

In order to solve the problems mentioned above, a number of new strategies have been developed in recent years to find effective solutions to optimization problems. Currently, methods based on the approaches of the Group Intelligence (GI) and partial search algorithms (PSA) have become widespread [32, 33]. They consist of mathematical models built as a result of behavioral studies of phenomena occurring in nature [34]. These methods are able to quickly find an approximate solution to the optimization problem in situations where the available initial data or available computational resources are severely limited. The main advantage of such algorithms, due to their versatile nature, is the possibility of application in any field [35–37]. Therefore, since the process of IC optimization can be represented as an optimization problem, the use of these approaches can contribute to the discovery of the best option for the ICs that are considered. Given the high sensitivity of the convergence rates of algorithms belonging to the GI family to the selected initial population, a number of new statistical methods for initial evaluation have also been developed to stimulate the search process [38]. Taking into account the characteristics and dimensionality of the search domain, they strive to distribute the number of available points as evenly throughout the available domain as possible. The goal of such algorithms is to exclude the omission of zones of the search domain containing effective solution options. In addition to the above, the optimization of ICs using GI involves multiple time-consuming calls to the simulation software, which is often not possible under timing constraints. For this reason, the introduction of artificial intelligence (AI) components, in particular, artificial neural network (ANN), into the optimization system is occurring [39]. Their goal is to replace the modeling tool for calculating output characteristics and necessary parameters, ensuring a high level of accuracy. The mentioned means are presented in more detail below.

Method for automating the selection of integrated circuit parameters using partial search algorithms based on GI:

In the approaches belonging to the family of PSAs that apply GI principles, the search process is evolutionary in nature and is carried out by simultaneously examining a set of solution candidates [40, 41]. Thus, in each subsequent evolution, the search points observed are obtained by performing certain transformation operations on the points available in the previous cycle. Different GI algorithms differ from each other in the types of transformation operations used, applied strategies, and random factors introduced into calculations [42–44]. It is the combination of such approaches that ensures examination of a wide search area in GIs.

Perhaps the most well-known method of GI optimization is the genetic algorithm (GA) [45], which aims to imitate the process of biological evolution. It consists of three basic operations: selection, crossover, and mutation, which are performed on the individuals in the current generation in each iteration. The goal of selection is to find the best individuals in the population and to pass on their traits, expressed in the form of genes, from generation to generation [46]. The objective function plays a crucial role in the evaluation process, which returns a numerical value describing the fitness of the individual. Crossover, performed with a certain probability, ensures the recombination of the candidate genes selected in the previous step, striving to identify solutions with a higher degree of fitness [47]. Mutation ensures the presence of random deviations in the genes, thereby guaranteeing the diversity of the population [48].

GA has also been adapted to solve the problem of IC optimization, incorporating a new implementation of the mutation operator. This variant of GA, called a design rule-guided genetic algorithm (DRG-GA), receives the data on the physical dimensions of the circuit components and the dependence of the output parameters from the designer as additional data [49]. Thus, after performing simulation, in case of detecting deviations from the proposed constraints, the mutation is performed only on the genes corresponding to those components that are " responsible " for the correction. The block diagram of the proposed approach is presented in Fig. 1.4. To assess the efficiency, the voltage generator (VG) (Fig. 1.5) was optimized, in which the generation frequency is determined by L inductance and C_0, C_{f_1}, C_{f_2} capacitances, and R_1, R_2 resistors are connected to the current mirror for noise reduction. The objective function is calculated by the following formula [49]:

$$\mathrm{Cost}\left(f\left(x\right)\right)=\sum_{i=1}^{2}W_i\cdot\exp\left(\frac{\left|y_i-y_{\mathrm{cons}_i}\right|}{y_{\mathrm{cons}_i}}\right)+\sum_{i=3}^{4}W_i\cdot\exp\left(\frac{y_i-y_{\mathrm{cons}_i}}{y_{\mathrm{cons}_i}}\right)$$

$$\text{satisfying}\, V_{\mathrm{out}}, V_{\mathrm{out}}\ \mathrm{MAD}>0.75V\,@\,30\,\mathrm{ns\,constraints}, \tag{1.2}$$

where $W_i\,(i=1,\ldots,4)$ are the values of the transistor widths $\left(W_{M_1}=W_{M_2}, W_{M_3}=W_{M_4}\right)$, y_i and y_{cons_i} $\left(i=1,\ldots,4\right)$ are correspondingly the measured value, obtained through simulation, and constraints applied to the i-th parameter. The maximum amplitude deviation (MAD) value of the output signals is set to 0.75 V.

The parametric optimization of the observed VG circuit was also performed using conventional GA and the obtained comparative results are summarized in

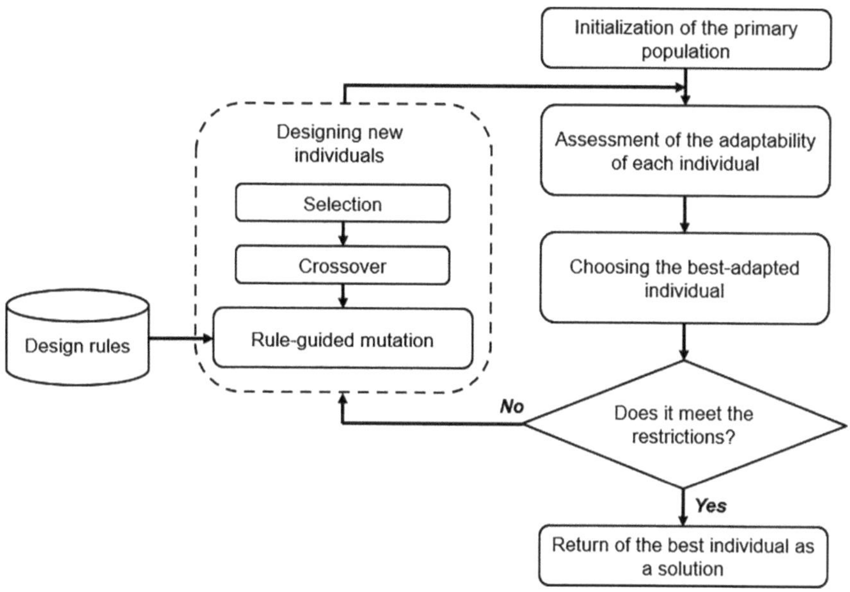

Fig. 1.4 The process of implementing the DRG-GA

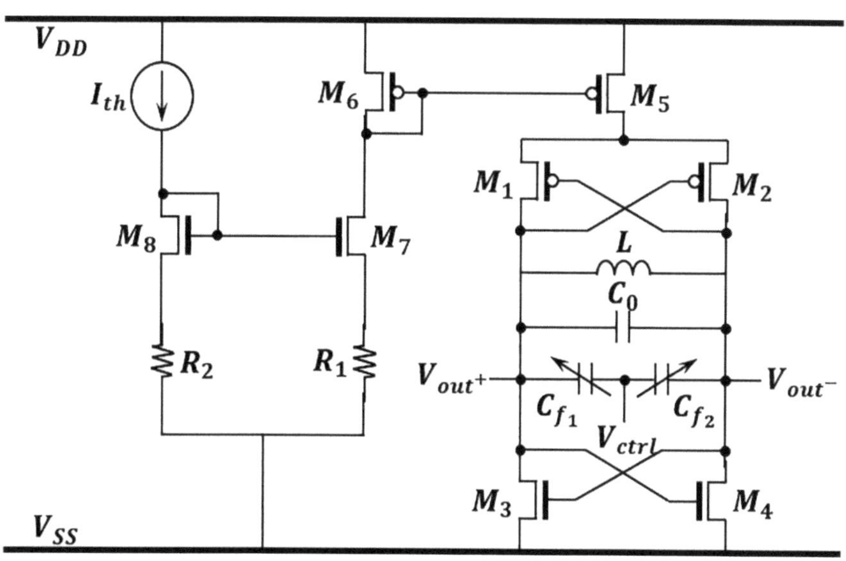

Fig. 1.5 Circuit of a VG optimized using the DRG-GA

Table. 1.1. The main advantage of the proposed new mutation operator lies in the increased probability of finding efficient solutions earlier, in the case of the VG example, about three iterations out of 10 [49].

Table 1.1 Comparative results of optimization of VG using GA and DRG-GA

Parameter name (unit)	Restriction cost	Value of the received GA	Resulting value of DRG-GA
Generation frequency (GHz)	2.5	2.56	2.46
Gain (MHz/V)	85	86.3	84.5
Phase noise (dB/Hz)	≤−115	−120.4	−120.3
Average current (mA)	≤1.6	1.52	1.49

Unlike GA, the basis of the OA called "Particle Swarm" (PSOA) is the principle of collective intelligent behavior of a group of animals [50]. Thus, each member of the group, cooperating with others and based on its own experience, periodically updates the trajectory chosen for the search for "food," changing the speed and direction of movement. The parameters that are pre-defined for the normal operation of the algorithm are:

- t_{max}—the number of iterations of the algorithm execution,
- n—the number of particles in the population,
- c_1, c_2—acceleration coefficients,
- P_{LB}, P_{UB}—minimum and maximum values of the allowable coordinate values of each particle, respectively,
- V_{max}—the maximum allowable speed for any particle,
- w_{min}, w_{max}—the limit values of inertial weight.

The procedure for implementing the algorithm is presented below (Fig. 1.6).

The principles used in the PSOA for an arbitrary particle to find the optimum in prospective search regions are as follows:

- Dynamic regulation of the movement speed, taking into account the best positions found so far for a given particle and the entire group (p_{best}, g_{best}),
- Calculating the position for the next iteration, given the current coordinates and velocity as well as the distance to p_{best} and g_{best}.

The computational part of the algorithm is implemented within the equation in the velocity update step, where the new value of the particle's velocity is estimated. The first of the addends is responsible for diversifying the search trajectory used by the current particle, by moving the particle to unexplored regions. It also seeks to prevent premature convergence to the local optimum, increasing the speed and avoiding "holes" in the plane. In addition, the inertial weight, which is part of the addend, also performs balancing between global (exploration) and so far found best solution-based (exploitation) search phases, depending on the values of t_{max} and the current region. The second addend provides the effect of its own experience, in other words, the cognitive component, which guides the particle to its own best position for exploring the nearby region. The last part of the equation provides the presence of a social component, introducing the effect of the best position found by the whole group into the calculation of the velocity. Due to this, the particle also considers

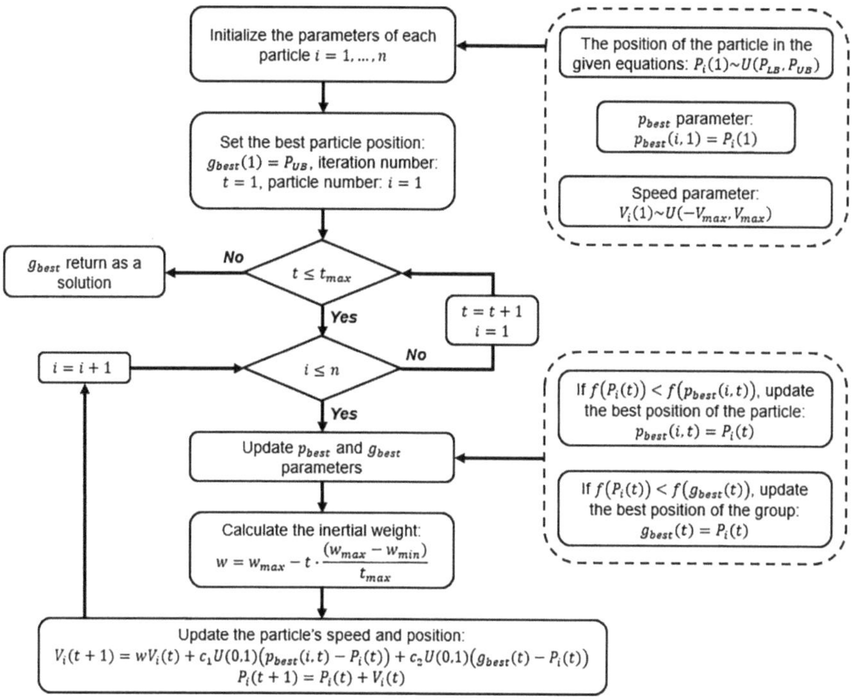

Fig. 1.6 The process of implementing the PSOA

regions partially explored by the group, in order to find solutions with higher fitness value [50].

Although there is no explicit separation of the strategies for exploration and exploitation phases in the PSOA, in terms of its qualitative criteria, the algorithm in some cases surpasses a number of more complex algorithms, including "Simulated Annelaing" optimization algorithm, "Ant Swarm," "Artificial Bee Swarm," "Sine-Cosine," "Differential Evolution" (DEOA) and a number of other methods [51]. The PSOA has also been adapted for parametric optimization of a differential amplifier (DA) with 65 nm technological models, the schematic structure of which is shown in Fig. 1.7 [52]. Here, the input cascade consists of n-MOS transistors to which a p-MOS current mirror is connected as a load. A current mirror made up of M_5, M_6 gates is used for the bias of the amplifier.

To facilitate the optimization process, the following constraints were imposed on the transistor sizes: $W_1 = W_2$, $W_3 = W_4$, $W_5 = W_6$, and the following values were chosen for the algorithm parameters: $c_1 = c_2 = 1.7$, $w_{\min} = 0.4$, $w_{\max} = 0.8$, $t_{\max} = 100$, $n = 20$. The results obtained for V_{in} with common mode range (CMR) of 0.75 V value are summarized in Table 1.2 [52].

In contrast to the generalization-based PSOA, one of the newest members of the family of GI algorithms, the Whale OA (WOA), models the hunting behavior of a specific animal, the humpback whale [53]. It distinguishes three different strategies

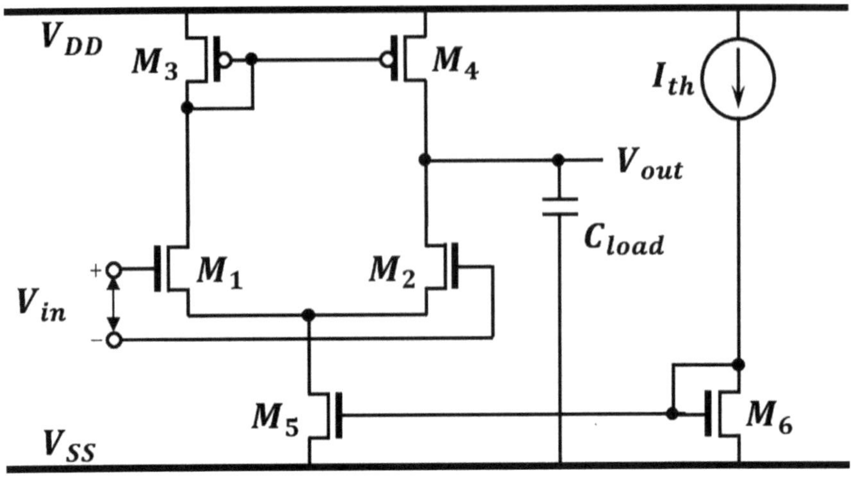

Fig. 1.7 DA circuit, optimized using the PGOA

Table 1.2 Results of DA circuit optimization, using the PSOA

Parameter name (unit)	Cost of the restriction	Value of the received PSOA
Gain (dB)	≥10	12.1
Cutoff frequency (MHz)	≥10	21.6
Frequency bandwidth (MHz)	≥80	84.8
Phase margin (PM) (°)	≥60	104
Power consumption (mkW)	≤100	76.1
Rise time rate (V μs^{-1})	≥100	104

for exploration and exploitation search phases. The principle of their application is presented below (Fig. 1.8):

For the algorithm to work properly, the user only needs to provide the population size (n), maximum number of iterations (t_{max}), and search plane boundaries as parameters (X_{LB}, X_{UB}), and the author suggests setting the degree of the exponential component to 1. The rationale for the strategies used to update the whale's position can be formulated as follows:

- During the step a) operation, the existing whales are grouped around the best solution, and a group search strategy is used to find a more favorable solution (otherwise called a bubble net attack).
- The b) step involves updating the whale's position toward the best solution found so far, with decreasing intensity over time (otherwise called prey encircling),
- The c) operator is responsible for exploring the observation plane within the permissible limits, randomly selecting the hunting position, and directing the current whale to new areas.

During the iterations, the effective balancing of exploration and exploitation search phases is carried out using a weighting factor, which provides a gradually

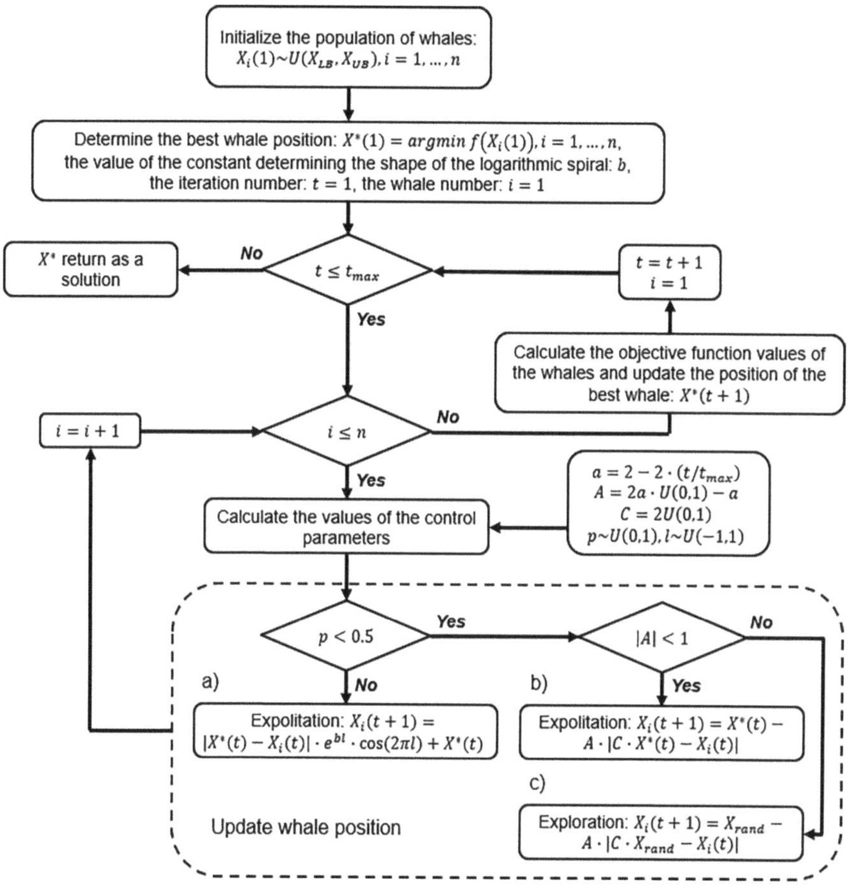

Fig. 1.8 The process of implementing the WOA

decreasing value of the amplitude of change [54]. Due to the variety of implemented search approaches, the WOA has been effectively adapted to solve problems in a number of areas [55–57]. The field of IC design is no exception, within the framework of which a two-stage amplifier operating at low voltage levels has been optimized using a 180 nm technological process, the structure of which is depicted in Fig. 1.9 [58]. The input of the circuit consists of a differential cascade of nMOS with a current mirror load, which provides the largest input voltage CLR value without additional parallel or cascode connections. M_6, M_7 transistors are biased so that the voltage between the sources and drains of M_3, M_4 transistors is sufficient to keep them in saturation. The output currents of the differential cascade, passing through M_6, M_7 transistors, are converted in the output cascade by a current mirror (M_8, M_9), and an amplifier with Miller compensation is used in the output cascade. The advantage of the resulting circuit is the input cascade with balanced loads and a high degree of amplification.

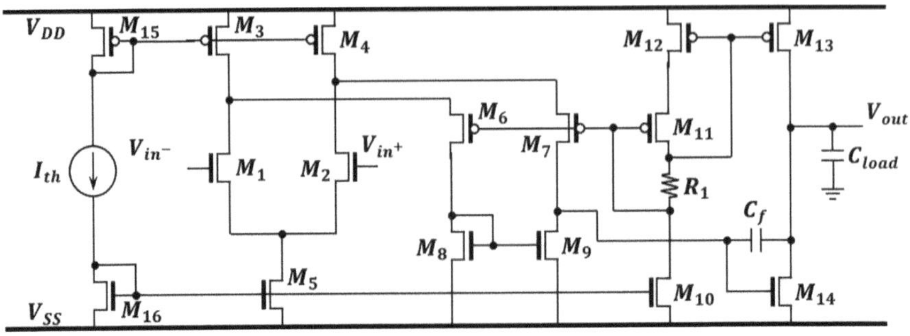

Fig. 1.9 Schematic view of a two-stage amplifier optimized using WOA

Table 1.3 Results of the optimization of a two-stage amplifier using WOA

	1st test		2nd test	
Parameter name (unit)	Constraint value	Measured value	Constraint value	Measured value
Gain (dB)	Maximum	89.36	≥70	–
Frequency bandwidth (MHz)	≥15	–	Maximum	25.5
Rise time rate (V μs⁻¹)	≥10	–	≥10	–
PM (°)	≥60	–	≥60	–

During the optimization, the following constraints were imposed on the transistor sizes $M_1 \equiv M_2$, $M_3 \equiv M_4 \equiv M_{15}$, $M_6 \equiv M_7$, $M_8 \equiv M_9$, $M_5 \equiv M_{10}$, $M_{11} \equiv M_{12}$. Two single-objective experiments with three additional constraints were performed, the results of which are presented in Table 1.3 [58].

Method of Initializing the Primary Population to Increase the Convergence Rate of Partial Search Algorithms

The search for the optimal solution of any GI PSA starts from the points included in the provided initial population. The more evenly distributed these points are over the search regions and include information about all the dimensions considered, the higher the probability of finding the optimum. The opposite is also true: if the initial points are concentrated only in certain regions, then regardless of the search strategies of the GI algorithm, avoiding premature convergence to the local optimum will be quite difficult [59–61].

Perhaps the most popular initial estimation method is random sampling (RS), which does not assume the presence of any information about the observed problem [62]. In RS, given the number of planes and the values of the upper and lower limits of each of them, coordinates are arbitrarily selected, which form the initial population points of the GI algorithm. It is the simplest approach from the point of view of implementation, but it does not guarantee that the output data will correspond to any distribution function and is very difficult to reproduce the results. For this reason, a number of methods have been developed that rely on statistical approaches for sampling and strive to reflect in the output the data distribution of the underlying problem [63].

One such algorithm is Latin Hypercube Sampling (LHS) [64]. This sampling technique is based on dividing the range of values of each input variable into intervals of equal length. In this way, coordinates are randomly selected from each of the intervals so that no more than one element is in the same row or column of the output point matrix. This process results in points that are equidistant from each other and evenly distributed along the range of available values of the input variables.

For the LHS to work, it is necessary to have the number of sampling points (n), the number of planes present in the point (D, in other words, the dimensionality of the problem), and the upper and lower limits of sampling ranges for each plane (LB_i, UB_i, $i = 1, \ldots, D$). The procedure for implementing the algorithm is presented below (Fig. 1.10).

The LHS has demonstrated its effectiveness in a number of areas, including computationally intensive computer simulations [65], environmental sciences [66], etc. This was possible due to the high degree of coverage of the input plane with a small number of sampling points (Fig. 1.11) [67].

Another sampling method is chaotic initialization (CI), which relies on the use of chaotic functions (CFs) to ensure the diversity of the initial population being formed [68]. CFs are mathematical functions that are characterized by chaotic behavior; in other words, small changes in the initial conditions lead to dramatically different

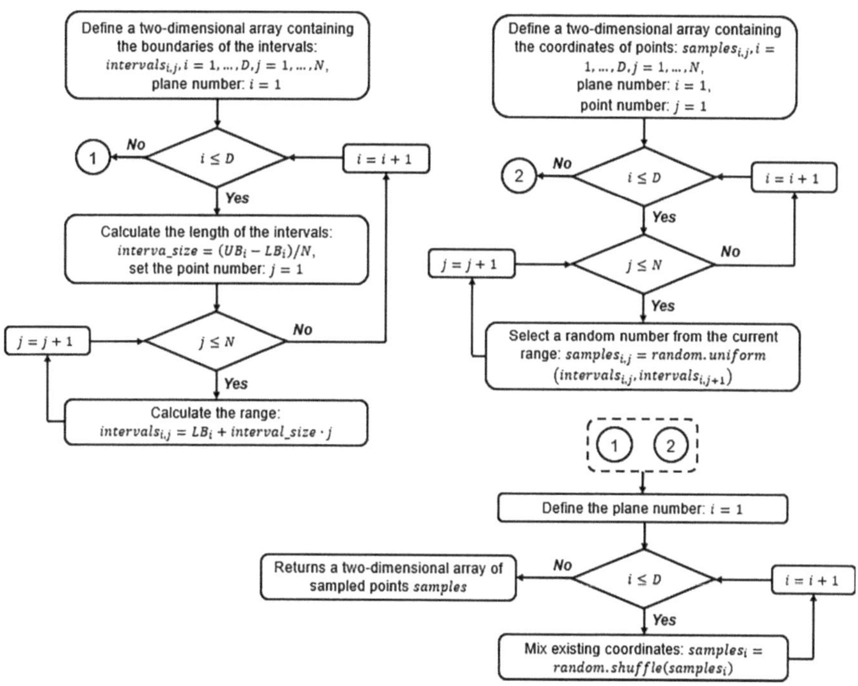

Fig. 1.10 Procedure of implementing the LHS

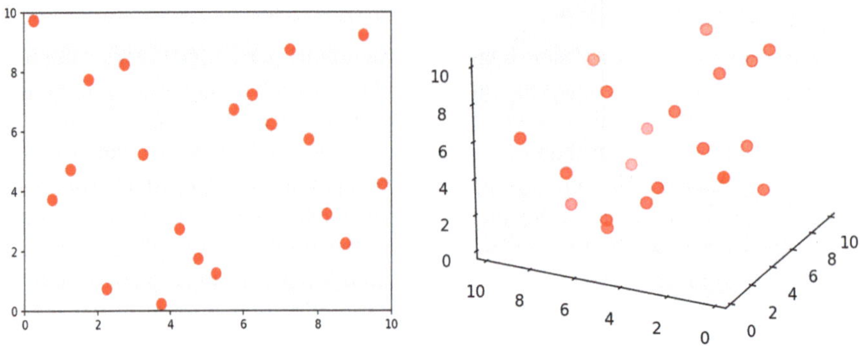

Fig. 1.11 Points initialized by the LHS in two-dimensional and three-dimensional planes

Table 1.4 Examples of CFs used by the CI

Chaotic function	Calculation formula	Number of planes	Range
Chebyshev	$x_{k+1} = \cos(k \bullet \cos^{-1}(x_k))$	1	$(0 \div 1)$
Gaussian	$x_{k+1} = \begin{cases} 0, \ if \ x_k = 0 \\ \frac{1}{x_k} \bmod(1), \text{otherwise} \end{cases}$	1	$(0 \div 1)$
Logistics	$x_{k+1} = ax_k(1 - x_k)$ $a \sim (3,75 \ldots 4)$	1	$(0 \div 1)$
Circular	$x_{k+1} = \left\{ x_k + b - \left(\frac{a}{2\pi}\right)\sin(2\pi x_k) \right\} \bmod(1)$	1	$(0 \div 1)$
Singer	$x_{k+1} = a\sin(x_k) + b\sin\left(\frac{x_k}{c}\right)$	1	$(-\infty \div \infty)$
Tent-shaped	$x_{k+1} = \begin{cases} \mu x_k, \ if \ x_k < 0.5 \\ \mu(1 - x_k), \text{otherwise} \end{cases}$	1	$(0 \div 1)$

output values. The advantage of implementing such functions is the possibility of generating pseudo-random values with the necessary statistical properties [69].

The same data is required for the implementation of the CI as in the case of the LHS, only the additional parameters necessary for CF's operation get added (Table 1.4). The work of the CI begins with the selection of a random point within the given limits for the current plane, after which this point is passed as input data to the CF. With a predetermined number of iterations, the output of the CF is again

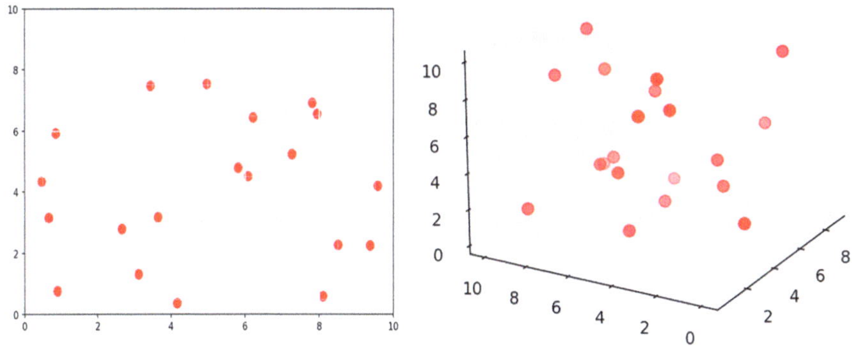

Fig. 1.12 Points, initialized by CI, in two-dimensional and three-dimensional planes using a tent-shaped CF

directed to the input in order to ensure the diversity of values. By repeating these operations a number of times corresponding to the number of points given for each plane, the necessary primary population is formed (Fig. 1.12).

A number of studies have been conducted on the improvement of convergence rates in various metaheuristic algorithms using points provided by the CI. In particular, the PSOA [70], the GA [71], the DEOA [72], and others have been considered, concluding that the CI improves the accuracy and convergence speed of these algorithms, especially for complex and multi-objective optimization problems.

The next observed method, opposition-based learning (OBL), unlike the previous ones, takes into account the specifics of the problem at hand during the calculation of the initial population [73]. It introduces the concept of opposite points and, given n, D, LB, UB data and the objective function $F(X)$, is implemented in several steps as follows:

- Initialization of given number of points using any elementary method (PN, LHN, CS, etc.). $P = [X(j)], j = 1, \ldots, n$,
- Calculation of the corresponding opposite points for the members of the resulting population: $X_{< \to i}(j) = \mathrm{UB}_i + \mathrm{LB}_i - X_i(j), i = 1, \ldots, D, j = 1, \ldots, n$,
- Calculation of adaptability values for members of each of the two resulting populations,
- Separating n points with the best value $\{F(X(j)) \cup F(X_{<}(j))\}, j = 1, \ldots, n$ from the population and returning them as the primary population.

The essence of this approach lies in the observation that simultaneous consideration of the positive and negative aspects of the existing search range leads to the search algorithm considering a wider range of solutions. The introduction of the OBL has made it possible to improve the rate of finding effective solutions to problems by a number of OA [74–76].

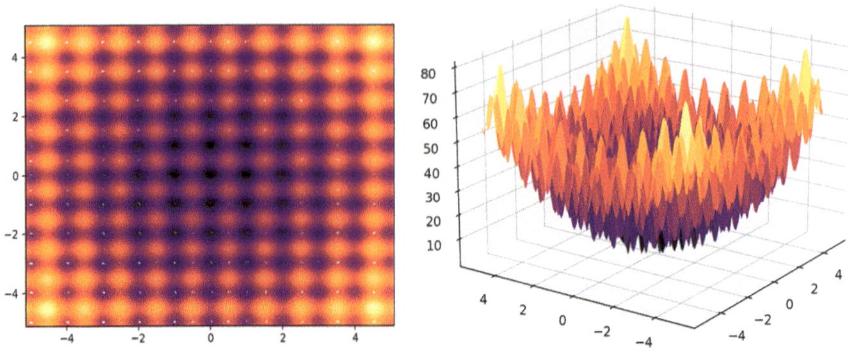

Fig. 1.13 Contour and surface appearance of the Rastrigin function for two input variables

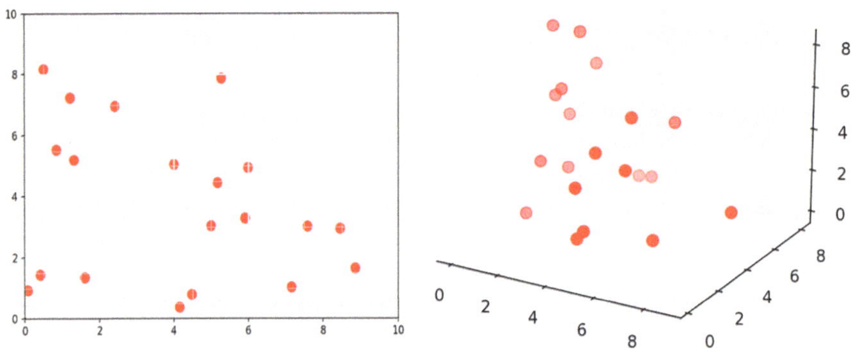

Fig. 1.14 Points, initialized by OBL, in two-dimensional and three-dimensional planes for the Rastrigin function

As an example, Fig. 1.13 shows the contour and surface appearance of the Rastrigin function [77] with multiple local optima for two input variables, and Fig. 1.14 shows the obtained initial evaluation points of the OBL.

Method of Replacing the Integrated Circuit Modeling Mean with Artificial Neural Networks During Parametric Optimization

The GI OAs considered above rely on modeling software tools to obtain accurate data on the behavior of the circuit during the search. Thus, to obtain the value of the objective function of any individual, a simulation is performed, within the framework of which the necessary output parameters are calculated, through which a numerical estimate of the adaptability is obtained. Depending on the efficiency of the search strategies used, the number of these calls can reach very large numbers before any acceptable solution is found. Since the implementation of these calls requires a large amount of time and computational resources, attempts have been made to replace the software tool modeling the circuits with machine learning and AI techniques. In particular, *k*-nearest neighbors [78], decision trees [79], random forests [80] and support vector machines [81] algorithms have been considered. However, the dependencies of the input and output parameters of the circuits were more accurately detected by ANNs, the efficiency of which could be further improved by increasing the number of hidden layers used (Fig. 1.15) [39].

However, before replacing the calls of the modeling tool with the appropriate ANN, it is necessary to select the ANN structure and further train it. In this case, the number of neurons in the input layer of the ANN will correspond to the number of independent design parameters in the circuit, and the output layer will contain neurons determined by the number of necessary measurements. There is no computational method for the optimal selection of the number of hidden layers as well as the number of neurons in each of them. They are selected based on the data obtained as a result of training, validation and testing of various ANN structures. Training, in turn, is based on the results of previously simulated versions of the circuit and is carried out using an error backpropagation algorithm. Within its framework, each group of input data is passed through the ANN, and the predicted values of the parameters of interest are obtained in the output layer. Next, the output error is calculated by some method and propagated to the connections of the previous layers using a special OA, changing their values (Fig. 1.16) [82]. Here it is necessary to choose the function for calculating the errors in the output, as well as the algorithm by which the values of the connections and deviations between neurons will be averaged during each training iteration. Currently, the OAs of ANNs that are widely used are stochastic gradient descent (SGD), adaptive gradient descent (AdaGrad), root mean square (RMS) propagation, AdaDelta, which solves the problem of constantly decreasing training step size, as well as the Adam algorithm, which is a hybrid of AdaGrad and RMS propagation approaches [83].

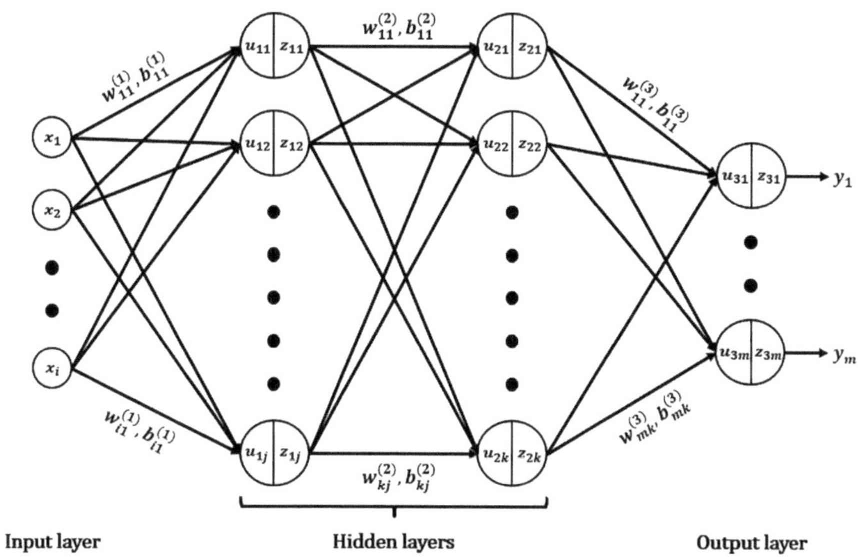

Fig. 1.15 Structure of an ANN with two hidden layers

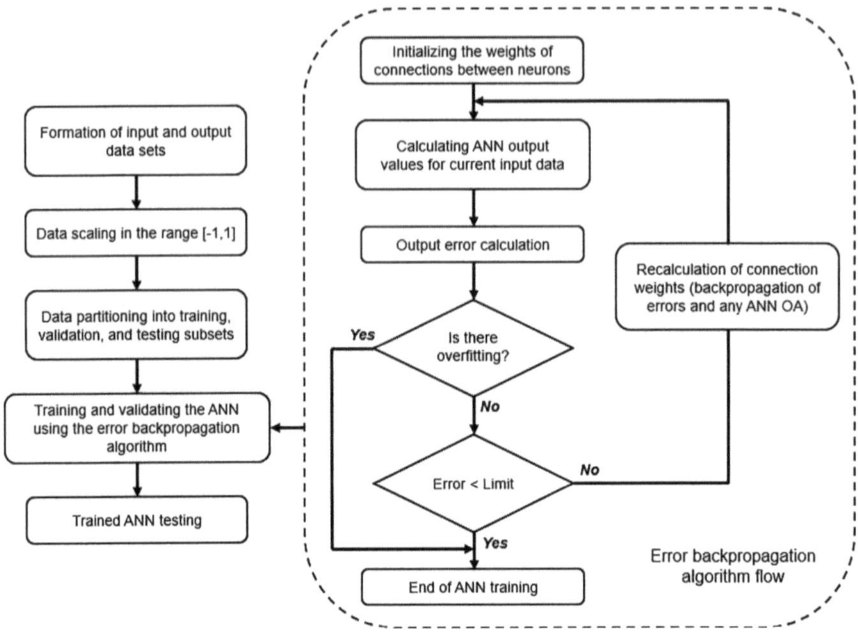

Fig. 1.16 The process of training an ANN using error backpropagation

In addition, the collection of reference data for training should be carried out using a minimum number of simulations, while maintaining an even distribution of data along the surface of the set of possible solutions, including boundary cases. This is necessary so that in practical applications, the ANN can generalize the "trained" dependencies with high accuracy to data that has not yet been encountered.

Simulation tool call substitution experiments were conducted using the ANN to predict the output characteristics of a 90 nm Miller OpAmp (Fig. 1.17) [84]. M_1, M_3, M_5, M_7 transistors and C_{load} capacitor sizes were used as independent design parameters, and the gain, PM, constant voltage noise suppression ratio (NSR), supply bus noise, input impedance, output impedance, and circuit area were selected as output characteristics. As a result, the input layer of the ANN consists of five neurons, and the output layer consists of seven neurons. A number of configurations were studied for the structure of hidden layers, and the best results were obtained in case of 3 layers and 1800 neurons.

During training, the AdaDelta algorithm was used as the ANN OA, and the output error was calculated according to the RMS deviation.

$$\text{Error} = \sqrt{\frac{\sum_{i=1}^{N}\left(x_i - \hat{x}_i\right)^2}{N}} \tag{1.3}$$

where N is the number of output parameters, x_i is the measured value of the i-th characteristic as a result of the simulation performed for the current input data, $\hat{x}_i$

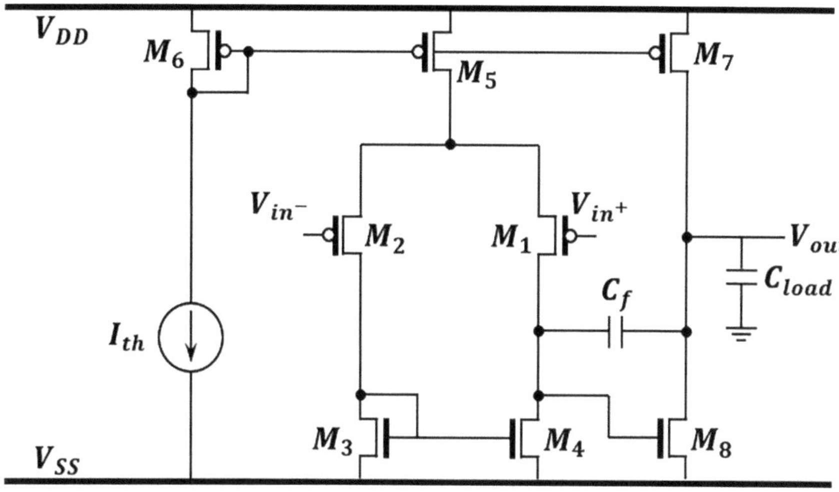

Fig. 1.17 Miller's OpAmp circuit for predicting output parameters using ANN

Table 1.5. RMS deviations of the predicted values of the output parameters of Miller's OpAmp through the ANN

Parameter name	RMS deviation	Parameter name	RMS deviation
Gain	0.17	Input resistance	0.14
PM	0.39	Output resistance	1.49
Input voltage NSR	0.16	Area	1.67
Power rails NSR	0.53		

is the predicted value of the same parameter by the ANN. The number of simulated variants of the circuit is 10,000, and the effective deviations of each output parameter are summarized in Table 1.5 [84]. As it can be seen from the results, the prediction accuracy for the critical parameters is high enough to be used in the further optimization process.

The ANN has also been embedded within the framework of an optimization system based on GA [85]. Unlike the previous approach, however, here the ANN is used only to predict the output parameters of circuit variants located in the local search area. In particular, a sampling of points is performed near the best individual found as a result of each iteration of the GA, with their subsequent modeling. Based on the obtained modeling results, the ANN is constructed and trained, after which all possible solution variants of the considered domain area are evaluated using the resulting model. The best candidate found is simulated to obtain accurate results, and if it does not satisfy the proposed design constraints, then a transition is made to the next iteration (Fig. 1.18). Under time constraints, periodic collection of ANN training data during each cycle of the system's operation is possible due to parallelization of simulations. This is possible due to the fact that after the formation of

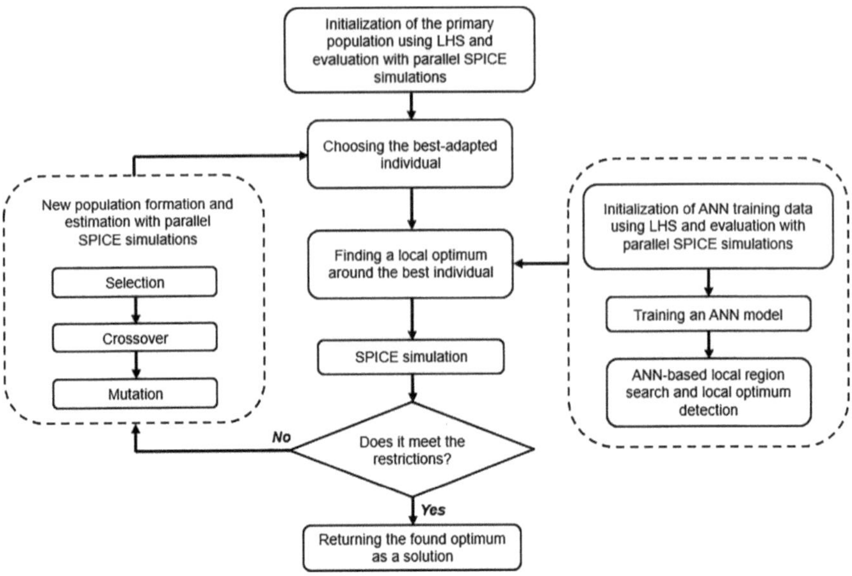

Fig. 1.18 The process of implementing GA with introduction of ANN for local search

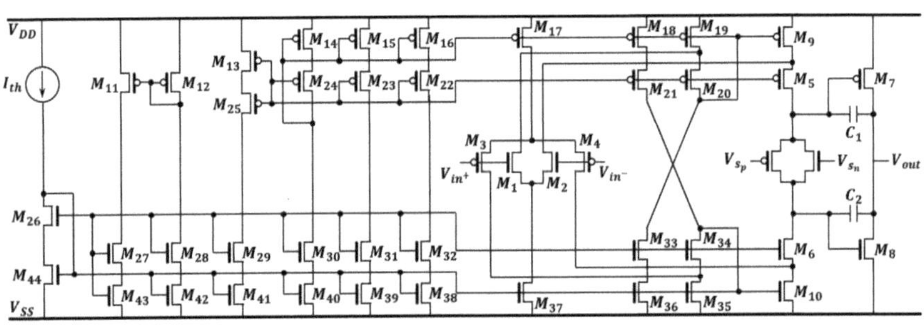

Fig. 1.19 Circuit of a two-stage rail-to-rail amplifier optimized using a GA with an embedded ANN

the population, the individuals of the GA can be observed independently of each other.

This system was used to optimize the two-stage rail-to-rail amplifier (Fig. 1.19), within which the observed parameters are gain, bandwidth, unity gain frequency (UGF) and PM. The sizes of M_1, M_3, M_5, M_6, M_7, M_8, M_9, M_{10} transistors and C_1/C_2 similar compensation capacitors have been considered as independent circuit parameters. The sizes of the remaining transistors are calculated, depending on the sizes of specified components, for example, M_{11}–M_{19}, M_{20}–M_{25}, M_{26}–M_{34}, M_{35}–M_{44} depend on the corresponding M_9, M_5, M_6, M_{10} values based on the matching requirements.

The constructed ANN consists of nine input and four output neurons, and the only hidden layer contains eight neurons, for the activation of which the hyperbolic

Table 1.6 Comparative results of optimization of a two-stage rail-to-rail OpAmp using GA and ANN-GA

Parameter name (unit)	Constraint value	Value obtained from the GA	Value obtained from the ANN-GA
Gain (dB)	Maximum	90,765	98,760
Capacity (Hz)	Maximum	548,455	543,308
UGF (MHz)	Maximum	17,314	37,199
PM (°)	≥60	65,692	61,494

tangent function of the sigmoid was used. The Levenberg-Marquardt method was used for training the model in combination with the mean relative error, and the amount of data for retraining in each iteration was a set of 632 input/output groups, taking into account the 79 computers available for parallel calculations. The comparative results of the optimization in combination with a conventional GA are summarized in Table 1.6 [85].

1.1.3 Problems in Design Mean for Automated Parametric Optimization of Digital Integrated Circuits

Existing tools of design means for automated parametric optimization of digital integrated circuits transform the optimization process into a single-objective or multi-objective optimization problem and solve it based on stochastic approaches. However, the development of the technological process, the associated complication of identifying relationships between the input and output parameters of ICs as well as the huge number of ICs being designed simultaneously, create challenges and difficulties, the solution of which by existing methods may require an unacceptable amount of time and computational resources. The shortcomings and main problems of existing means for automated optimization of ICs are presented below.

Problems in the Automation Method of Selecting Parameters of Integrated Circuits Using Partial Search Algorithms Based on Group Intelligence

The above section presented three main types of local optimization algorithms [45, 50, 53] which seek to find the optimum for a problem by applying radically different strategies. However, there is a drawback that is characteristic of all local optimization approaches, namely early convergence. This occurs when the algorithm falls into a "hole" of the local optimum during the search, and in subsequent iterations focuses only on exploring the vicinity of that region (Fig. 1.20) [86, 87]. It can occur for a number of reasons:

- Lack of diversity, which occurs when an algorithm focuses too much on a certain region of the search domain and more often leans toward exploitation search. The reason may be inefficient population formation mechanisms for the given

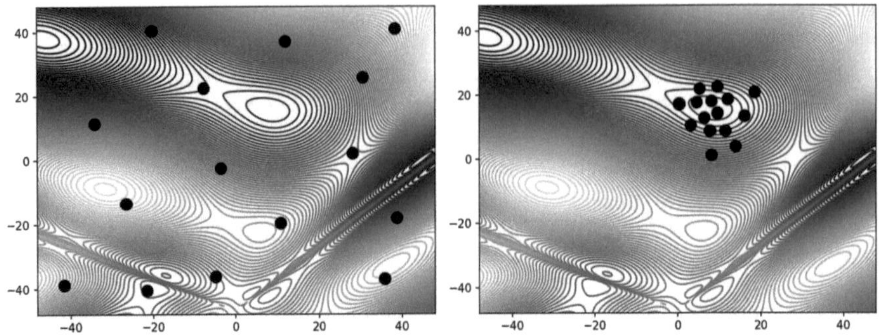

Fig. 1.20 Scenario of early convergence of an optimization algorithm to a local minimum

problem, as well as improperly balanced transitions between the applied strategies,
- Wrong choice of initial parameters, such as high probability of selection and mutation of individuals, inappropriate initial particle velocities, etc.,
- Features of the surface of the chosen objective function, in particular the variety of rough and tricky domains in which local and global optima follow each other.

The problem becomes more complicated when considered in the context of IC optimization, where there are no fixed approaches for choosing algorithm parameters and objective functions. Mostly, preference is given to default values and linear or exponential expressions connecting the output parameters obtained as a result of modeling. However, even in these cases, the probability of encountering a local optimum is high, taking into account the number of independent parameters present in the design and the sizes of the ranges of their permissible values, which can include tens of thousands of solution options.

In such situations, it is necessary to keep in mind that all evolutionary algorithms are, first of all, random methods that never guarantee the discovery of the optimum. However, depending on the approaches used, the probability of discovery can be increased. Therefore, taking into account the above shortcomings, there is a need to develop new hybrid methods that, by combining the advantages of different approaches, will more effectively implement the global search as well as balancing between the strategies used.

Problems in the Special Initial Estimation Method of the Primary Population to Increase the Convergence Rate of Partial Search Algorithms

The main difficulty in practical application of the stochastic methods of initial population's estimation [62, 64, 68, 73], considered in subsection above, in the optimization of ICs lies in the reproduction of the results obtained by them and in ensuring the proportional distribution of the points sampled in the experiments conducted separately from each other. Although all the considered approaches use various strategies for selecting the positions of individuals in the population, the proportional representation of all design parameters in the points obtained in the output is not ensured [88]. In addition, each of the methods has its own individual shortcomings.

For the effective operation of the CI, it is necessary to correctly select the CF and its corresponding initial parameters. However, there is still no clear explanation of why specific functions give more favorable results in some cases. For example, in [89], out of ten CFs used to solve the problem considered, the best result was obtained when choosing the logistic function. In the scenario of [90], the same ten candidates obtained optimal results when using the tent-shaped function. For example, the generated points of the functions given in Table 1.4 (Fig. 1.21) show how many free spaces can arise in different cases and how unevenly distributed the data can be [88]. In addition, the initial conditions also have a significant impact on the formation of the output results of the CI, and even minor changes in them can lead to a sequence of points that differ drastically. Moreover, the CI method was originally developed to solve problems with a small number of parameters, and it does not generalize well as the number of parameters increases, even at the expense of increasing the number of sample points.

The main drawback of the OBL becomes especially apparent in the context of IC optimization, in particular, when after obtaining the inverse points, it is necessary to perform an assessment of their adaptability [73]. Since each of these assessments is interpreted as a call of the modeling software tool, this method requires twice as many simulations to form the same population. This, depending on the size of the chosen population, can amount to hundreds of additional calls.

Considering the shortcomings of the above-mentioned methods, there is a need to develop new deterministic mechanisms for sampling members of the primary population, which will ensure the reproducibility of experiments and improve the convergence rates of OA, while not requiring additional parameter selection and additional simulations of the circuit.

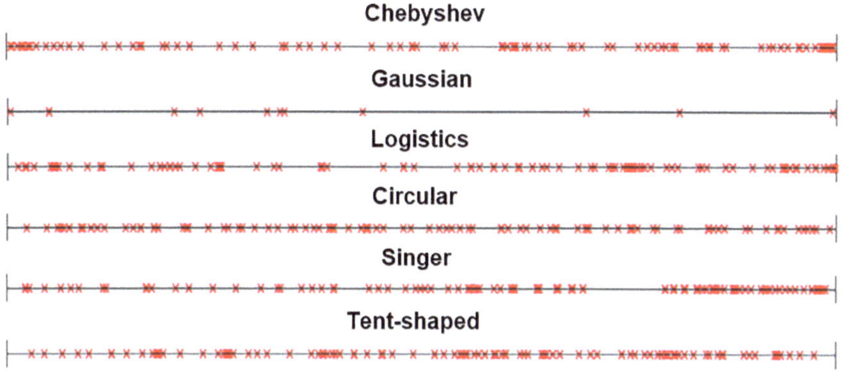

Fig. 1.21 Appearance of 100 points generated by six CFs of the CI on a one-dimensional plane

Problems in the Method of Replacing Integrated Circuit Modeling Software with Artificial Neural Networks During Parametric Optimization
The main difficulty in the method of replacing the software tool for simulating circuits with an ANN in the process of optimizing ICs is the collection of the simulation data necessary for training the ANN [84, 85]. The problem is also complicated by the fact that an additional set of data is needed for testing the trained ANN, which does not participate in the training in any way, but requires additional resources for simulations. In the case of ICs, due to the complexity of the circuit and the sets of permissible values of variable parameters, hundreds of calls to the simulation tool may be required.

It should be noted that so far there is no method for calculating how much data is sufficient for effective training of the ANN [91]. Therefore, any clear assessment of the quality of the data can be obtained only during testing or practical application of the trained ANN. The selection of points in the simulation data set also becomes more difficult as the size of the observed circuits increases and technological processes become smaller. Thus, the greater the influence of secondary factors on circuit operation, the more random the relationships between the input and output parameters become. This complicates the discovery of the behavioral rules of the circuit for ANN and their generalization to data not observed during training. Therefore, if certain parts of the set of all possible solutions are underrepresented in

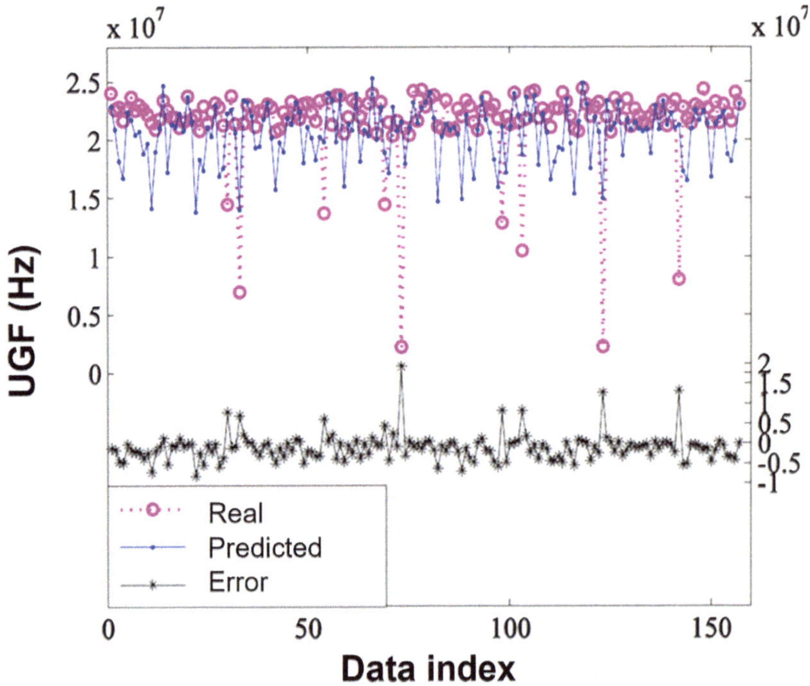

Fig. 1.22 Predicted and actual simulation values of ANN for a two-stage rail-to-rail OpAmp UGF

Table 1.7 Comparative results of training global and local ANNs

Model	Design parameters range	Model parameters	Number of training data	Average relative error (%)
Global ANN	[0.7 ÷ 1.3]	[32, 32, 32, 32, 32]	45,583	30.2
Local ANN	[0.95 ÷ 1.05]	[16]	632	0.758

the training data, the predicted values of the ANN may differ significantly from the actual values obtained as a result of simulation (Fig. 1.22) [85]. As a result, a situation arises where, for sufficiently complex circuits, even training based on datasets containing tens of thousands of simulation results may not provide the required accuracy of the predicted data (Table 1.7) [85].

Based on the above, the use of only one ANN model for predicting parameters over the entire range of possible solutions is quite problematic and practically impossible. The use of the ANN model for predictions only in the local domain provides higher accuracy results, but in this case, it is necessary to carry out a sampling of 632 points for each iteration, with their subsequent modeling and retraining of the ANN [85]. In practice, such an approach can be used only in the presence of a large number of computational resources. Thus, 80 computers were used during the experiments, 79 of which were used only for the parallel execution of the simulations.

Taking all this into account, there is a need to develop new methods and tools for collecting data for training the ANN, which will ensure high accuracy of the parameters predicted by the ANN as a result of training while at the same time requiring fewer calls to the modeling software tool.

1.1.4 Proposed Principles to Increase the Efficiency of Design Means for Automated Parametric Optimization of Digital Integrated Circuits

Considering the increasing amount of constraints imposed on the selection of design parameters for ICs, the time and computational resource requirements for automating optimization, and the likelihood of returning a suboptimal solution in existing methods due to inefficient initial populations, incorrectly balanced search strategies, and excessive simulations, a need arises to develop new solutions and methods aimed at increasing the diversity of the population and the rate of convergence of algorithms to the optimum. The following principles are proposed within the framework:

- Applying a hybrid GI approach based on the "Harris Hawks" optimization algorithm (HHOA) and GA to ensure higher search domain coverage during optimization.

Considering the dependence of the described methods on the selected operators and initial values of the parameters, which can lead to inefficient implementation of the exploration or exploitation-based searches, it is proposed to apply the principles of GA for global search of the solution and the approaches of HHOA for exploitation-based search for the selection of IC parameters. The latter is one of the newest members of the family of GI PSAs and has not yet been tested in the IC optimization process. Its main advantage lies in the diversity of strategies implemented for the exploitation search phase, which, combined with correctly composed selection and crossover operators of GA, can provide the capability to consider wider sections of the search domain.

- A deterministic method for selecting members of the primary population, based on sampling the edge points of embedded cubes, which does not require additional parameter settings.

The methods described above are stochastic in nature and cannot ensure stability of the quality of values between different tests. In addition, this depends on the parameters selected by the user and, in some cases, requires additional computational resources. For these reasons, it is proposed to perform the initialization of the primary population of the GI algorithm using the developed method called "Double Diagonal Uniform Initialization" (DDUI), which does not contain adjustable parameters and performs only the minimum number of necessary operations from the calculations point of view, excluding the underrepresentation of the input components in the output.

- Introduction of the generative adversarial network (GAN) into the ANN training data collection system.

Within the framework of the book, it is proposed to implement the generation of additional amounts of training data for the ANN based on the small pool of initial data samples, thanks to the introduction of the GAN. This will make it possible to obtain the intended number of training data sets, while simultaneously carrying out a smaller number of simulations. To increase the efficiency of the ANN, it is also proposed to introduce an additional guidance subsystem based on the Spearman correlation coefficients (SCC).

Conclusions

- Along with the development of the technological process, the influence of secondary factors on integrated circuits has significantly increased, which complicates the detection of connections between input and output parameters during optimization. From the point of view of a single-objective or multi-objective optimization problem, this circumstance translates into additional obstacles and irregularities on the search plane and leads to significant complications in finding the optimum. Therefore, the development of automated parametric optimization means for integrated circuits has become an extremely important challenge at present.

- Existing automated parametric optimization means for integrated circuits are based on partial search approaches using group intelligence and perform numerous calls to the modeling software tool to obtain data on the behavior of the circuit. Even in the case of introducing artificial neural networks, a significant number of data sets are required to make predictions with the necessary accuracy. This indicates that the latter do not meet modern requirements, and there is a need to develop new solutions for organizing the search process.
- Automated parametric optimization approaches for integrated circuits have been proposed that will meet modern requirements and significantly increase the convergence rate to the optimum by introducing additional, computationally non-time-consuming components into the system, while making fewer calls to the modeling software tool.

1.2 Proposed Solutions for Design Means for Automated Parametric Optimization of Digital Integrated Circuits

The problems described in Sect. 1.1 indicate the shortcomings of modern means of automated parametric optimization of ICs. The parallel design of a number of ICs corresponding to various technological processes, the sharp increase in the influence of often unpredictable secondary factors on circuit operation as well as their widespread use in vital areas have led to the tightening restrictions imposed on the means of design automation. Often, designers need to carry out additional adjustment of circuit parameters after the operation of the OA or develop new optimization approaches. The proposed changes to the means of automated parametric optimization of ICs are presented below, which solve the shortcomings of the existing methods, while complying with the proposed restrictions.

1.2.1 Method to Increase the Efficiency of Partial Search Algorithms for Automated Selection of Parameters for Integrated Circuits

As mentioned earlier, the widespread use of the tested GI PSAs for the automated selection of the sizes of IC components is hindered by the non-explicit separation of the implemented exploration and exploitation search strategies, or their limited number. For this reason, there is a tendency to get stuck in the local optimum domain and insufficient examination of other search regions. To overcome this, it is proposed to use a hybrid algorithmic approach, which will apply different approaches to find more favorable solutions in the best solution-based and global search stages. In particular, it is proposed to use GA to implement the global search process, and to use the corresponding strategies of the HHOA for best solution-based search (Fig. 1.23) [92].

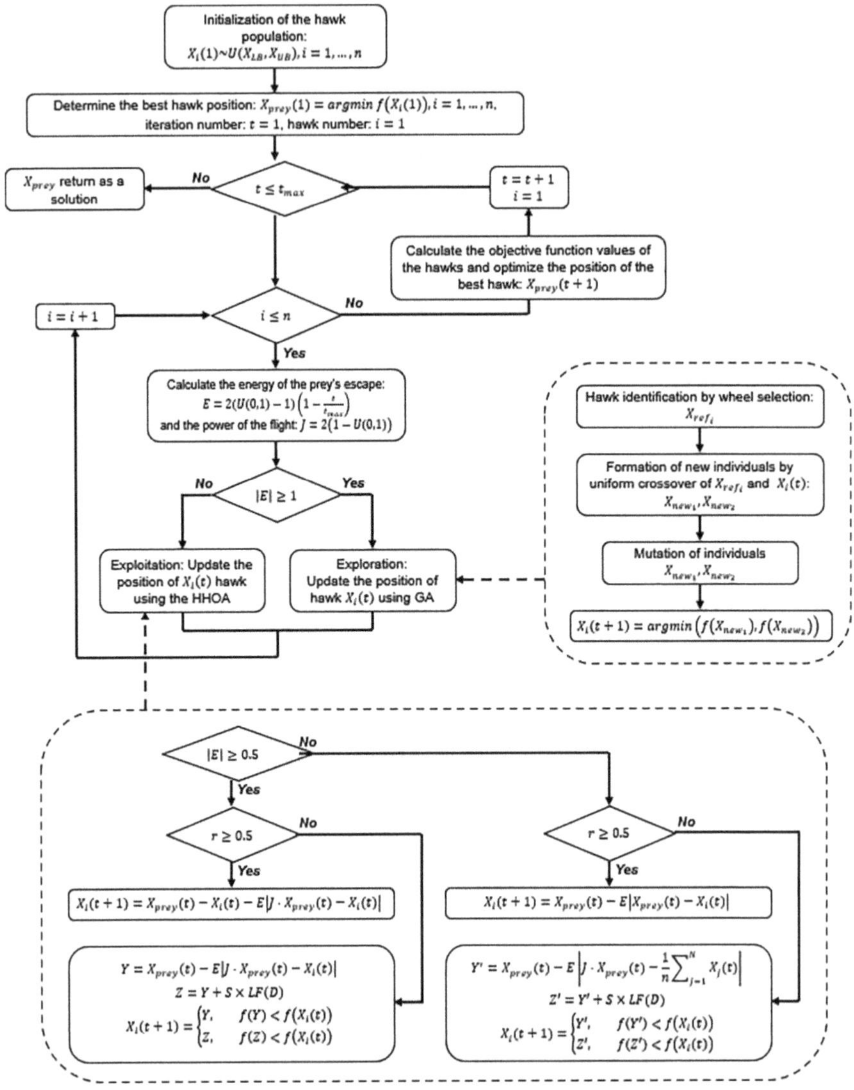

Fig. 1.23. The process of implementing the GA-HHOA

Consisting of only three main functions, GA has proven itself to be particularly effective in global search [93] and, when combined with the right types of selection/crossover/mutation operators, it can provide a high level of coverage of wide areas of the search domain. Based on experiments, the best results have been obtained under the conditions of combining the wheel selection, uniform crossover and multipoint mutation approaches [92]. In particular, wheel selection, based on adaptability values, assigns proportional probability values to all individuals in the population and makes a selection based on them. The advantage of this approach is that the probability of selecting any individual is not zero and only the " most suitable " solution options are not always selected. In case of uniform crossover, instead of splitting individuals, the genes of each parent are considered separately and can be passed on to the offspring with equal probability, ensuring the diversity of the data in the offspring. In the proposed approach, since in each iteration of the algorithm all individuals of the population are successively considered, only one reference candidate is selected by means of the wheel selection $X_i(t)$ for further crossbreeding with the current individual: X_{ref_i}. Next, by performing a uniform crossover, two new individuals are formed: X_{new_1} and X_{new_2}. By performing the mutation on them, the final versions of the candidates are obtained, from which the most favorable adaptation passes to the next generation, by assigning the value to the current individual. These operations define the global search or exploration procedure (Fig. 1.23) [92].

The HHOA is one of the newest members of the family of GI algorithms [94], the main feature of which is the variety of integrated search strategies. In particular, the primary version includes two exploration and four exploitation approaches, the choice between which during an arbitrary iteration of the algorithm is carried out using a special computational parameter, called the prey escape energy. The algorithm is based on mathematical models of various functions of prey detection and encircling by Harris's hawks, which are:

- Exploration (prey detection)—a surprise attack based on the position of an arbitrary member of the population or on the average coordinates of all members,
- Exploitation (prey encircling)—soft or hard besiege, with the introduction of levy flights if necessary [94].

Like a number of other GI PSA algorithms, the HHOA algorithm has a tendency to settle in the local optimum region, and the global search strategies available in it do not always satisfy the conditions of the imposed problems. On the other hand, the exploitation approaches are quite effective in terms of exploring small surrounding domains [95]. For this reason, only the exploitation methods of the HHOA algorithm are used in the proposed algorithm (Fig. 1.23) [92].

In addition to the above, it is also proposed to introduce a system for archiving calculated fitness values into the system, which will allow avoiding additional simulations of the circuit in case of obtaining identical candidates. This problem becomes especially evident in conditions of a small number of ICs or a small possible set of values of IC components due to technological process limitations, when as a result of arbitrary actions, individuals with identical values can be obtained [92]. The procedure for implementing the proposed GA-HHOA algorithm with all the introduced changes is presented in Fig. 1.23.

To evaluate the effectiveness of the proposed method, in the first stage, it was applied, together with five other algorithms, to solve the problem of minimization of ten test functions (Tables 1.8 and 1.9). The considered functions are: weighted sphere, Mikalevitz, Ackley, Iasom, Grievank, Rastrigin, Rosenbrock, Schwefel, Stiblinsky-Tang and Zakharov [77], whose mathematical expressions, minimum and maximum values of the solution search domain, as well as the optimum points

Table 1.8 Data on testing functions (part 1)

f_n	Name	Function	Search domain boundaries	Optimum point
f_1	Weighted ball	$f(x)=\sum_{i=1}^{D} i*x_i^2$	[−5.12 ÷ 5.12]	0
f_2	Mikalevits	$f(x)=-\sum_{i=1}^{D}\sin(x_i)\sin^{2m}\left(\frac{ix_i^2}{\pi}\right)$	[0 ÷ π]	~−99.56129 when $D = 100$
f_3	Ackley	$f(x)=20-20\exp\left(-0{,}2\sqrt{\frac{1}{D}\sum_{i=1}^{D}x_i^2}\right)$ $+e-\exp\left(\frac{1}{D}\sum_{i=1}^{D}\cos(2x_i\pi)\right)$	[−32,768 ÷ 32,768]	0
f_4	Iasom	$f(x)=\left(-\prod_{i=1}^{D}\cos(x_i)\right)$ $*\exp\left(-\sum_{i=1}^{D}(x_i-\pi)^2\right)$	[−100 ÷ 100]	−1
f_5	Grievank	$f(x)=\frac{1}{4000}\sum_{i=1}^{D}x_i^2$ $-\prod_{i=1}^{D}\cos\left(\frac{x_i}{\sqrt{i}}\right)+1$	[−600 ÷ 600]	0

Table 1.9 Data on testing functions (part 2)

f_n	Name	Function	Search domain boundaries	Optimum point
f_6	Rastrigin	$f(x)=\sum_{i=1}^{D}\left(x_i^2-10\cos(2\pi x_i)\right)+10D$	[−5.12 ÷ 5.12]	0
f_7	Rosenbrock	$f(x)=\sum_{i=1}^{D-1}\left(100\left(x_{i+1}-x_i^2\right)^2+(x_i-1)^2\right)$	[−5 ÷ 5]	0
f_8	Schwefel	$f(x) = 418,9829 * D$ $-\sum_{i=1}^{D}x_i\sin\left(\sqrt{\lvert x_i\rvert}\right)$	[−500 ÷ 500]	0
f_9	Stiblinski-Tang	$f_9(x)=\frac{1}{2}\sum_{i=1}^{D}\left(x_i^4-16x_i^2+5x_i\right)$	[−5 ÷ 5]	−39.166165 * D
f_{10}	Zakharov	$f(x)=\sum_{i=1}^{D}x_i^2+\left(\frac{1}{2}\sum_{i=1}^{D}ix_i\right)^2$ $+\left(\frac{1}{2}\sum_{i=1}^{D}x_i^4\right)^2$	[−100 ÷ 100]	0

are given in Figs. 1.23 and 1.24. The other considered GA algorithms are: GA, HHOA, WOA, PSOA, and DEOA.

In order to collect realistic statistical data, the minimization process was performed 100 times for each of the functions using an arbitrary algorithm, with the following parameters:

- The number of independent parameters available is 100,
- The number of individuals in the population is equal to 100,
- The number of iterations for performing the optimization is equal to 200.

To exclude the dependence of the results on the population initialization method, all algorithms were given the same initial population before the consecutive minimization process, which was generated using the uniform RS. As a performance criterion, the best (minimum), worst (maximum), and average values obtained for all methods were considered, which are summarized in Tables 1.10 and 1.11 [92].

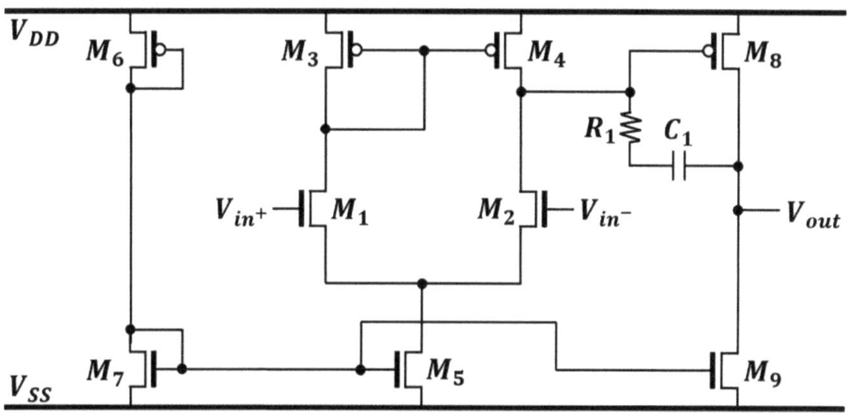

Fig. 1.24 Circuit of an optimized two-stage OpAmp

Table 1.10 Results of minimization of test functions (GA-HHOA check) (part 1)

F.	The cost	GA-HHOA	GA	HHOA	WOA	PSOA	DEOA
f_1	Best	1.50E-57	8933,789	1.39E-44	7.29E-43	3402,737	5558,368
	Worst	2.28E-48	13245.67	2.60E-35	1.18E-37	7558,163	8558,669
	Medium	1.18E-49	11367.34	1.94E-36	2.76E-38	5148,686	7407,124
f_2	Best	−41.4853	−36.1461	−29.6152	−29.8026	−22.5705	−24.0593
	Worst	−31.1347	−28.5344	−12.8581	−18.1181	−13.8896	−21.0733
	Medium	−33.4469	−32,1958	−22.6650	−24.4258	−17,1933	−22.3287
f_3	Best	0	7.320686	0	3.55E-15	3.641858	7.426574
	Worst	0	8.576761	0	1.42E-14	4.719050	8.407669
	Medium	0	7.879382	0	5.51E-15	4.288697	7.909273
f_4	Best	−0.99981	0	−0.99998	0	0	0
	Worst	−0.87398	0	−0.97856	0	0	0
	Medium	−0.91651	0	−0.99335	0	0	0
f_5	Best	0	1.070556	0	0	0.320891	1.072712
	Worst	0	1.127869	0	0.414168	0.638235	1.095862
	Medium	0	1.094295	0	0.037352	0.487318	1.080772

Table 1.11 Results of minimization of testing functions (GA-HHOA check) (part 2)

F.	The cost	GA-HHOA	GA	HHOA	WOA	PGOA	DEOA
f_6	Best	0	890.2924	0	0	763.2385	1125,158
	Worst	0	1077,726	0	0	1010,287	1254,704
	Medium	0	965.3466	0	0	874.5329	1193,619
f_7	Best	7.06E-09	119602.4	0.000268	97.22799	15978.83	123151.3
	Worst	14.22185	249029.4	0.084789	98.11171	154901.9	224891.6
	Medium	2.080795	165122.2	0.014989	97.84475	54362.79	165555.7
f_8	Best	0	22995.09	0	9.39	40606.02	36002.13
	Worst	26.44028	26789.99	31.52147	12904.59	41257.71	37731.77
	Medium	9.389623	24718.89	12.68531	6492,891	41016.48	36893.77
f_9	Best	−3916.61	−2560.25	−3916.62	−3916.61	−3051.45	−1768.91
	Worst	−3916.52	−2137.72	−3916.48	−3097.97	−2696.82	−1414.35
	Medium	−3916.57	−2357.45	−3916.60	−3849.23	−2903.45	−1573.38
f_{10}	Best	1.59E-35	528.3849	1.01E-19	793.9915	623.2383	814,117
	Worst	6.57E-15	850.3978	880.9389	1019,348	1255,925	2202,962
	Medium	3.29E-16	712.4746	55.68902	872.9994	955.5419	1247,804

Table 1.12 Limitations imposed on the optimization process of a two-stage OpAmp

Parameter name (unit)	Constraint value	Parameter name	Constraint value
Gain (dB)	≥60 and ≤90	Gate length	[0.4 ÷ 1.2]
PM (°)	≥30 and ≤90	Diffusion width	[0.3 ÷ 2.4]
Gain margin (dB) (GM)	≥−50 and ≤−10	Number of parallel connection gates	[1 ÷ 4]
Bandwidth (MHz)	≥350 and ≤550		

The results show that the proposed hybrid GA-HHOA method outperforms all the considered algorithms in the problem of minimum detection for 9 out of 10 functions, being able to either find the coordinates of the minimum point or come very close to it. The only function in which the necessary point could not be found is Iasom. In its contour form, along almost the entire search domain, it represents a flat surface, in which the influence of random factors on the movement of individuals is greater. In the case of that function, the HHOA was more "successful," but the GA-HHOA, unlike the other four algorithms, was able to approach the optimum. As a result, the proposed approach can be effectively applied to optimization problems.

Next, to evaluate the performance of the method, it was used to solve the parametric optimization problem of three ICs. The first IC under consideration is a two-stage OpAmp, consisting of differential and common-source amplifiers, the schematic structure of which is shown in Fig. 1.24. The circuit was implemented using models corresponding to the CMOS technological process, and the constraints imposed on the optimization are given in Table 1.12. As an additional design constraint, the same dimensions were used for M1 and M2 transistors during simulations, and the size of the resistance component R was kept constant. As a result, there are eight components in the circuit, which correspond to 24 independent

Table 1.13 Optimization results of a two-stage OpAmp (GA-HHOA check)

Optimization algorithm	Solution	Gain	PM	GM	Bandwidth	Adaptability
GA-HHOA	Best	82,795	65,626	−23,209	549.1	10,322
	Worst	74,354	57,958	−49,425	470.8	12,598
	Medium	–	–	–	–	11,884
GA	Best	73,784	50,413	−19,424	521.8	10,859
	Worst	54,183	65,485	−35,706	452.6	394,819
	Medium	–	–	–	–	46,013
HHOA	Best	74,958	63,472	−25.27	508.7	10.65
	Worst	64,759	37,458	-41,355	350.2	12,701
	Medium	–	–	–	–	12,551
WOA	Best	72,227	65,226	−25,165	529.8	10,621
	Worst	66,121	55,169	−49,052	366.1	13,148
	Medium	–	–	–	–	12,057
PSOA	Best	81,963	66,927	−25,804	498.0	10.49
	Worst	80,873	55,057	−49,833	415.5	12,894
	Medium	–	–	–	–	11,953

design parameters. To obtain statistical data, the circuit was optimized 100 times using the six optimization algorithms considered above. The iteration limit is set to 20, the population consists of 200 solution candidates, and all algorithms are given the same starting points to begin the search process.

During optimization, the following formula for calculating the adaptability value was used as a qualitative criterion for individuals:

$$\text{Adaptability} = \begin{cases} \sum_{i=1}^{N} w * e^{-\frac{2V_i}{V_i + \text{LB}_i}}, \text{ if } V_i \leq \text{UB}_i \\ \sum_{i=1}^{N} w * e^{-\frac{\text{UB}_i + V_i}{2V_i}}, \text{ if } V_i > \text{UB}_i \end{cases}, \tag{1.4}$$

where LB_i, UB_i represent lower and upper permissible values of the i-th measured parameter in the constraints, respectively, and w is the additional penalty coefficient. As a result, the obtained results of the optimization of the two-stage OpAmp are summarized in Table 1.13 [92].

As seen from the data, GA-HHOA outperformed the other methods, being able to detect the individual with the lowest fitness value in all the benchmark classes, showing an average of 27.76% higher efficiency.

The second circuit considered is the high-speed decoder (HSD), whose schematic view is shown in Fig. 1.25 [96]. It consists of a logic NOR circuit, an inverter, a transfer switch, and a dual-voltage circuit.

For optimization, the IC components were implemented using the CMOS models. The proposed constraints are summarized in Table 1.14, and the design constraints associated with the components are as follows:

$$\text{M}_3 \equiv \text{M}_4, \text{M}_5 \equiv \text{M}_6, \text{M}_3 = [1{,}0 \div 3{,}0] * \text{M}_5,$$

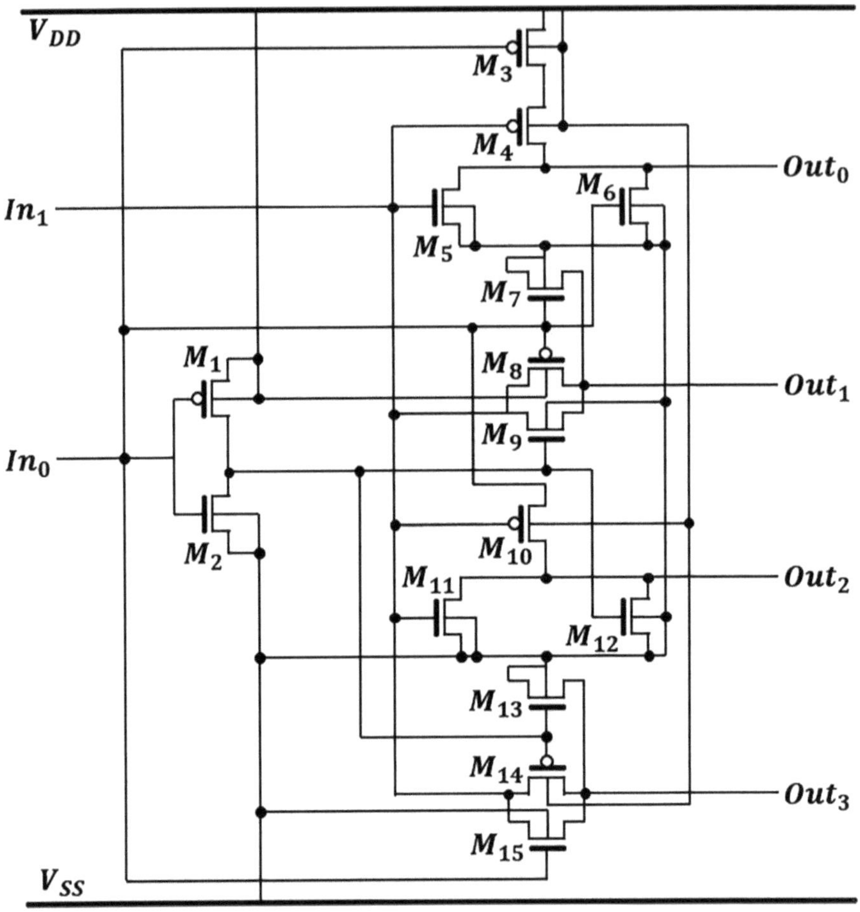

Fig. 1.25 Schematic view of the optimized HSD

$$M_8 \equiv M_9, M_{14} \equiv M_{15},$$

$M_{11} \equiv M_{12}$, $M_{10} = [1,0 \div 3,0] * M_{11}$, $M_1 = [1,0 \div 3,0] * M_2$:

As a result, seven IC components and three coefficients are involved in the optimization process, bringing the number of independent design parameters to 17, in conjunction with 16 measured parameters. The same GI OAs were considered for optimization as in the previous case, using expression (1.4) to assess the adaptability of individuals. The results of optimization performed 100 times by each algorithm under the conditions of 20 iterations and a population consisting of 200 individuals are summarized in Table 1.15.

Based on the data, it can be argued that the proposed method managed to outperform the other considered approaches in all cases in finding a solution with minimal

Table 1.14 Limitations on the HSD optimization process

Parameter name (unit)	Constraint value (10% ÷ 90%)	Parameter name (unit)	Constraint value (10% ÷ 90%)
Out_0 T_{rise} (ps) (1)	≥29 and ≤35	Out_2 T_{rise} (ps) (1)	≥29 and ≤35
Out_0 T_{fall} (ps) (2)	≥29 and ≤35	Out_2 T_{fall} (ps) (2)	≥29 and ≤35
Out_0 Price (ps) (3)	≥0 and ≤10	Out_2 Prise (ps) (3)	≥0 and ≤10
Out_0 P_{fall} (ps) (4)	≥0 and ≤10	Out_2 P_{fall} (pv) (4)	≥0 and ≤10
Out_1 T_{rise} (ps) (1)	≥29 and ≤35	Out_3 T_{rise} (pv) (1)	≥29 and ≤35
Out_1 T_{fall} (ps) (2)	≥29 and ≤35	Out_3 T_{fall} (ps) (2)	≥29 and ≤35
Out_1 Price (ps) (3)	≥0 and ≤10	Out_3 Price (ps) (3)	≥ 0 and ≤10
Out_1 P_{fall} (ps) (4)	≥0 and ≤10	Out_3 P_{fall} (ps) (4)	≥ 0 and ≤10
Parameter name		Limit value	
Gate length		0.03	
Diffusion width		[0.3 ÷ 2.4]	
Number of parallel connection gates		[1 ÷ 4]	

fitness. Overall, the GA-HHOA demonstrated 44.09% higher productivity on average.

As a final test circuit for evaluating the method, an embedded cascode OpAmp was considered, the schematic of which is shown in Fig. 1.26 [92]. In it, the differential pair acts as an input cascade, the outputs of which are connected to transistors of common gate chains of opposite polarity. For the purpose of optimization, the circuit was implemented using FinFET technological process, within which, for each of the transistors, under the conditions of a fixed gate length (L), three parameters were considered: the number of fins (Nfin), the number of parallel connection gates (Nfin Dim) and the number of vertical connection gates (L Dim). Its purpose is to ensure obtaining larger working dimensions of a transistor under the conditions of a constant L and a limited Nfin. Within the framework of optimization, this is achieved by replacing the current transistor with a chain of identical transistors connected in parallel with Nfin Dim and vertically with L Dim (Fig. 1.27).

The technical and technological constraints imposed on the optimization are given in Table 1.16, and the design rules related to matching the transistor sizes are as follows:

$$M_1 \equiv M_2, M_3 \equiv M_4, M_5 \equiv M_6, M_7 \equiv M_8$$

$$M_3 = 2 * M_{11}, M_5 = 2 * M_{12}$$

$$M_9 = 4 * M_{10}, M_{10} = 4 * M_{15}, M_{17} = 2 * M_{15}$$

The number of effective independent parameters becomes 24, corresponding to eight transistors. The process was carried out according to the same principles in the case of the previously considered circuits, and the results are summarized in Table 1.17. Based on them, a conclusion is formed that for ICs implemented using FinFET technology, the proposed method is able to find the solution with the

Table 1.15 Results of HSD optimization (GA-HHOA check)

Optimization algorithm	Solution	(1); (2) (3); (4)	(5); (6) (7); (8)	(9); (10) (11); (12)	(13); (14) (15); (16)	Adaptability
GA-HHOA	Best Worst Medium	34.24; 34.93 9.97; 9.96/ 32.46; 35.65 9.69; 9.65/ –	34.67; 30.38 6.31; 7.18/ 30.85; 29.08 7.45; 5.12/ –	29.91; 34.63 9.26; 9.97/ 29.36; 29.19 8.57; 8.58/ –	32.67; 33.66 6.76; 8.78/ 32.87; 29.02 8.55; 9.57/ –	40,718 75,485 42,453
GA	Best Worst Medium	29.52; 33.02 9.59; 9.7/ 30.2; 33.24 9.34; 8.41/ –	34.86; 33.35 6.37; 8.44/ 34.59; 34.7 6.04; 9.44/ –	31.83; 34.4 9.83; 9.96/ 32.12; 32.35 9.25; 8.95/ –	32.67; 34.91 6.95; 6.94/ 32.68; 43.11 6.89; 9.91/ –	40.75 77,771 47,642
HHOA	Best Worst Medium	29.26; 32.9 9.57; 9.02/ 29.14; 33.06 9.49; 9.11/ –	34.78; 29.44 8.96; 8.33/ 41.38; 32.42 9.98; 9.94/ –	34.9; 29.83 9.53; 8.48/ 30.37; 30.05 7.52; 7.67/ –	29.26; 34.14 9.38; 6.46/ −82.7; 32.82 9.94; 6.49/ –	41,236 1063,507 124,636
WOA	Best Worst Medium	29.87; 32.99 9.97; 9.96/ 29.33; 33.49 9.48; 9.09/ –	34.82; 33.34 5.52; 8.36/ 34.81; 29.57 9.17; 7.03/ –	34.84; 29.77 8.72; 9.77/ 34.67; 30.71 8.63; 8.42/ –	32.68; 33.75 5.87; 8.35/ 32.66; 27.17 9.51; 6.19/ –	40,956 417,664 87,534
PGOA	Best Worst Medium	29.14; 32.74 9.7; 9.56/ 29.79; 32.61 9.18; 8.79/ –	33.63; 32.83 7.74; 9.83/ 33.8; 34.01 3.81; 8.73/ –	29.3; 29.09 9.42; 8.31/ 34.14; 31.26 9.94; 8.02/ –	32.68; 34.83 8.15; 9.78/ 32.68; 966.3 5.95; 9.87/ –	41.15 97.25 45.27

necessary minimum adaptability, while ensuring the satisfaction of the technical specifications' constraints. In general, compared to other algorithms, GA-HHOA demonstrated an average of 72.51% higher performance.

As a result, based on the combination of the results of minimization of observed testing functions and optimization of ICs, it can be argued that the proposed hybrid GA-HHOA approach, based on the implementation of GA computational processes, in particular, the combination of wheel selection, uniform crossover, as well as multipoint mutation, and the combination of HHOA with the exploitation-based search strategies, provides an average of 48.11% higher efficiency than other existing popular GI OA.

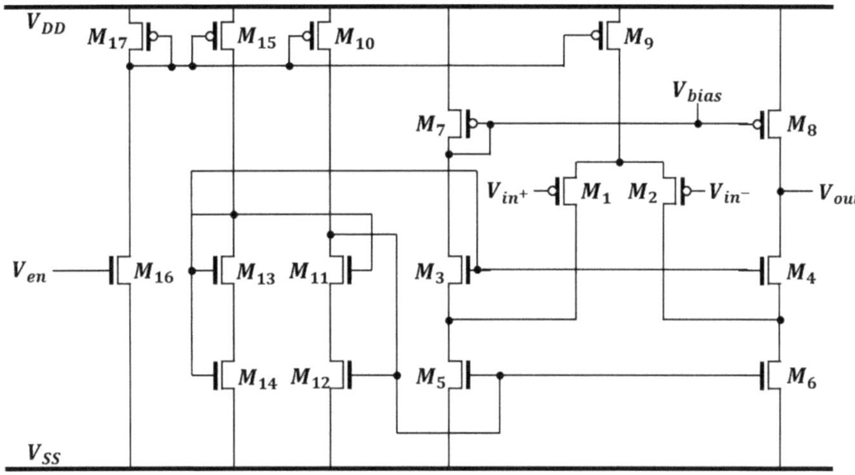

Fig. 1.26 Optimized embedded cascode OpAmp circuit

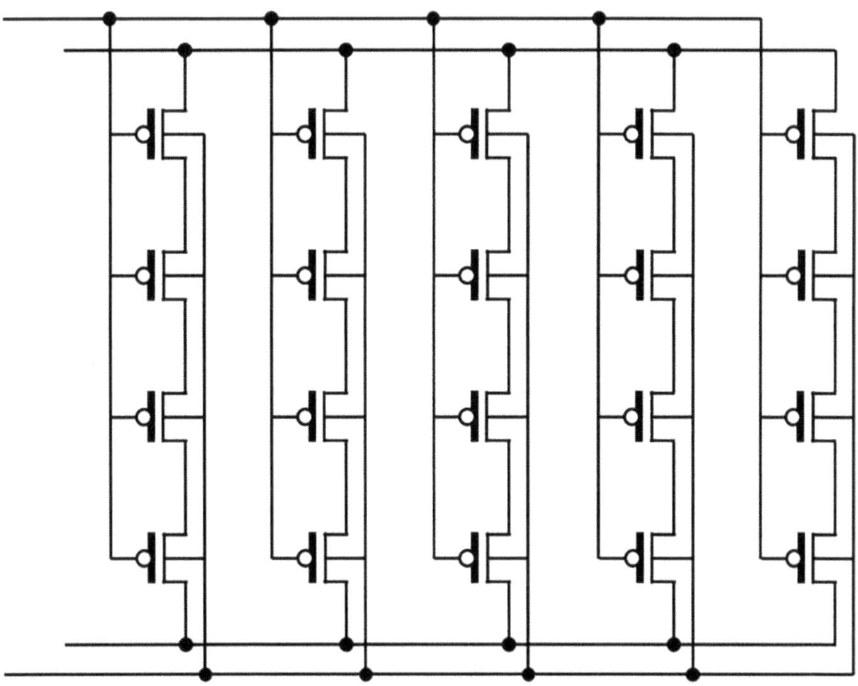

Fig. 1.27 Circuit with vertical and parallel connections of transistors to increase L and Nfin of transistors in the FinFET technological process

Table 1.16 Limitations imposed on the optimization process of embedded cascode OpAmp

Parameter name (unit)	Constraint value	Parameter name	Constraint value
Gain (dB)	≥70 and ≤90	Valve length	0.014
PM (°)	≥70 and ≤90	Number of fins	[4 ÷ 8]
Gain margin (dB) (GM)	≥−50 and ≤−20	Number of parallel connection gates	[1 ÷ 40]
Bandwidth (MHz)	≥ 40 and ≤500	Number of vertical connection gates	[1 ÷ 40]

Table 1.17 Optimization results of embedded cascode OpAmp (GA-HHOA check)

Optimization algorithm	Solution	Gain	PM	Gain margin (GM)	Bandwidth	Adaptability
GA-HHOA	Best	71,812	77.25	−25,184	89,006	11,294
	Worst	70,549	73,535	−20,334	18,634	538,469
	Medium	–	–	–	–	78,675
GA	Best	76,659	72,199	−49,071	40,006	14,034
	Worst	70,422	87,611	−32,539	6246	772,002
	Medium	–	–	–	–	437,624
HHOA	Best	70.06	74,889	−49,251	49,264	13,842
	Worst	56,036	89,786	−20,258	5879	1189,744
	Medium	–	–	–	–	505,437
WOA	Best	80,751	70,529	−46,041	40,109	13,236
	Worst	70,247	86,901	−49,267	2865	885,138
	Medium	–	–	–	–	571,349
PSOA	Best	70,184	70,347	−47,685	40,089	13,803
	Worst	76,766	83,018	−46,666	3252	869,849
	Medium	–	–	–	–	537,742

1.2.2 *Initialization Method of Primary Population of Partial Search Algorithms*

To solve the problems of the dependence of the above-presented methods on random factors and the need to perform additional simulations, it is proposed to use a deterministic approach that does not involve additional computational resources. In particular, a new initial population formation strategy is proposed, which is called DDUI [97]. Its essence lies in dividing the entire search domain into cubes embedded at equal distances, with the subsequent sampling of their peripheral points. In other words, the primary (main) and secondary diagonals of the domain are simultaneously considered, which are divided into proportional segments, and the positions of the necessary points are obtained. Thus, to calculate the coordinates of N points, each plane of the search domain is divided into $[N/2]$ equal segments. In the first stage, the points of the main diagonal are sampled, which are located from each other at a distance of:

$$dx = \frac{UB_j - LB_j}{N/2 - 1} \tag{1.5}$$

where UB_j and LB_j indicate the upper and lower limits of the j-th domain. As a result, in the D-plane problem, the j-th coordinate of i-th point will be calculated by the following expression:

$$P_{i,j} = LB_j + i * dx, j = 1, \ldots, D \tag{1.6}$$

Next, the secondary diagonal is considered, in which the distance between the points is obtained by the same formula (1.5), the coordinates of the i–th individual in the population are calculated as follows:

$$P_{i,j} = \begin{cases} LB_j + i * dx, & \text{if } i\%2 = 0 \\ UB_j - i * dx, & \text{otherwise} \end{cases} \tag{1.7}$$

As a result, [N/2] points are taken from each diagonal, the aggregation of which constitutes the primary population. However, if the number of necessary N points is an odd number, then the midpoint of the diagonals, which is also their intersection point, is sampled twice. In this case, to avoid data duplication, that point is simply removed from one of the diagonals. The positions of the points sampled in the two-dimensional and three-dimensional domains using the algorithm are given in Fig. 1.28, and the general implementation process is summarized in Fig. 1.29 [97].

In addition to the absence of random phenomena and the simplicity of the operations involved, the main advantage of the proposed approach is that the generated equidistant points carry information about each plane. This allows avoiding unwanted groupings of points in certain planes, which is especially noticeable when solving large-scale problems. In addition, in most recent OAs, linear iteration is

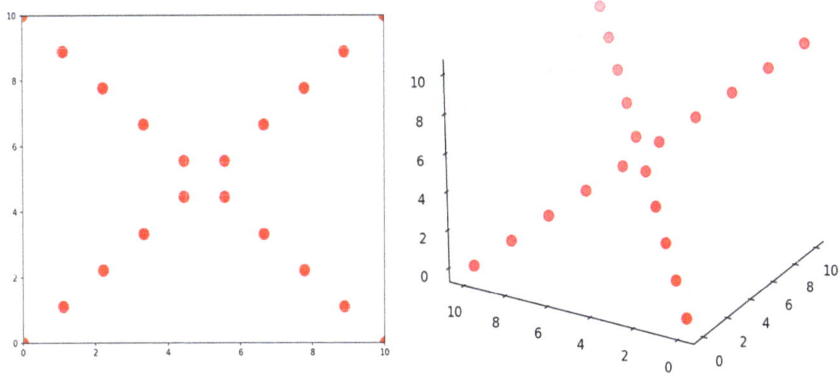

Fig. 1.28 Dots initialized by DDUI in two-dimensional and three-dimensional planes

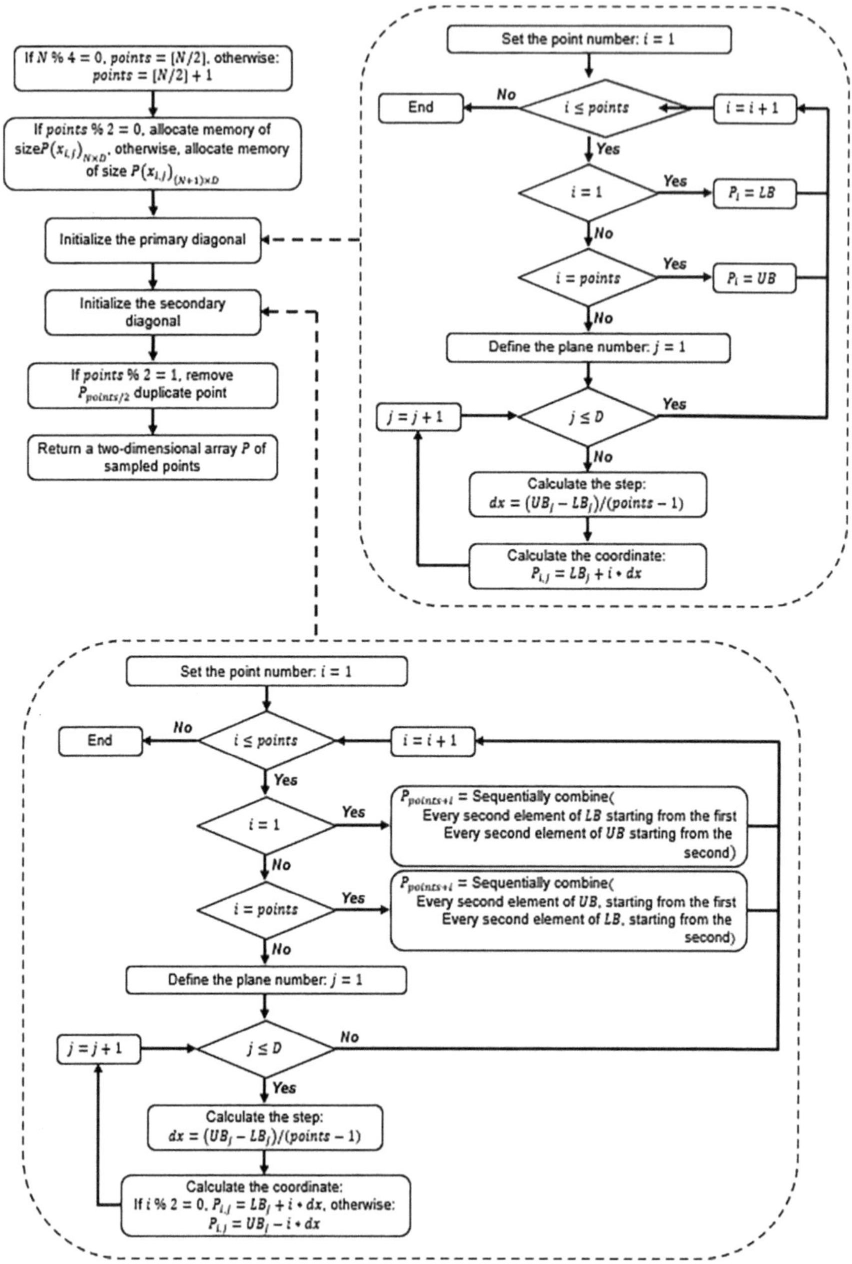

Fig. 1.29 The process of implementing the DDUI

used to update the position of the individuals. This, combined with the information available at the equidistant points in each plane and the non-zero iteration coefficient, allows obtaining coordinates for an arbitrary point in the considered domain. The main limitation of the OA is that, in order to distribute the points evenly along both diagonals, it cannot generate an arbitrary number of points. The number of sampling points should be $4 * i$ or $4 * i + 1$, where $i = 1, 2, 3, \ldots$, however, this includes commonly used population sizes such as 100, 200, 400, etc. [97].

To evaluate the effectiveness of the proposed strategy, it was first applied to find the minimum values of eight testing functions, which are: Ackley (f_1), Iasom (f_2), Grievank (f_3), Rastrigin (f_4), Rosenbrock (f_5), Schwefel (f_6), Stiblinski-Tang (f_7), and Zakharov (f_8) [77] (Tables 1.8 and 1.9 f_3–f_{10}). The methods of RS, LHS, OBL, and CI with tent-shaped CF were considered as alternatives, and the minimization process was carried out using the WOA algorithm. In order to obtain statistical data, the minimization of f_1–f_8 functions with each initialization method was carried out 100 times, with populations consisting of 100 individuals and within 200 iterations. To assess the performance dependence on the dimensionality of the problem, the process was carried out for search domains consisting of 10, 50, and 100 dimensions, the results of which are summarized in Tables 1.18, 1.19, and 1.20, respectively [97]. It should be noted that the search for the absolute minimum of such functions is usually carried out for over 400–500 iterations, but since the current experiments

Table 1.18 Results of minimization of test functions (DDUI check) ($D = 10$)

F.	The cost	DDUI	RS	LHS	OBL	CI
f_1	Best	0	0	0	0	3.55E-15
	Worst	1.07E-14	1.42E-14	1.42E-14	1.42E-14	1.42E-14
	Medium	5.51E-15	5.86E-15	6.22E-15	6.39E-15	6.04E-15
f_2	Best	−1	0	0	0	−1
	Worst	−0.995393	0	0	0	0
	Medium	−0.998617	0	0	0	−0.42688
f_3	Best	0	0	0	0	0
	Worst	0	0.474477	0.447371	0.248930	0.74061
	Medium	0	0.115378	0.061806	0.055661	0.11250
f_4	Best	0	0	0	0	0
	Worst	1.42E-14	0	38.40678	59.52248	6.20447
	Medium	7.11E-16	0	3.165975	6.439556	0.31022
f_5	Best	5.29E-09	5.156798	0.006609	0.006203	0.00043
	Worst	0.007103	7.165505	7.180006	7.206342	7.19182
	Medium	0.000926	6.488363	6.474888	6.155187	5.64577
f_6	Best	−4189,828	−4189,748	−4189,828	−4189,818	−4189.82
	Worst	−4189,598	−2650,022	−2662,835	−2664,699	−2345.75
	Medium	−4189,793	−3411,158	−3531,479	−3597,201	−3798.93
f_7	Best	−391.6616	−391.6587	−391.6614	−391.6557	−391,659
	Worst	−391.6549	−320.6448	−334.3338	−333.9817	−391,619
	Medium	−391.6604	−370.3126	−383.0364	−378.7673	−391,647
f_8	Best	4.499357	10103.10	8090,126	9829,655	45.7919
	Worst	8.584559	35568.96	42038.54	34923.42	22127.7
	Medium	6.389910	19309.57	23440.08	22512.79	5920.59

Table 1.19 Results of minimization of test functions (DDUI check) ($D = 50$)

F.	The cost	DDUI	RS	LHS	OBL	CI
f_1	Best	3.55E-15	0	0	3.55E-15	3.55E-15
	Worst	7.11E-15	1.42E-14	1.42E-14	1.42E-14	1.42E-14
	Medium	5.15E-15	6.04E-15	6.39E-15	8.70E-15	7.46E-15
f_2	Best	−1	0	0	0	−1
	Worst	−0.215439	0	0	0	0
	Medium	−0.855592	0	0	0	−0.10789
f_3	Best	0	0	0	0	0
	Worst	0	0.312606	0.224252	0	0.25278
	Medium	0	0.015631	0.011213	0	0.01264
f_4	Best	0	0	0	0	0
	Worst	0	5.68E-14	0	5.68E-14	5.68E-14
	Medium	0	2.84E-15	0	2.84E-15	2.84E-15
f_5	Best	1.21E-05 0.363177	46.94658	47.25961	47.57934 48.52713	4.90902
	Worst	0.022933	48.48067	48.51746	47.89927	48.5143
	Medium		47.94278	47.87552		45.7317
f_6	Best	−20949.14	−20945.48	−20947.02	−20947.64	−20948.5
	Worst	−20946.24	−12764.78	−14170.39	−14634.68	−14999.6
	Medium	−20948.70	−17584.73	−18355.21	−18933.11	−18982.1
f_7	Best	−1958,308	−1958,258	−1958,181	−1958,084	−1958.28
	Worst	−1958,237	−1457,922	−1591,349	−1552,201	−1956,58
	Medium	−1958,294	−1915,232	−1886,188	−1834,910	−1957.87
f_8	Best	201.2862	129088.2	149936.4	133597.9	23044.3
	Worst	215.3145	187324.9	223126.8	210587.4	159037
	Medium	208.6043	156493.7	176492.5	168902.9	52760.3

are aimed only at assessing the effect of the initial estimation method, the selected 200 iterations are sufficient.

Starting from the case of ten dimensions, as a result of minimizing the functions 100 times, using the DDUI, it was possible to find the minimum value of six out of eight functions. And in case of the remaining f_5 and $f_8$functions, it was possible to find a solution tens of times closer to the optimum using the DDUI than using all other approaches. The graphs of the general convergence rates of all methods for the case $D = 10$ are given in Fig. 1.30, which shows that optimal solutions are also found much earlier using the DDUI, and in fact the search can be stopped without wasting additional computational resources. The same statements regarding efficiency are also applicable to the worst and average results. Next, in case of 50 dimemsions of the plane, complications arise with regard to detecting the minimum of the functions f_5 and f_8. However, the solutions obtained by the DDUI surpass the other methods in effectiveness.

It should be noted that in the case of $D = 50$, the global minimum for the Ackley function was not found, but considering the worst and average results, the performance of the DDUI again exceeds that of the others. In the last 100-dimensional plane optimization problem, the results replicate those of the previous cases.

As a result, in all the considered scenarios, the proposed DDUI method outperforms other approaches and can be applied in the context of the IC parametric

Table 1.20 Results of minimization of test functions (DDUI check) ($D = 100$)

F.	The cost	DDUI	RS	LHS	OBL	CI
f_1	Best	3.55E−15	3.55E-15	0	0	3.55E-15
	Worst	1.42E-14	1.42E-14	1.42E-14	1.42E-14	7.11E-15
	Medium	7.11E-15	7.28E-15	5.33E-15	6.04E-15	5.86E-15
f_2	Best	-1	0	0	0	−0.99997
	Worst	−0.003422	0	0	0	0
	Medium	−0.682719	0	0	0	−0.39638
f_3	Best	0	0	0	0	0
	Worst	0	0	0	0	0
	Medium	0	0	0	0	0
f_4	Best	0	0	0	0	0
	Worst	0	2.27E-13	1.14E-13	1.14E-13	0
	Medium	0	1.14E-14	5.68E-15	5.68E-15	0
f_5	Best	1.45E-07	97.34662	97.80314	97.31989	97.2842
	Worst	2.525692	98.15001	98.12555	98.12313	98.1099
	Medium	0.131511	97.83444	97.96993	97.89779	97.8350
f_6	Best	−41898.28	−41897.52	−41892.49	−41894.94	−41898.2
	Worst	−41895.67	−28071.92	−24079.93	−23955.85	−30007.9
	Medium	−41897.71	−38405.96	−34462.54	−36111.84	−38064.6
f_7	Best	−3916,616	−3916,149	−3916,026	−3916,548	−3916.37
	Worst	−3916,493	−2983,068	−3506,109	−3005,931	−3912.43
	Medium	−3916,594	−3756,602	−3865,889	−3742,011	−3915.02
f_8	Best	408.2037	282207.7	276021.2	277208.6	39730.8
	Worst	428.5733	394584.2	376304.1	394964.3	590361
	Medium	417.9449	331750.8	333909.9	338027.5	150988

optimization problem. In particular, the circuits presented in Figs. 1.24, 1.25, and 1.26 were optimized, and the imposed constraints are the same as in the case of the algorithms of the previous subsection (Tables 1.12, 1.14, and 1.16).

For all three circuits (two-stage OpAmp, HSD, OpAmp with embedded cascode), the optimization was carried out 100 times, using the WOA algorithm and the objective function (1.4), using populations containing 200 individuals within 20 iterations (Tables 1.21, 1.22, and 1.23). The same initial evaluation approaches as in the case of function minimization were considered as DDUI "competitors." It should be noted that the total number of simulations performed in the case of OBL exceeds the number of other approaches by 200, due to the evaluation of the opposing points.

The results of the optimization of the two-cascade OpAmp are summarized in Fig. 1.21, from which it can be seen that the efficiency of the DDUI exceeded the other approaches by 2.64% on average, 2.17% in the best case, and 6.06% in the worst case; therefore, in the generalized case, it showed 3.62% higher productivity. The parametric optimization data of HSD are presented in Fig. 1.22, where the efficiency increase for the above criteria is 109.05%, 1.01%, and 131.02%, respectively, which gives a high effective probability of detecting an individual with low fitness of 69.96%.

The optimization results of the last testing circuit, the OpAmp with embedded cascode, are presented in Fig. 1.23, where the proposed method, albeit slightly, showed itself with higher efficiency. In particular, the DDUI was able to find the

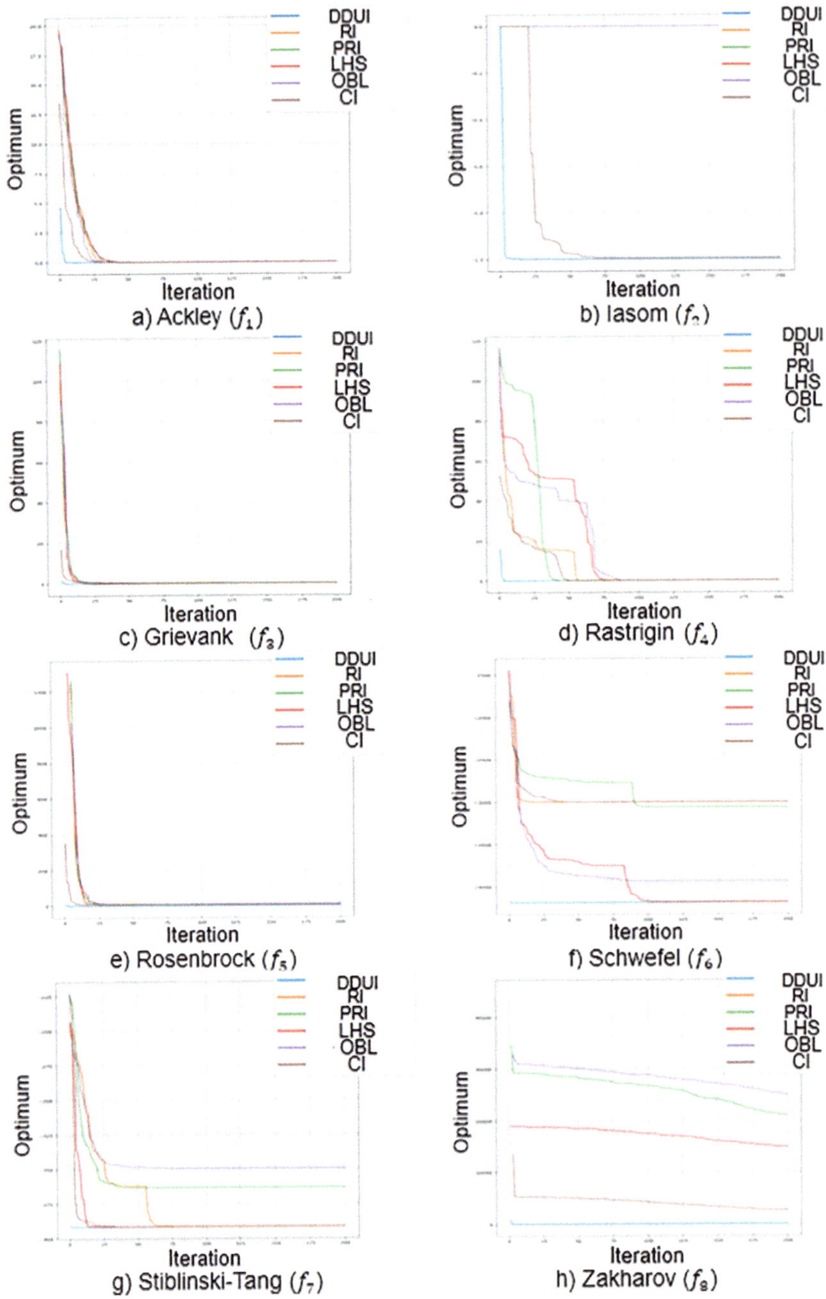

Fig. 1.30 Convergence rates of minimization of testing functions for case $D = 10$

Table 1.21 Optimization results of a two-stage OpAmp (DDUI check)

Optimization algorithm	Solution	Gain	PM	Gain Margin (GM)	Bandwidth	Adaptability
DDUI	Best	83.82	63,998	−22,418	546.0	10.33
	Worst	62,581	66.0	−26,176	440.8	11.15
	Medium	–	–	–	–	10,717
RS	Best	74,434	69,175	−28,159	435.8	10.85
	Worst	64,011	53,531	−21,968	350.4	11,704
	Medium	–	–	–	–	11,431
LHS	Best	73,063	67,532	−23.67	549.8	10,493
	Worst	61,861	49,477	−49,583	500.1	13,087
	Medium	–	–	–	–	10,846
OBL	Best	77,102	65,021	−22.16	548.8	10,433
	Worst	62,603	68,732	−31,031	362.1	11,524
	Medium	–	–	–	–	10.78
CI	Best	78,303	62,392	−22,277	549.0	10,455
	Worst	63,225	67,993	−31.26	446.6	11,164
	Medium	–	–	–	–	10.97

best solution among the considered methods, as well as obtain the best average adaptability, but its worst result from the ten optimizations performed is inferior to the OBL. Despite this, in this case, the DDUI result is preferable, since, as mentioned, the OBL uses 200 more simulations, which may be unacceptable under certain limitations. As a result, the DDUI in case of the given IC, in the generalized case, showed 37.34% higher efficiency.

As a result, taking into account the results of function minimization and IC optimization, the proposed approach to initializing the OA of the GI outperforms the other considered algorithms by an average of 36.97%, while not depending on random factors and not requiring additional computational resources. The main limitation of the algorithm is the lack of the ability to sample an arbitrary number of points, which can be an obstacle in problems requiring a strictly defined number of points.

1.2.3 Method to Reduce the Number of Calls to the Simulation Tool Implemented within the Framework of Parametric Optimization

The main drawback of the first of the systems presented above was the low accuracy of the predicted data of the ANN when considering the entirety of the search domain. Very often, even the data of thousands of simulations are not enough for the ANN to form a complete picture of the behavior of the observed IC. On the other hand, the second system considered uses the ANN to predict only the local neighborhood of the best individual of the current iteration. However, in this case, hundreds of points are sampled and simulated in the vicinity of the individual at each stage, which is realistically possible only with the availability of a huge amount of computational

Table 1.22 Results of HSD optimization (DDUI check)

Optimization algorithm	Solution	(1); (2) (3); (4)	(5); (6) (7); (8)	(9); (10) (11); (12)	(13); (14) (15); (16)	Adaptability
DDUI	Best Worst Medium	29.06; 31.91 9.63; 9.01/ 29.18; 31.68 9.63; 9.01/ –	34.95; 31.28 8.48; 8.59/ 33.41; 30.64 7.67; 7.8/ –	34.94; 32.99 9.16; 9.95/ 36.0; 30.36 9.98; 9.09/ –	32.69; 32.89 8.78; 8.05/ 32.66; 31.47 8.17; 6.91/ –	40,799 75,209 48,734
RS	Best Worst Medium	29.43; 32.3 9.8; 9.4/ 30.06; 33.91 9.73; 9.59/ –	34.67; 31.96 6.27; 7.85/ 39.11; 33.57 9.87; 9.42/ –	29.38; 34.59 9.78; 9.98/ 32.58; 33.88 9.76; 9.58/ –	29.27; 33.93 7.07; 7.54/ −82.6; 31.35 9.29; 7.48/ –	41,251 1061,954 278,152
LHS	Best Worst Medium	29.16; 33.07 9.82; 9.84/ 29.47; 33.56 9.82; 9.83/ –	34.98; 29.14 5.69; 5.86/ 42.56; 30.34 9.92; 9.21/ –	31.31; 29.02 9.98; 9.95/ 31.09; 29.17 7.36; 7.62/ –	29.2; 34.94 6.08; 9.7/ −82.7; 34.14 9.95; 6.27/ –	41,559 1064,002 326,717
OBL	Best Worst Medium	29.79; 33.31 9.88; 9.82/ 29.52; 32.85 9.63; 9.67/ –	33.38; 31.48 8.68; 8.45/ 31.24; 34.92 6.96; 9.97/ –	34.69; 29.07 9.67; 7.52/ 29.25; 31.42 7.72; 9.94/ –	32.69; 34.35 9.19; 9.31/ 32.7; 42.17 7.78; 9.12/ –	40,891 77,761 53,944
CI	Best Worst Medium	29.15; 33.38 9.93; 9.96/ 29.2; 32.62 9.51; 7.86/ –	34.74; 33.65 6.28; 8.23/ 43.38; 29.61 9.93; 9.09/ –	34.78; 30.51 9.08; 8.19/ 29.91; 29.3 7.34; 8.42/ –	29.3; 32.5 6.92; 7.4/ −82.6; 29.39 9.97; 8.56/ –	41,159 1064,691 147,625

resources. For this reason, it is proposed to introduce a GAN system in the training data collection phase of the ANN, in conjunction with the SCC, the purpose of which is to automatically generate the remaining necessary data, having only a certain amount of already simulated data [98].

A GAN [99] consists of two competing multilayer neural networks, one of which is called a generator (G) which aims to generate new realistic data, and the other one is called a discriminator (D) which evaluates the generated data using some scalar output value. Both components of the GAN are trained in parallel on a limited set of data x. During this time, the generator tries to modify its internal p_g distribution so that when applied to the input $p_z(z)$ noise (in other words, the hidden domain), the

Table 1.23 Optimization results of OpAmp with embedded cascode (DDUI check)

Optimization Algorithm	Solution	Gain	PM	Gain Margin (GM)	Bandwidth	Adaptability
DDUI	Best	75,611	84,317	−31,506	83,373	11.21
	Worst	31,772	89.6	−48,673	57,176	545,005
	Medium	–	–	–	–	148,549
RS	Best	70,349	72,458	−45,762	62,367	12,674
	Worst	−63,854	178,218	0.002	0.004	1561,039
	Medium	–	–	–	–	412,427
LHS	Best	73,862	79,184	−43.98	40,991	12,904
	Worst	72,397	79,018	−31,114	17,787	549,116
	Medium	–	–	–	–	247,681
OBL	Best	71,181	86,282	−35,256	63,696	11,698
	Worst	70,031	78,955	−23.01	20,756	513,846
	Medium	–	–	–	–	257,634
CI	Best	70,149	88,884	−42,617	126,624	11,248
	Worst	80,367	87,398	−40,275	5793	785,163
	Medium	–	–	–	–	347,534

output will have $p_{\text{data}}(x)$ results, close to the distribution of the actual data. The discriminator, in turn, tries to maximize the chance of making a correct attribution by returning the probability of the received data being real at each time point using a sigmoid activation function. In other words, a two-player minimax game takes place between *D* and *G* with the following value function *V(G, D)*:

$$\min_{G}\max_{D} V(G,D) = E_{x\sim p_{\text{data}}(x)}\left[\log D(x)\right] + E_{z\sim p_z(z)}\left[\log\left(1 - D\left(G(z)\right)\right)\right] \quad (1.8)$$

Ideally, the networks are trained until $p_g = p_{\text{data}}$ condition is met, i.e. *G* has no more room for improvement and *D* is unable to distinguish the input data: $D(x) = 1/2$. The networks are trained using the error backpropagation algorithm, using the following loss functions:

$$L_D = \nabla_{\theta_d} \frac{1}{m}\sum_{i=1}^{m}\left[\log D\left(x^{(i)}\right) + \log\left(1 - D\left(G\left(z^{(i)}\right)\right)\right)\right] \quad (1.9)$$

$$L_G = \nabla_{\theta_g} \frac{1}{m}\sum_{i=1}^{m}\log\left(1 - D\left(G\left(z^{(i)}\right)\right)\right) \quad (1.10)$$

The described GAN system has some significant problems, the main ones being the following [100]:

- The optimal discriminator returns efficient data for training the generator, but if the latter's "learning" rate lags behind, p_g begins to differ too much from p_{data}, this leads to gradient values approaching zero and the generator stops "learning" anything.

- The existing generator loss function has large deviations in the range of gradient values, which can lead to unstable behavior of the entire model.

For this reason, in the automated optimization system of ICs, it is proposed to use Wasserstein GAN with an embedded gradient penalty (WGAN-GP) [98], [100, 101]. The main difference lies in the calculation of the loss, where the Wasserstein distance is used, which has a smoother gradient everywhere. Due to this, the overall system continues to "learn" regardless of the generator's productivity, and the discriminator, instead of the probability, gives a special numerical estimate at the output using a "linear" activation function, becoming a "judging" ANN. To stabilize the system's behavior, a gradient penalty component is added to the calculations, which "punishes" the model if its gradient norm moves from the target value of 1. As a result, combining all the factors, the discriminator's loss function looks like this:

$$L_D = \nabla_w \frac{1}{m} \sum_{i=1}^{m} \left[D_w\left(G_\theta(z)\right) - D_w(x) + \lambda\left(\left\|\nabla_{\hat{x}} D_w(\hat{x})\right\|_2 - 1\right)^2 \right] \tag{1.11}$$

where $\hat{x} \leftarrow \epsilon x + (1-\epsilon)G_\theta(z)$, in which ϵ is a parameter uniformly sampled from the interval [0,1], w and θ are the trained parameters of generator and discriminator networks respectively, $x \sim P_r$ is a sample of real data and $z \sim p(z)$ is a point taken from the hidden domain, λ is the penalty coefficient, the value of which is usually assigned to 10.

The loss function of generator ANN is calculated on $\left\{z^{(i)}\right\}_{i=1}^{m} \sim p(z)$ group of hidden points using the following formula:

$$L_G = \nabla_\theta \frac{1}{m} \sum_{i=1}^{m} -D_w\left(G_\theta(z)\right) \tag{1.12}$$

In addition to the above, it is proposed to integrate a mechanism for calculating the correlation coefficient matrix into the additional simulation data generation system [98]. Its purpose is to conduct a sensitivity analysis of parameters by examining the available simulated data. This will allow identifying positive or negative dependencies between parameters under conditions of changing values, expressed as a two-dimensional array of coefficients. These coefficients, by processing the data obtained sequentially at the output of the generator, will eliminate unrealistic candidates. In particular, it is proposed to use the SCC method, since it has proven to be more effective when processing small amounts of data [101]. In SCC, the values are calculated using the following expression:

$$r_s = \rho_{R(X),R(Y)} = \frac{\operatorname{cov}\left(R(X),R(Y)\right)}{\sigma_{R(X)}\sigma_{R(Y)}} \tag{1.13}$$

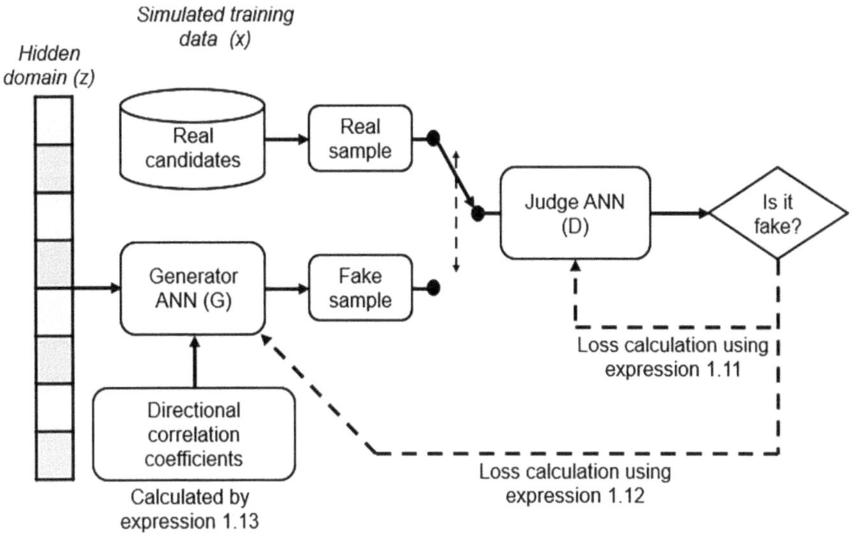

Fig. 1.31 The process of generating data using the WGAN-GT with embedded SCC

where $R(X)$, $R(Y)$ represent arbitrary (X_i, Y_i) pair rank as a result of gradual classification of n sampled and simulated data entries, $\mathrm{cov}(R(X), Y(X))$ corresponds to the covariance of those categorical variables, and $\sigma_{R(X)}$ and $\sigma_{R(Y)}$ show the standard deviation of those same values. The block diagram of an efficient system for generating additional training data by implementing a SCC in a WGAN-GP system is presented in Fig. 1.31 [98].

The developed system can be implemented both in the global domain together with the ANN which replaces the simulation software tool for searching, and in the local domain of an individual for the purpose of generating additional data. Figure 1.32 shows the appearance of the system with the proposed modifications within the framework of the problem of parametric optimization of ICs using GA [98].

Unlike the global search system using the ANN, the proposed components in the local search system presented in Fig. 1.33 are used in each iteration. During the next period, when the simulation is performed for the individuals sampled from the neighborhood of the current best solution, the SCC matrix is built based on the received data, and the GAN is trained. Then, if the required accuracy is ensured, new data is obtained using the generator ANN of GAN for training the predicting ANN. Following this, during the evaluation of the points of the local region of the current individual, depending on the size of the selected neighborhood and the limits of the design parameter values, the number of available coordinates can reach millions.

Evaluating all these points using the ANN is a rather time-consuming process, which can neutralize the time saved by using the GAN. For this reason, an algorithm for limiting the number of points considered is also implemented in the local search

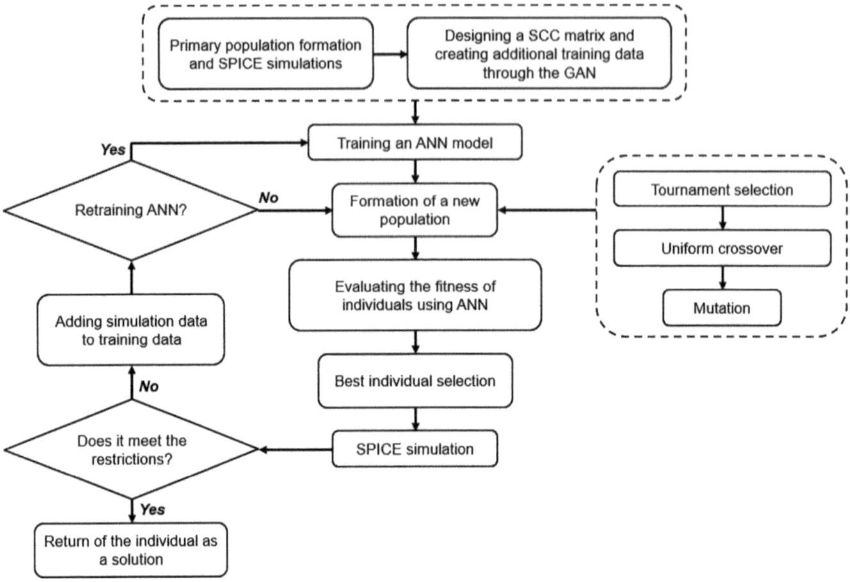

Fig. 1.32 Global search for ANN and GAN with the introduction of the implementation of GA process

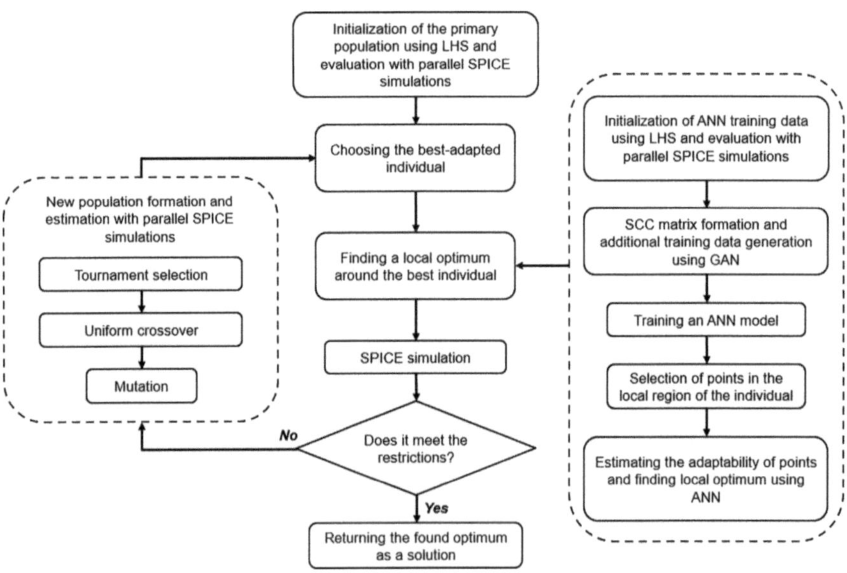

Fig. 1.33 Local search for ANN and GAN with the introduction of the implementation of the GA process

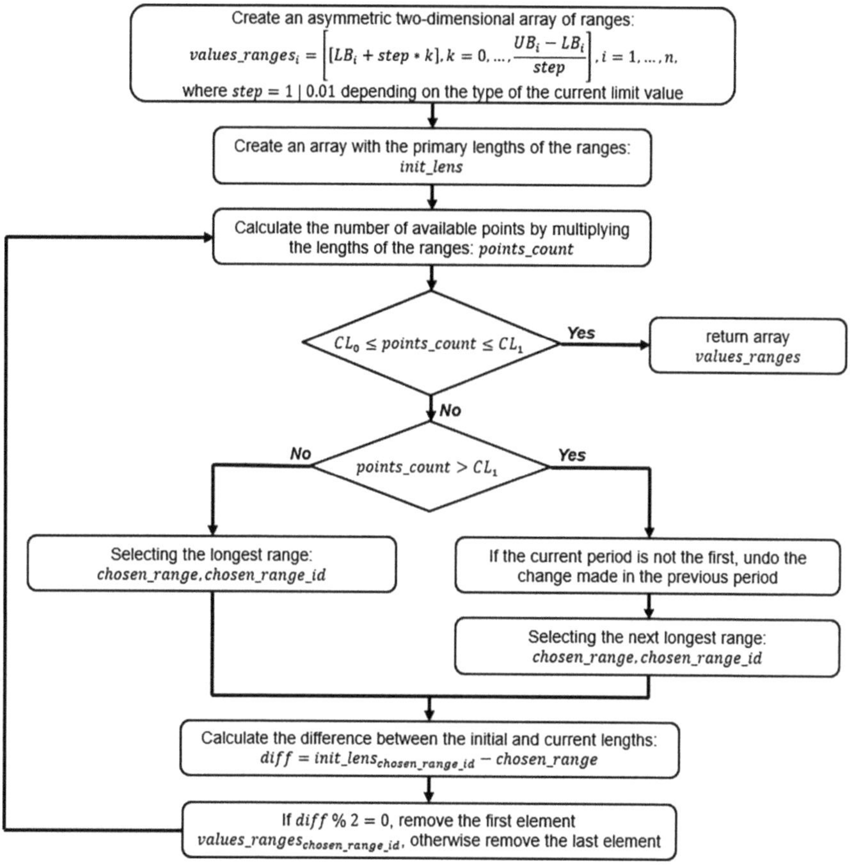

Fig. 1.34 The process of selecting points in the local region of an individual for the prediction of ANN

system, the principle of which is presented in Fig. 1.34. In essence, the approach used sequentially reduces the range of components with the largest number of possible values of IC until the total number of possible values falls within the given limits.

It should be noted that in the local search system, the GAN with its two ANNs and the predictive ANN are retrained from scratch in each iteration of the system operation. In the case of global search, only the predictive ANN is periodically retrained, and only on additional data, preserving the previously existing "knowledge." However, as noted in the previous section, this system is applicable only in cases where the allowable values of the design parameters are strictly limited. Otherwise, without simulating new data, it will not be possible to predict the behavior of individuals with the necessary accuracy, even using the GAN.

The efficiency assessment of the system presented in Fig. 1.33 with the introduction of the proposed WGAN-GP and SCC methods was carried out using the circuits presented in Figs. 1.24, 1.25, and 1.26, respectively, according to the constraints presented in Figs. 1.12, 1.14, and 1.16, and for the formation of individuals, the design rules related to the sizes of the components of the ICs were taken into account. Unlike the system in Fig. 1.32, the approach in Fig. 1.33 will allow, during iterations, through periodic retraining, to form a more generalized statistical picture of the productivity of the developed approach. For all ICs, GA was used as the OA, combining tournament selection, uniform crossover and multipoint mutation, and the optimization was carried out within 20 iterations. The structure of 3 ANNs used, depending on the design parameters and the number of measurements performed, is given in Table 1.24. The following formula for the root mean square logarithmic error was used to calculate the loss during training:

$$L(y,\hat{y}) = \frac{1}{N}\sum_{i=1}^{N}\left(\log(y_i+1)-\log(\hat{y}_i+1)\right)^2 \tag{1.14}$$

In the developed model, the generator ANN, using a sigmoid activation function, outputs data related to the dimensions of IC components, normalized in the range [0, 1]. It also provides the adaptive values obtained by the expression (1.14) of the corresponding measurements, instead of the measured values themselves, which makes the training process easier since the numbers returned by the neurons of the output layer of the ANN are located in the same range. A 5% error threshold was used as the criterion for completing the ANN training and starting the generation of additional data. For local search using the predictive ANN, a range consisting of ±4% of the coordinates was chosen as the boundary of the current individual's surroundings. To improve training efficiency, the predictive ANN receives normalized versions of the design parameter values as input, and in each iteration, the target number of training data entries is taken as ten times the number of independent parameters [85]. The predictive ANN also outputs the adaptive values of the measured parameters. In all three ANNs, the dropout regularization technique was used to avoid overfitting during training.

Table 1.24 Structural descriptions of ANNs used in the developed system

Parameter name	Predictive ANN	Generator ANN	Discriminator ANN
Number of neurons	[8, 18, 36, 14] * D	[10, 22, 40, 16] * D	[10, 22, 40, 16] * D
Inner layers' activation function	LeakyReLU	LeakyReLU, sigmoid + linear	LeakyReLU, sigmoid + linear
ANN OA	Adam	Adam	Adam
OA settings	0.001; 0.9; 0.999	0.0001; 0; 0.9	0.0001; 0; 0.9

Table 1.25 Optimization results of a two-stage OpAmp (WGAN-GT and SCC check)

Parameter name	Received value	Parameter name	Received value
Gain	60,032	Bandwidth	548.2
PM	64.23	Adaptability	10,883
Gain margin (GM)	−22,604		

Table 1.26 Results of HSD optimization (WGAN-GT and SCC check)

Parameter name	Received value	Parameter name	Received value
(1); (2)	30.06; 33.56	(11); (12)	9.97; 9.94
(3); (4)	9.97; 9.95	(13); (14)	32.67; 34.73
(5); (6)	34.95; 29.33	(15); (16)	8.43; 9.92
(7); (8)	8.06; 6.42	Adaptability	41,093
(9); (10)	30.06; 30.2		

Table 1.27 Optimization results of the OpAmp with embedded cascode (WGAN-GT and SCC check)

Parameter name	Received value	Parameter name	Received value
Gain	76.68	Bandwidth	40,085
PM	71,947	Adaptability	12,424
Gain margin (GM)	−30,927		

The optimization results of the two-stage OpAmp are summarized in Fig. 1.25. They show that the developed system was able to find an effective solution, and the GAN was able to identify the behavioral features of the circuit, providing additional training data with a maximum acceptable deviation of 4.92%. During all iterations of the optimization process, the GAN generated an average of 11.25% of the necessary training data, which allowed avoiding hundreds of additional simulations (Table 1.25).

The effective numbers for the parametric optimization of the HSD are given in Fig. 1.26. In this case, it was also possible to find a version of IC that satisfied the proposed constraints, which was also facilitated by the GAN, which generated an average of 10.48% of the training data during the iterations, while ensuring a maximum deviation of 4.75% from the reference data (Table 1.26).

In case of the OpAmp with embedded cascode, the corresponding data are given in Table 1.27. However, due to the complexity of simulating the structure of the circuit and the behavioral features, the use of the GAN has managed to reduce the number of simulations performed in all iterations by an average of only 4.87%, providing a maximum deviation of 5.02%. However, in terms of time saved, the advantages of implementing the GAN will be significant in case of a large number of similar circuits.

As a result, during the parametric optimization of three different ICs, the proposed system for generating additional data for training the predictive ANN, based on the combination of the WGAN-GP and SCC strategies, allowed reducing the number of necessary simulations by an average of 8.86%. However, it is necessary to take into account that with the increase in the number of independent components in the optimized IC, the process of training additional generator and discriminator ANNs also becomes more complicated, which may require structural changes in ANNs related to the number of hidden layers and neurons in them.

Conclusions

1. A hybrid approach based on strategies used in genetic and Harris hawk optimization algorithms was developed, which, using a combination of wheel selection, uniform crossover, and multipoint mutation for exploration, and mathematical models of four techniques adopted by Harris hawks during preying for exploitation phases, outperformed other currently widely used swarm intelligence optimization algorithms in the parametric optimization of three considered ICs, providing an average of 48.11% higher efficiency.
2. A deterministic approach to initializing the primary population of optimization algorithms using group intelligence has been developed, which, considering the existing search domain as a set of equidistant embedded cubes, sequentially samples the peripheral points, without using additional estimates of the objective function, was able to significantly increase the convergence rates of optimization algorithms in the case of optimizing the considered ICs, providing an average of 36.97% higher performance.
3. A system has been designed to replace calls to the simulation software tool during the optimization of integrated circuits with an artificial neural network, which, through the implementation of the Wasserstein generative adversarial network with an embedded gradient penalty and the computational processes of Spearman correlation coefficients, further reduces the number of simulations required to collect training data for the predictive neural network in each optimization iteration, generating 8.86% more data on average, with a maximum deviation of 5.02%.

References

1. R.R. Schaller, Moore's law: past, present and future. IEEE Spectr. **34**(6), 52–59 (1997). https://doi.org/10.1109/6.591665
2. T.P. Dash, S. Dey, E. Mohapatra, et al., Vertically-stacked silicon nanosheet field effect transistors at 3 nm technology nodes, in *2019 Devices for Integrated Circuit (DevIC)*, (IEEE, 2019), pp. 99–103. https://doi.org/10.1109/DEVIC.2019.8783300
3. J.R. Baker, *CMOS Circuit Design, Layout, and Simulation*, 3rd edn. (John Wiley & Sons, 2010), 920p
4. B. Razavi, *Design of Analog CMOS Integrated Circuits*, 2nd edn. (Mc Graw Hill, 2017), 625p

5. A. Boora, B.K. Thangarasu, K.S. Yeo, An ultra-low power 900 MHz intermediate frequency low noise amplifier for low-power RF receivers, in *2020 IEEE 33rd International System-on-Chip Conference (SOCC)*, (IEEE, 2020), pp. 163–167. https://doi.org/10.1109/SOCC49529.2020.9524753
6. A. Skillman, T. Edsö, A technical overview of Cortex-M55 and Ethos-U55: arm's most capable processors for endpoint AI, in *2020 IEEE Hot Chips 32 Symposium (HCS)*, (IEEE, 2020), pp. 1–20. https://doi.org/10.1109/HCS49909.2020.9220415
7. Y. Abdulrahman, Artificial Intelligence, 5G, and IoT, in *Intelligent Connectivity: AI, IoT, and 5G*, (Wiley, 2022), pp. 251–268. https://doi.org/10.1002/9781119685265.ch14
8. C. Zhang, Intelligent Internet of things service based on artificial intelligence technology, in *2021 IEEE 2nd International Conference on Big Data, Artificial Intelligence and Internet of Things Engineering (ICBAIE)*, (IEEE, 2021), pp. 731–734. https://doi.org/10.1109/ICBAIE52039.2021.9390061
9. B. Asmika, G. Mounika, P.S. Rani, Deep learning for vision and decision making in self driving cars-challenges with ethical decision making, in *2021 International Conference on Intelligent Technologies (CONIT)*, (IEEE, 2021), pp. 1–5. https://doi.org/10.1109/CONIT51480.2021.9498342
10. L. Bai, Z. Zhu, Z. Liu, L. Chen, et al., Equivalent circuit and parasitic parameter extraction of 3D comb-like redistribution layer applied for FOWLP, in *2022 23rd International Conference on Electronic Packaging Technology (ICEPT)*, (IEEE, 2022), pp. 1–5. https://doi.org/10.1109/ICEPT56209.2022.9873315
11. Y. Luo et al., Influence of parasitic capacitance and resistance on performance of 6T-SRAM for advanced CMOS circuits design, in *2022 China Semiconductor Technology International Conference (CSTIC)*, (IEEE, 2022), pp. 1–3. https://doi.org/10.1109/CSTIC55103.2022.9856731
12. Z. Abedi, S. Hemmady, T. Antonsen, E. Schamiloglu, P. Zarkesh-Ha, Electromagnetic compatibility in leakage current of CMOS integrated circuits, in *2019 International Symposium on Electromagnetic Compatibility*, (EMC EUROPE, 2019), pp. 765–768. https://doi.org/10.1109/EMCEurope.2019.8872123
13. A. Yadav, Optimizing leakage current and fluctuations at ground rail in 1-bit 8T full adder circuit using various MTCMOS techniques, in *2022 4th International Conference on Inventive Research in Computing Applications (ICIRCA)*, (IEEE, 2022), pp. 174–179. https://doi.org/10.1109/ICIRCA54612.2022.9985706
14. N. Ginzberg, T. Gidoni, D. Regev, E. Cohen, Transmit-receive cross-modulation distortion correction in a 5-6 GHz full duplex quadrature balanced CMOS RF front-end, in *2020 IEEE/MTT-S International Microwave Symposium (IMS)*, (IEEE, 2020), pp. 1211–1214. https://doi.org/10.1109/IMS30576.2020.9224029
15. J. Wei, D. Li, J. Qiao, L. Zhao, Analysis and a reduction method of temporal noise in the CMOS image sensor readout chain, in *2018 IEEE International Conference on Integrated Circuits, Technologies and Applications (ICTA)*, (IEEE, 2018), pp. 128–129. https://doi.org/10.1109/CICTA.2018.8705719
16. Z. Zhang, H. Qiu, J. Wang, Research on eliminating parasitic oscillation of inverter by using PI and repetitive control, in *2016 IEEE International Conference on Electronic Information and Communication Technology (ICEICT)*, (IEEE, 2016), pp. 321–324. https://doi.org/10.1109/ICEICT.2016.7879708
17. Y. Wen, J. Dean, B.A. Floyd, P.D. Franzon, High dimensional optimization for electronic design, in *2022 ACM/IEEE 4th Workshop on Machine Learning for CAD (MLCAD)*, (IEEE, 2022), pp. 153–157. https://doi.org/10.1109/MLCAD55463.2022.9900104
18. M. Stănescu et al., Multi-objective optimization algorithms for automated circuit sizing of analog/mixed-signal circuits, in *2021 International Semiconductor Conference (CAS)*, (IEEE, 2021), pp. 117–120. https://doi.org/10.1109/CAS52836.2021.9604128

19. A. Brik, L. Labrak, L. Carrel, et al., Fast extraction of predictive models for integrated circuits using n-performance Pareto fronts, in *2019 IFIP/IEEE 27th International Conference on Very Large Scale Integration (VLSI-SoC)*, (IEEE, 2019), pp. 281–286. https://doi.org/10.1109/VLSI-SoC.2019.8920305
20. L. Lamrini, M.C. Abounaima, M.T. Alaoui, et al., A Filtering approach used in a massive data context to reduce the set of choices in a multicriteria decision aid process: Pareto solutions, in *2020 IEEE 2nd International Conference on Electronics, Control, Optimization and Computer Science (ICECOCS)*, (IEEE, 2020), pp. 1–6. https://doi.org/10.1109/ICECOCS50124.2020.9314445
21. G.C. Da Silva, E.F. Wanner, L.C.T. Bezerra, T. Stützle, Revisiting Pareto-optimal multi- and many-objective reference fronts for continuous optimization, in *2021 IEEE Congress on Evolutionary Computation (CEC)*, (IEEE, 2021), pp. 1171–1178. https://doi.org/10.1109/CEC45853.2021.9504952
22. S. Tanaka, K. Takadama, H. Sato, Impacts of single-objective landscapes on multi-objective optimization, in *2022 IEEE Congress on Evolutionary Computation (CEC)*, (IEEE, 2022), pp. 1–8. https://doi.org/10.1109/CEC55065.2022.9870226
23. M.K. Afifi, H.M. Sharaf, M.M. Sayed, D.K. Ibrahim, Comparative study between single-objective and multi-objective optimization approaches for directional overcurrent relays coordination considering different fault locations, in *2019 IEEE Milan PowerTech*, (IEEE, 2019), pp. 1–6. https://doi.org/10.1109/PTC.2019.8810815
24. R.A. De Lima Moreto, C.E. Thomaz, S.P. Gimenez, Gaussian fitness functions for optimizing analog CMOS integrated circuits. IEEE Trans. Computer-Aided Des. Integr. Circ. Syst. **36**(10), 1620–1632 (2017). https://doi.org/10.1109/TCAD.2017.2661804
25. S. Jeloka et al., System technology co-optimization and design challenges for 3D IC, in *2022 IEEE Custom Integrated Circuits Conference (CICC)*, (IEEE, 2022), pp. 1–6. https://doi.org/10.1109/CICC53496.2022.9772831
26. М.М. Гурарий, М.М. Жаров, Н.Н. Прокопенко, et al., Применение методов оптимизации при проектировании аналоговых схем радиоэлектроники. Журнал радиоэлектроники [электронный журнал]. **4** (2018). https://doi.org/10.30898/1684-1719.2018.4.6
27. P. Gupta, S. Gourishetty, H. Mandadapu, Z. Abbas, PVT variations aware robust transistor sizing for power-delay optimal CMOS digital circuit design, in *2019 IEEE International Symposium on Circuits and Systems (ISCAS)*, (IEEE, 2019), pp. 1–5. https://doi.org/10.1109/ISCAS.2019.8702068
28. J. Tisza, M. Chauca, Strategies for optimization of time delay in nanoelectronics circuits with process variations effects, in *2019 IEEE Sciences and Humanities International Research Conference (SHIRCON)*, (IEEE, 2019), pp. 1–4. https://doi.org/10.1109/SHIRCON48091.2019.9024886
29. J. Liu et al., From specification to silicon: towards analog/mixed-signal design automation using surrogate NN models with transfer learning, in *2021 IEEE/ACM International Conference on Computer Aided Design (ICCAD)*, (IEEE, 2021), pp. 1–9. https://doi.org/10.1109/ICCAD51958.2021.9643445
30. Q. Li, S.-Y. Liu, X.-S. Yang, Influence of Initialization on the Performance of Metaheuristic Optimizers. arXiv:2003.03789. https://doi.org/10.48550/arXiv.2003.03789
31. M.A. Muñoz, S. Yuan, K. Michael, S.K. Halgamuge, Algorithm selection for black-box continuous optimization problems: a survey on methods and challenges. Inf. Sci. **317**, 224–245 (2015). https://doi.org/10.1016/j.ins.2015.05.010
32. M. Abdel-Basset, L. Abdel-Fatah, A.K. Sangaiah, et al., Metaheuristic algorithms: a comprehensive review, in *Computational Intelligence for Multimedia Big Data on the Cloud with Engineering Applications*, (Academic Press, 2018), pp. 185–231. https://doi.org/10.1016/B978-0-12-813314-9.00010-4

33. W. Wong, C.I. Ming, A review on metaheuristic algorithms: recent trends, benchmarking and applications, in *2019 7th International Conference on Smart Computing & Communications (ICSCC)*, (IEEE, 2019), pp. 1–5. https://doi.org/10.1109/ICSCC.2019.8843624
34. А.П. Карпенко, Современные алгоритмы поисковой оптимизации. Алгоритмы вдохновленные природой. 2-е издание. Московский государственный технический университет им. Н.Э. Баумана (национальный исследовательский университет, 2016), 448с
35. X. Chen, G. Liu, N. Xiong, et al., A survey of swarm intelligence techniques in VLSI routing problems. IEEE Access **8**, 26266–26292 (2020). https://doi.org/10.1109/ACCESS.2020.2971574
36. M.K. Hussein, M.H. Mousa, Efficient task offloading for IoT-based applications in fog computing using ant colony optimization. IEEE Access **8**, 37191–37201 (2020). https://doi.org/10.1109/ACCESS.2020.2975741
37. K.K. Arpan, Bio inspired computing – a review of algorithms and scope of applications. Expert Syst. Appl. **59**, 20–32 (2016). https://doi.org/10.1016/j.eswa.2016.04.018
38. M. Sarhani, S. Voß, R. Jovanovic, Initialization of metaheuristics: comprehensive review, critical analysis, and research directions. Int. Trans. Oper. Res. **30**, 3361–3397 (2022). https://doi.org/10.1111/itor.13237
39. M. Fayazi, Z. Colter, E. Afshari, R. Dreslinski, Applications of artificial intelligence on the modeling and optimization for analog and mixed-signal circuits: a review. IEEE Trans. Circ. Syst. I Regul. Pap. **68**(6), 2418–2431 (2021). https://doi.org/10.1109/TCSI.2021.3065332
40. A. Mohammadi, S.H. Zahiri, Analysis of swarm intelligence and evolutionary computation techniques in IIR digital filters design, in *2016 1st Conference on Swarm Intelligence and Evolutionary Computation (CSIEC)*, (2016), pp. 64–69. https://doi.org/10.1109/CSIEC.2016.7482117
41. T. Yerriswamy, G. Murtugudde, C.I. Ming, Study of evolutionary techniques in the field of network security, in *2020 International Conference on Smart Technologies in Computing, Electrical and Electronics (ICSTCEE)*, (IEEE, 2020), pp. 594–598. https://doi.org/10.1109/ICSTCEE49637.2020.9277082
42. P. Saha, S. Ahmed, H.S. Kalluru, Z. Abbas, Low power PVT-aware transistor sizing and approximate design generation for standard cells using swarm intelligence, in *2021 IEEE International Symposium on Circuits and Systems (ISCAS)*, (IEEE, 2021), pp. 1–5. https://doi.org/10.1109/ISCAS51556.2021.9401794
43. A. Chordia, S. Hemaram, J.N. Tripathi, A swarm intelligence based automated framework for variability analysis, in *2020 IEEE MTT-S Latin America Microwave Conference (LAMC 2020)*, (IEEE, 2021), pp. 1–3. https://doi.org/10.1109/LAMC50424.2021.9602347
44. A. Ahmid, T.-M. Dao, L.Ê. Ngan, Comparison study of discrete optimization problem using meta-heuristic approaches: a case study. Int. J. Ind. Eng. Oper. Manag. (2019). https://doi.org/10.46254/j.ieom.20190201
45. Ш.Р. Давронов, Обзор современных генетических алгоритмов и их применение на практике. Молодой ученый. **36**(483), 15–18 (2023). https://moluch.ru/archive/483/105878/
46. A. Shukla, H.M. Pandey, D. Mehrotra, Comparative review of selection techniques in genetic algorithm, in *2015 International Conference on Futuristic Trends on Computational Analysis and Knowledge Management (ABLAZE)*, (IEEE, 2015), pp. 515–519. https://doi.org/10.1109/ABLAZE.2015.7154916
47. P. Kora, P. Yadlapalli, Crossover operators in genetic algorithms: a review. Int. J. Comput. Appl. (2017). https://doi.org/10.5120/ijca2017913370
48. A. Hassanat, K. Almohammadi, E. Alkafaween, E. Abunawas, et al., Choosing mutation and crossover ratios for genetic algorithms—a review with a new dynamic approach. Information **10**(12), 390 (2019). https://doi.org/10.3390/info10120390

49. R. Zhou, P. Poechmueller, Y. Wang, An analog circuit design and optimization system with rule-guided genetic algorithm. IEEE Trans. Computer-Aided Des. Integr. Circ. Syst. **41**(12), 5182–5192 (2022). https://doi.org/10.1109/TCAD.2022.3166637
50. Z. Yudong, W. Shuihua, J. Genlin, A comprehensive survey on particle swarm optimization algorithm and its applications. Math. Prob. Eng. **2015**, 931256 (2015). https://doi.org/10.1155/2015/931256
51. S. Yarat, S. Senan, Z. Orman, A comparative study on PSO with other metaheuristic methods, in *Applying Particle Swarm Optimization. International Series in Operations Research & Management Science*, vol. 306, (Springer, Cham, 2021), pp. 49–72. https://doi.org/10.1007/978-3-030-70281-6_4
52. R. Rashid, N. Nambath, Hybrid particle swarm optimization algorithm for area minimization in 65 nm technology, in *2021 IEEE International Symposium on Circuits and Systems (ISCAS)*, (IEEE, 2021), pp. 1–5. https://doi.org/10.1109/ISCAS51556.2021.9401139
53. S. Mirjalili, A. Lewis, The whale optimization algorithm. Adv. Eng. Softw. **95**, 51–57 (2016). https://doi.org/10.1016/j.advengsoft.2016.01.008
54. F. Wantao, Convergence analysis of whale optimization algorithm. J. Phys. Conf. Ser. **1757**, 012008 (2021). https://doi.org/10.1088/1742-6596/1757/1/012008
55. F.S. Gharehchopogh, H. Gholizadeh, A comprehensive survey: whale optimization algorithm and its applications. Swarm Evolut. Comput. **48**, 1–24 (2019). https://doi.org/10.1016/j.swevo.2019.03.004
56. K. Wang, X. Jiang, Y. Jiang, Sensor selection method for target tracking based on hybrid binary whale optimization algorithm in wireless sensor networks, in *2019 IEEE 5th International Conference on Computer and Communications (ICCC)*, (IEEE, 2019), pp. 592–596. https://doi.org/10.1109/ICCC47050.2019.9064434
57. F. Hasan, F. Ahmad, M. Shahid, et al., Solving portfolio selection problem using whale optimization algorithm, in *2022 3rd International Conference on Computation, Automation and Knowledge Management (ICCAKM)*, (IEEE, 2022), pp. 1–5. https://doi.org/10.1109/ICCAKM54721.2022.9990079
58. A. Lberni, M.A. Marktani, A. Ahaitouf, A. Ahaitouf, Adaptation of the whale optimization algorithm to the optimal sizing of analog integrated circuit: low voltage amplifier performances, in *2020 IEEE 2nd International Conference on Electronics, Control, Optimization and Computer Science (ICECOCS)*, (IEEE, 2020), pp. 1–6. https://doi.org/10.1109/ICECOCS50124.2020.9314428
59. K. Devika, G. Jeyakumar, Solving multi-objective optimization problems using differential evolution algorithm with different population initialization techniques, in *2018 International Conference on Advances in Computing, Communications and Informatics (ICACCI)*, (IEEE, 2018), pp. 1–5. https://doi.org/10.1109/ICACCI.2018.8554861
60. V. De Albuquerque Torreão, R. Vimieiro, Effects of population initialization on evolutionary techniques for subgroup discovery in high dimensional datasets, in *2018 7th Brazilian Conference on Intelligent Systems (BRACIS)*, (IEEE, 2018), pp. 25–30. https://doi.org/10.1109/BRACIS.2018.00013
61. A. Ashraf et al., Studying the impact of initialization for population-based algorithms with low-discrepancy sequences. Appl. Sci. **11**(17), 8190 (2021). https://doi.org/10.3390/app11178190
62. Z. Zhao et al., Random sampling over joins revisited, in *Proceedings of the 2018 International Conference on Management of Data*, (IEEE, 2018), pp. 1525–1539. https://doi.org/10.1145/3183713.3183739
63. J.O. Agushaka, A.E. Ezugwu, L. Abualigah, et al., Efficient initialization methods for population-based metaheuristic algorithms: a comparative study. Arch. Computat. Methods Eng. **30**, 1727–1787 (2023). https://doi.org/10.1007/s11831-022-09850-4
64. M.D. Shields, J. Zhang, The generalization of Latin hypercube sampling. Reliab. Eng. Syst. Saf. **148**, 96–108 (2017). https://doi.org/10.1016/j.ress.2015.12.002

65. L. Aibin, L. Wenyi, Reliability evaluation of distribution network with distributed generation based on Latin hypercube sequential sampling, in *2020 3rd International Conference on Electron Device and Mechanical Engineering (ICEDME)*, (IEEE, 2020), pp. 97–99. https://doi.org/10.1109/ICEDME50972.2020.00028
66. R. Sheikholeslami, S. Razavi, Progressive Latin hypercube sampling: an efficient approach for robust sampling-based analysis of environmental models. Environ. Model Softw. **93**, 109–126 (2017). https://doi.org/10.1016/j.envsoft.2017.03.010
67. D. Donovan, K. Burrage, P. Burrage, T.A. McCourt, et al., Estimates of the coverage of parameter space by Latin hypercube and orthogonal array-based sampling. Appl. Math. Model. **57**, 553–564 (2018). https://doi.org/10.1016/j.apm.2017.11.036
68. T. Dongping, Particle swarm optimization with chaos-based initialization for numerical optimization. Intell. Autom. Soft Comput. (2017). https://doi.org/10.1080/10798587.2017.1293881
69. D. Yang, Z. Liu, J. Zhou, Chaos optimization algorithms based on chaotic maps with different probability distribution and search speed for global optimization. Commun. Nonlinear Sci. Numer. Simul. **19**, 1229–1246 (2014). https://doi.org/10.1016/j.cnsns.2013.08.017
70. H. Djellali, N. Ghoualmi, Improved chaotic initialization of particle swarm applied to feature selection, in *2019 International Conference on Networking and Advanced Systems (ICNAS)*, (IEEE, 2019), pp. 1–5. https://doi.org/10.1109/ICNAS.2019.8807837
71. A.M. Algelany, M.A. El-Shorbagy, Chaotic enhanced genetic algorithm for solving the nonlinear system of equations. Comput. Intell. Neurosci. **2022**, 1376479 (2022). https://doi.org/10.1155/2022/1376479
72. C. Min-Yuan, T. Duc-Hoc, C. Minh-Tu, Chaotic initialized multiple objective differential evolution with adaptive mutation strategy (CA-MODE) for construction project time-cost-quality trade-off. J. Civ. Eng. Manag. **22**(2), 210–223 (2016). https://doi.org/10.3846/13923730.2014.897972
73. S. Mahdavi, S. Rahnamayan, K. Deb, Opposition based learning: a literature review. Swarm Evolut. Comput. **39**, 1–23 (2018). https://doi.org/10.1016/j.swevo.2017.09.010
74. T. Bezdan, A. Petrovic, M. Zivkovic, I. Strumberger, et al., Current best opposition-based learning salp swarm algorithm for global numerical optimization, in *2021 Zooming Innovation in Consumer Technologies Conference (ZINC)*, (IEEE, 2021), pp. 5–10. https://doi.org/10.1109/ZINC52049.2021.9499275
75. W. Deng, H. Ni, Y. Liu, et al., An adaptive differential evolution algorithm based on belief space and generalized opposition-based learning for resource allocation. Appl. Soft Comput. **127** (2022). https://doi.org/10.1016/j.asoc.2022.109419
76. S. Shalini, S. Akash, Development and applications of an intelligent crow search algorithm based on opposition based learning. ISA Trans. **129**, 210–230 (2020). https://doi.org/10.1016/j.isatra.2019.09.004
77. https://www.sfu.ca/~ssurjano/optimization.html
78. X. Tang, A. Xu, Multi-class classification using kernel density estimation on K-nearest neighbours. Electron. Lett. **52**(8), 600–602 (2016). https://doi.org/10.1049/el.2015.4437
79. J. Shi, Q. He, Z. Wang, GMM clustering-based decision trees considering fault rate and cluster validity for analog circuit fault diagnosis. IEEE Access **7**, 140637–140650 (2019). https://doi.org/10.1109/ACCESS.2019.2943380
80. T.-S. Wu, C. Alkan, T.W. Chen, Complexity reduction for analog circuit performance models using random forests, in *2009 17th IFIP International Conference on Very Large Scale Integration (VLSI-SoC)*, (IEEE, 2009), pp. 29–34. https://doi.org/10.1109/VLSISOC.2009.6041326
81. R. Zhang, N. Uetake, T. Nakada, Y. Nakashima, Design of programmable analog calculation unit by implementing support vector regression for approximate computing. IEEE Micro. **38**(6), 73–82 (2018). https://doi.org/10.1109/MM.2018.2873953

82. К.И. Фаустова, Нейронные сети: применение сегодня и перспективы развития. Территория науки. **4**, 83–87 (2017)
83. A. Zohrevand, Z. Imani, An empirical study of the performance of different optimizers in the deep neural networks, in *2022 International Conference on Machine Vision and Image Processing (MVIP)*, (IEEE, 2022), pp. 1–5. https://doi.org/10.1109/MVIP53647.2022.9738743
84. A. Elsiginy, E. Azab, M. Elmahdy, Comparative study of evolutionary algorithms for a hybrid analog design optimization with the use of deep neural networks, in *2020 32nd International Conference on Microelectronics (ICM)*, (IEEE, 2020), pp. 1–4. https://doi.org/10.1109/ICM50269.2020.9331809
85. Y. Li, Y. Wang, Y. Li, et al., An artificial neural network assisted optimization system for analog design space exploration. IEEE Trans. Comput.-Aided Des. Integr. Circuits Syst. **39**(10), 2640–2653 (2020). https://doi.org/10.1109/TCAD.2019.2961322
86. D. Kavaliauskas, L. Sakalauskas, Study of convergence in metaheuristics algorithms. Baltic J. Mod. Comput. **7**(3), 436–443 (2019). https://doi.org/10.22364/bjmc.2019.7.3.10
87. A. Izzah, B.A. Nugroho, W.F. Mahmudy, F.A. Bachtiar, et al., Convergence analysis in swarm intelligence for city tour optimization, in *2019 International Seminar on Research of Information Technology and Intelligent Systems (ISRITI)*, (IEEE, 2019), pp. 109–113. https://doi.org/10.1109/ISRITI48646.2019.9034656
88. A. Tharwat, W. Schenck, Population initialization techniques for evolutionary algorithms for single-objective constrained optimization problems: deterministic vs. stochastic techniques. Swarm Evol. Comput. **67** (2021). https://doi.org/10.1016/j.swevo.2021.100952/
89. A. Tharwat, A.E. Hassanien, Chaotic antlion algorithm for parameter optimization of support vector machine. Appl. Intell. **48**, 670–686 (2018). https://doi.org/10.1007/s10489-017-0994-0
90. M. Hossam, E. Emary, C. Grosan, Feature selection via chaotic antlion optimization. PLoS One **11**(3), e0150652 (2016). https://doi.org/10.1371/journal.pone.0150652
91. D. Soekhoe, P. Van der Putten, A. Plaat, On the impact of data set size in transfer learning using deep neural networks. Adv. Intell. Data Anal. XV **9897**, 50–60 (2016). https://doi.org/10.1007/978-3-319-46349-0_5
92. A.V. Vardumyan, H.T. Kostanyan, A Harris Hawks optimizer based hybrid algorithm for optimal design of analog integrated circuits, in *2022 International Conference on Frontiers of Communications, Information System and Data Science (CISDS)*, (IEEE, 2022), pp. 143–147. https://doi.org/10.1109/CISDS57597.2022.00031
93. K. Sourabh, C.S. Singh, K. Vijay, A review on genetic algorithm: past, present, and future. Multimed. Tools Appl. **80**, 8091–8126 (2021). https://doi.org/10.1007/s11042-020-10139-6
94. A.A. Heidari, S. Mirjalili, H. Faris, I. Aljarah, et al., Harris hawks optimization: algorithm and applications. Future Gener. Comput. Syst. **97**, 849–872 (2019). https://doi.org/10.1016/j.future.2019.02.028
95. P. Lemin, C. Zhennao, H. Ali Asghar, et al., Hierarchical Harris hawks optimizer for feature selection. J. Adv. Res. (2023). https://doi.org/10.1016/j.jare.2023.01.014
96. C.S. Savalam, K. Prasanti, A.S. Haranath, Design and implementation of high performance and low power mixed logic line decoders. Int. J. Innov. Technol. Explor. Eng. **8**, 635–640 (2019)
97. A.V. Vardumyan, An efficient primary population initialization method for metaheuristic algorithms. Proc. RA NAS NPUA. Ser. Tech. Sci. **75**(3), 431–442 (2022). https://doi.org/10.53297/0002306X-2022.v75.3-431
98. V.S. Melikyan, A.V. Vardumyan, A.G. Harutyunyan, N.A. Asatryan, et al., System for automated IC design based on generative adversarial and artificial deep neural networks. Proc. Univ. Electron. **28**(5), 612–620 (2023). https://doi.org/10.24151/1561-5405-2023-28-5-612-620

99. I.J. Goodfellow, J. Pouget-Abadie, M. Mirza, B. Xu, et al., Generative adversarial networks. arXiv:1406.2661 (2014)
100. M. Arjovsky, S. Chintala, L. Bottou, Wasserstein generative adversarial networks. arXiv:1701.07875 (2017)
101. P. Radhakrishnan, B. Vignesh, A note on rank correlation and semi-supervised machine learning based measure, in *2017 Innovations in Power and Advanced Computing Technologies (i-PACT)*, (IEEE, 2017), pp. 1–8. https://doi.org/10.1109/IPACT.2017.8245035

Chapter 2
Development of Means to Design Integrated Circuits Taking into Account Aging Phenomena

2.1 General Issues of Developing Means to Design Integrated Circuits, Taking into Account Aging Phenomena

2.1.1 Need for Developing Means for Designing Integrated Circuits, Taking into Account Aging Phenomena

The continued scaling of modern submicron technologies, which have enabled the manufacturing of metal-oxide-semiconductor (MOS) transistors with channel lengths of up to 3 nm, has led to challenges in the design of integrated circuits (ICs) [1–4]. The dramatic increase in the impact of short channel phenomena [5–10] inherent to 14 nm and below technologies has led to a complexity in the design process [11–17].

In ICs, the above-mentioned phenomena lead to deterioration in reliability of the nodes [15–20], an increase in the design time, the number of additional pins in case of introducing control elements, as well as an increase in their occupied area and power consumption, and other negative consequences [21–26]. In general case, not taking into account the above-mentioned factors leads to decrease in the percentage of yield ICs, and therefore, decrease in cost [27, 28].

One of the most important factors affecting the reliability of ICs is the aging phenomenon [29–33], as a result of which circuit parameters deteriorate over time and can even lead to circuit failure [34–36]. Therefore, it is necessary to identify the expected problems at the design stage through the necessary tests and apply preventive actions.

IC aging tests are performed using specific models provided by semiconductor manufacturers and under clearly defined operating conditions, which are +10% deviation from the nominal supply voltage and +125 °C operating temperature [37]. IC aging tests are performed using simulation tools, such as HSPICE [38] and

V. Melikyan, K. Li, *Parametric Optimization Techniques of Integrated Circuits*,
https://doi.org/10.1007/978-3-032-20400-4_2

FineSim [39] developed by Synopsys, LTspice [40] by Analog Devices, Spectre [41] and PSpice [42] by Cadence, and ELDO [43] by Siemens.

In order to solve the identified problems, designers are researching and implementing various methods. There are various ways to avoid the effects of aging phenomena: the effect can be reduced or neutralized by implementing circuit-level solutions [44–47]. In order to neutralize the negative effects of technological deviations, the occurrence of the aging phenomenon can be detected, and the consequences of its impact can be compensated for by means of adjusting pins available in a number of modern ICs [48–51]. However, the addition of each pin is associated with high costs. Therefore, there is a need to design monitoring, detecting, and compensation circuits included in ICs [51–55].

Thus, given the importance of modern ICs in systems such as self-driving cars and ICs implanted in human bodies for medical reasons, failures due to aging can cause irreversible consequences [56, 57]. Therefore, there is a need to develop solutions to address the problems arising from the effects of aging phenomena that will enable ICs to operate uninterruptedly for more than 10 years, ensuring the safety of human life.

2.1.2 Key Factors Determining the Aging of Integrated Circuits

IC aging is defined as a change in transistor parameters that changes in magnitude over time. Types of IC degradation mechanisms that fit this definition include hot carrier injection (HCI) [29–31, 58–64], bias temperature instability (BTI) [29–31, 58–65], time-dependent dielectric breakdown (TDDB) [29–31, 66], and electromigration (EM) [9, 29–31]. Each of these phenomena has different modeling approaches and affects transistor parameters differently, but their physical origins are often similar or even the same. The most common physical degradation mechanisms are deep oxide traps, generation of interface states, and trapping of electrons and holes [60]. The main aging phenomena discussed above are presented below.

HCI Phenomenon. In modern complementary metal-oxide-semiconductor (CMOS) technologies, the scaling of supply voltages (VDD) and threshold voltages (V_{th}) has slowed down or stopped altogether due to the non-scalability of the subthreshold slope, while the gate length (L) and dielectric thickness (t_{ox}) continue to scale [58–64]. This leads to an increase in the effective electric fields along the channel from the drain to the source horizontally, and from the gate to the substrate vertically. This aggressive, non-steady scaling results in the phenomenon of hot charge carriers. Under these conditions, charge carriers can acquire large kinetic energy when they pass through regions of high electric field. When the energy of a charge carrier is significantly greater than that in thermal equilibrium, it is called "hot" [58–61]. These hot carriers can gain enough energy to inject themselves into the gate dielectric, damage the environment, or cause leakage currents between the gate and the substrate, thus causing instabilities in the device characteristics. In general, the hot carrier injection (HCI) phenomenon leads to a decrease in the output current (I_d) of the transistor and an increase in V_{th}. The change in I_d is described by the following formula [58]:

$$\Delta I_d = I_d B V_{ds}^{p} 5^{\frac{\Delta E}{kT}} L^{\delta} t^{q}, \tag{2.1}$$

where B, p, and δ are experimentally determined parameters, V_{ds} is the drain–source voltage, ΔE is the thermal energy, k is the Boltzmann constant, T is the temperature, L is the channel length, t is the stress time, and q is the time exponent, which varies in the range 0.25–0.45.

HCI mechanisms are of four types: channel hot electron (CHE) injection [59–61], drain avalanche hot carrier (DAHC) injection [59–61], secondary generated hot electron (SGHE) injection [59–61], and substrate hot electron (SHE) injection [59–61], which are presented below.

- **CHE.** The gate and drain voltages V_g and V_d must be equal (Fig. 2.1) [60]. In this state, only the "lucky" electrons are able to overcome the gate dielectric and leave the channel. The lucky electrons are hot charge carriers that have not collided with particles in the channel and, therefore, can continuously gain kinetic energy along the channel. CHE leads to the flow of gate current (I_g) or eventually to the formation of oxide traps [59–61]. Since holes have a larger effective mass than electrons, NMOS transistors suffer more from CHE.

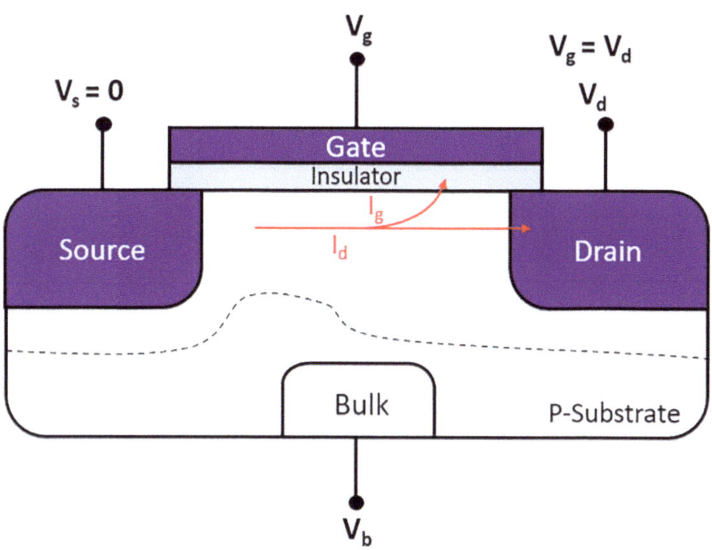

Fig. 2.1 Illustration of the hot electron injection phenomenon in the channel

- **DAHC.** When the transistor is in a strong saturation state such that V_d is greater than V_g, impact ionization occurs on the output side of the channel (Fig. 2.2) [60]. The resulting electron-hole pairs begin to move under the influence of the electric field inside the channel. Most of the charge carriers move toward the gate, causing further ionization or the generation of interface states [59]. The remaining particles are accelerated along the channel depending on their charge. Only a small fraction are attracted to the body, contributing to the body current (Ib) [59–61].
- **SGHE:** Charge carriers produced by high-energy collisions, as in the case of DAHC, are also capable of further impact ionization. In this case, the particles produced are called secondary hot charge carriers (Fig. 2.3) [59–61].
- **SHE:** When the bulk voltage (V_b) is positive or negative, depending on the type of transistor, the charge carriers gain energy in the channel and are likely to overcome the barrier of the gate dielectric (Fig. 2.4) [59–61]. The mechanism is independent of the source and drain voltages and is therefore uniformly distributed along the channel. SHE occurs mainly in circuits where the bulk is subjected to a non-zero voltage. NMOS transistors suffer from the injection of hot electrons of substrate, while PMOS transistors suffer from the injection of holes of substrate [59–61].

BTI Phenomenon. The BTI phenomenon [58–61] causes the generation of interface states and oxide traps on the Si/SiO_2 surface under the influence of a vertical electric field across the gate ($V_g > |V_{th}|$) and at high temperatures (100–150 °C) [62–65]. The greatest effect of BTI is observed in p-type MOS (PMOS) transistors when stressed with VDD and a large negative gate voltage at high temperatures

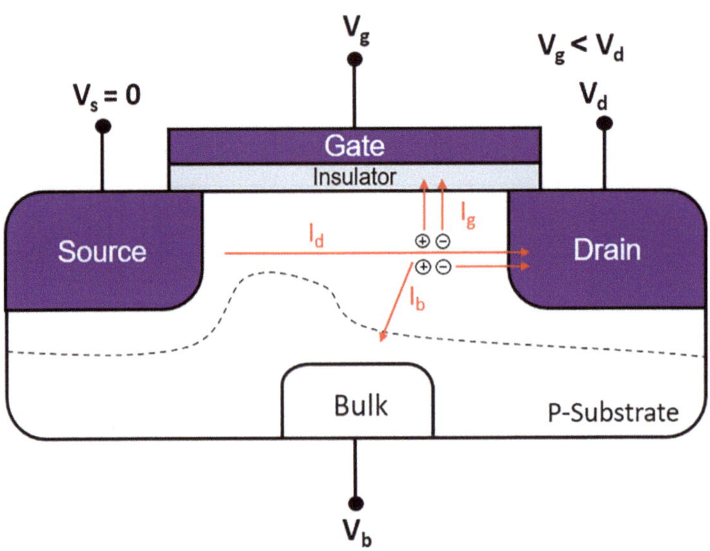

Fig. 2.2 Illustration of the injection phenomenon of DAHC

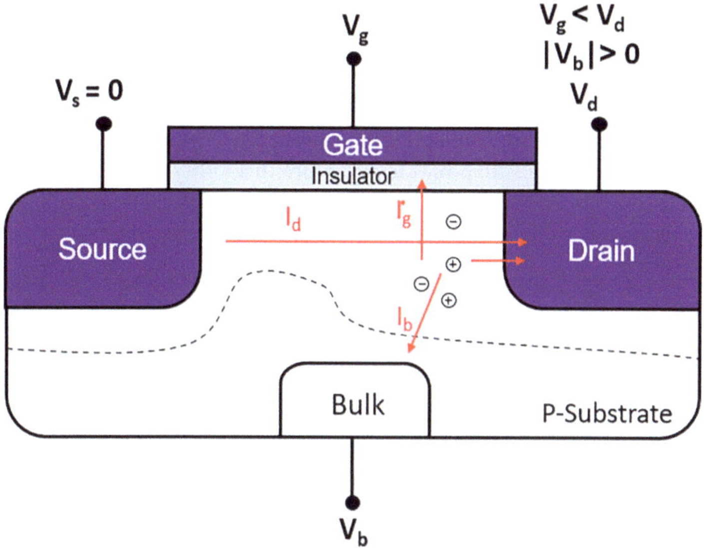

Fig. 2.3 Illustration of the injection phenomenon of SGHE

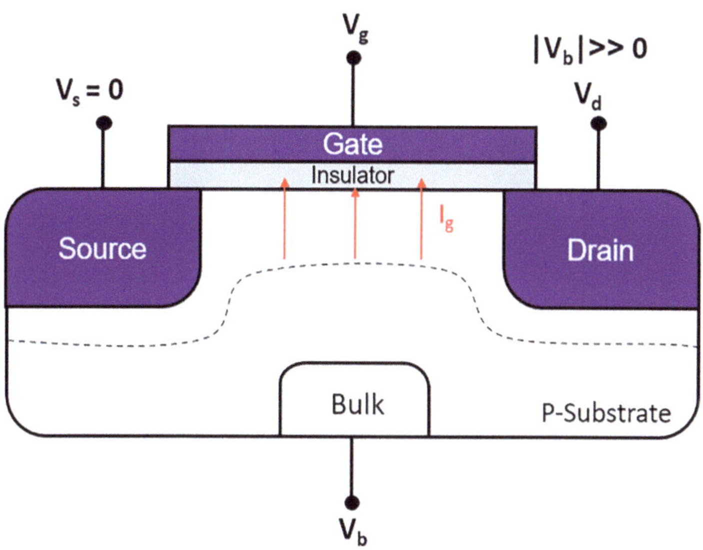

Fig. 2.4 Illustration of the injection phenomenon of SHE

(100–150 °C). This phenomenon is called negative-bias temperature instability (NBTI) because of the negative difference between the gate and source voltages. The holes in the PMOS channel interact with the passivated hydrogen bonds in the dielectric, generating interface states and oxide traps. This leads to an increase in the absolute value of V_{th} [58–65]. Starting with the 45 nm CMOS technology, high-K materials have been used in semiconductor fabrication to overcome the increase in leakage current due to the scaling of the gate oxide thickness. These high-K materials have lower leakage currents compared to SiO_2 dielectrics due to their high dielectric constant (k) [58]. With the introduction of high-K materials, the BTI phenomenon has also become more prominent in n-type MOS (NMOS) transistors, which is called positive-bias temperature instability (PBTI) due to the positive difference between the gate and source voltages. Currently, BTI is one of the most important reliability issues for both digital and analog circuits.

The BTI effect begins to recover after the stress is removed (Fig. 2.5) [58]. The V_{th} recovery phenomenon consists of long transient recovery processes. The recovery component occurs after the stress phase, during the removal of the stress signal, due to the escape of charges from the trap [58]. This recovery phenomenon is beneficial for the long-term use of the device, but it also makes the characterization of the BTI quite difficult.
V_{th} for NMOS and PMOS transistors due to PBTI and NBTI is modeled by the following formula [58]:

$$\Delta V_{\text{th}} = A\left(\frac{V_{\text{g}}}{t_{\text{ox}}}\right)^{m} e^{\frac{\Delta E}{kT}} L^{\alpha} W^{\beta} t^{n}, \tag{2.2}$$

where A, m, α, β are experimentally determined parameters, V_{g} is the gate voltage, ΔE is the thermal energy, k is the Boltzmann constant, T is the temperature, L is the channel length, W is the width, t is the stress time, and n is the time index, which varies in the range 0.19–0.26.

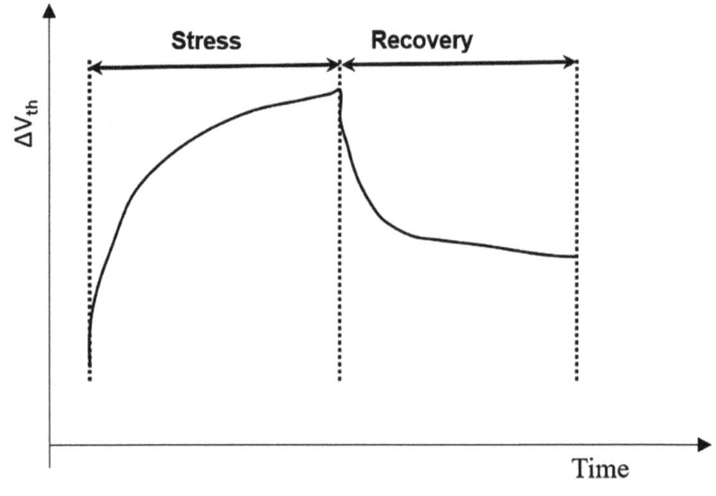

Fig. 2.5 Stress and recovery phases during the lifecycle of a transistor

TDDB Phenomenon. The dielectric between the gate and the channel has the role of separating the charges accumulated on their surfaces. Each dielectric can withstand a certain maximum electric field. When the dielectric is exposed to high fields, dielectric breakdown occurs [59]. It occurs as follows: a large number of environmental states and traps are created in the dielectric. Due to the increase in the number of traps, an electric current begins to flow through them due to tunneling [60]. If the number of traps continues to increase so much that the traps connect to each other, a large current begins to flow through this connection between the gate and the substrate, which leads to dielectric breakdown (Fig. 2.6) [59, 60, 66].

TDDB is a statistical process. This process can be modeled using the Weibull probability distribution [60]:

$$F\left(t_{\mathrm{bd}}\right)=1-\exp\left[-\left(\frac{t_{\mathrm{bd}}}{\alpha_{\mathrm{bd}}}\right)^{\beta_{\mathrm{bd}}}\right], \tag{2.3}$$

where $F(t_{bd})$ is the cumulative density function for the time-to-breakdown event, t_{bd} is the stress time, α_{bd} and β_{bd} are technology-dependent parameters.

EM Phenomenon. The phenomenon of material migration in solids occurs due to various processes. The most common are chemical diffusion due to concentration gradients, material migration due to temperature, material migration due to mechanical stress, and material migration due to high electric fields. The last one is called electromigration and is the main factor in material migration in ICs [9, 60]. Most semiconductor ICs use aluminum (Al) or copper (Cu) metal wires to connect them

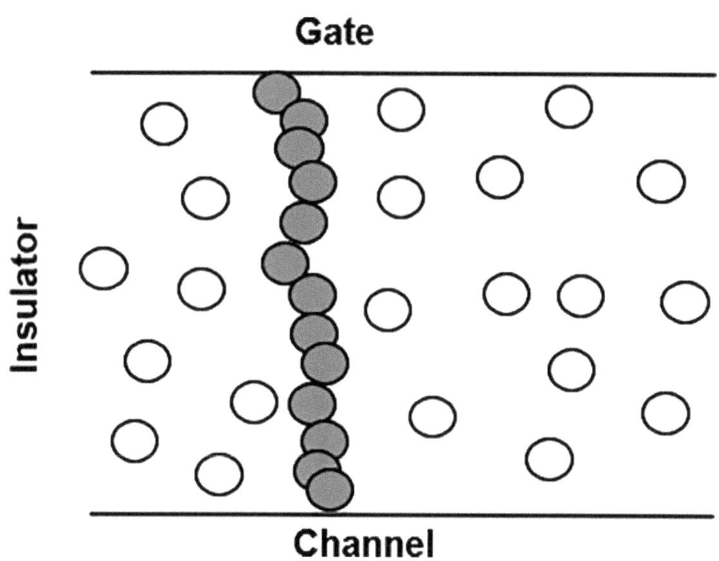

Fig. 2.6 Illustration of the TDDB phenomenon

in and out of ICs, or from one part of an IC to another [59]. EM mainly occurs at the edges of crystal structures, where the crystal is inhomogeneously distributed. At high currents, charge carriers interact with atoms in the crystal lattice and move some of them in the direction of the current, forming the so-called hills and holes [59]. This can cause the resistance of the wire to change, or at some point it may break altogether. It is characterized by the mean time to failure (MTTF), which for EM is modeled by Black's law [60]:

$$\mathrm{MTTF} = \frac{A}{J^n}\exp\left(\frac{E_\mathrm{a}}{kT}\right), \tag{2.4}$$

where A is a constant depending on the cross-sectional area of the conducting wires, J is the current density, E_a is the activation energy, k is the Boltzmann constant, T is the temperature, and n is a scaling factor. Electromigration is highly dependent on the physical design.

The mentioned aging phenomena are modeled by companies, and special models are created that are later used during the design of ICs to take aging phenomena into account. The most famous models are MOSRA [67] developed by Synopsys and AgeMOS [68, 69] by Cadence.

The Impact of Aging Phenomena on Analog and Mixed-signal Integrated Circuits

Analog and mixed-signal systems play a major role in modern large-scale integrated circuits. They are generally the most sensitive circuits in the entire IC. And very often they are the ones that are affected by aging phenomena, which can lead to serious failures.

Nowadays, the main challenges of ICs are reliability, area, functionality, power consumption, and speed. Ensuring the last one has become a particularly serious challenge, given the already existing high frequencies and reliability issues. One of the most important circuits for ensuring high frequencies in ICs is the voltage-controlled oscillator (VCO) [5, 70–74].

VCO. The main function of a VCO is to generate a high-frequency periodic signal, the frequency of which depends on the voltage applied to its control input. The two most important parameters for the design of a VCO are high linearity and frequency [5]. Due to the effects of aging, the frequency of a VCO can decrease due to an increase in V_th [70, 71]. One of the most widespread VCO circuits consists of a ring oscillator (RO) with an odd number of inverters and a voltage-to-current converter, and a CMOS inverter is connected at the output of the circuit to rectify the edges (Fig. 2.7) [5, 70]. The output frequency of the VCO (f_out) depends on the control voltage (V_ctrl) applied to its input (Fig. 2.8) [5, 70]. It is expressed in the following formula [5]:

$$f_\mathrm{out} = f_0 + K_\mathrm{VCO}V_\mathrm{ctrl}, \tag{2.5}$$

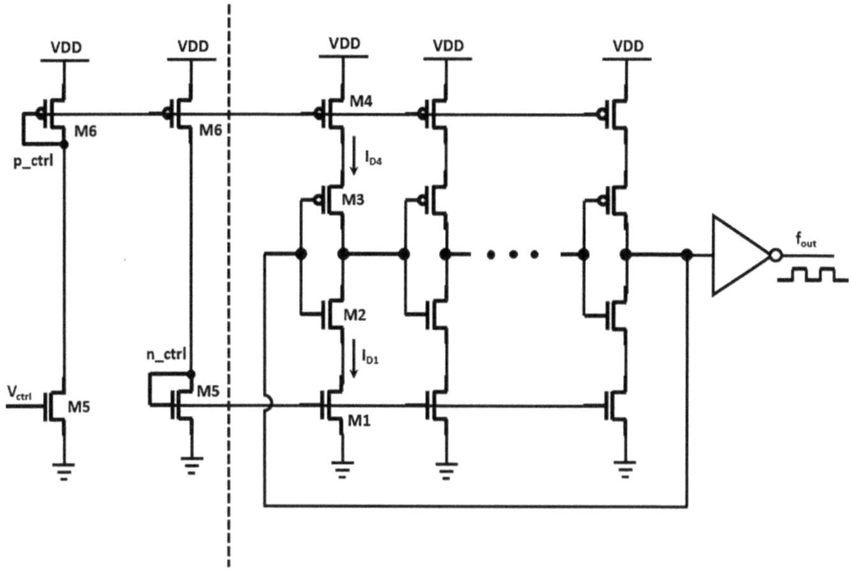

Fig. 2.7 VCO circuit

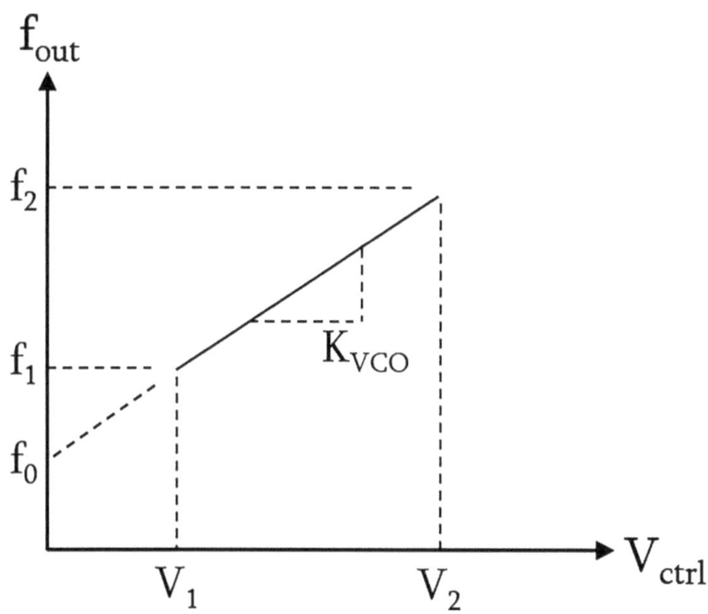

Fig. 2.8 The dependence of VCO f_{out} on V_{ctrl}

where f_0 is the minimum output frequency of the VCO, and K_{VCO} is the gain.

The above VCO circuit was designed for the SAED 14 nm FinFET [75] process technology, and a timing analysis was performed using HSPICE simulation (Fig. 2.9) [70], from which it can be seen that the VCO output frequency is 3.22 GHz.

To study the aging of the designed VCO, an HSPICE simulation was performed, aging the circuit for 10 years. As a result of aging, the threshold voltages of both M1-M4 and M5 and M6 transistors increase, which results in a decrease in the currents I_{D4} and I_{D1}, which leads to a decrease in the frequency (Fig. 2.10) [70].

From the results obtained, it becomes clear that the output frequency of the VCO decreased by 6.5% due to aging.

An HSPICE simulation was also performed for 1–10 years to understand how the frequency changes over time (Fig. 2.11) [70].

The results show that the largest decrease occurred during the first year, which means that the circuit will generate a low-frequency signal starting from the first year. The last one further increases the importance of the problem and the need for solutions.

Foundries for 3 nm and below technologies have encountered difficulties in manufacturing transistors with thick oxide gate layers [76]. This problem forces the replacement of all transistors with thick oxide gate layers with thin oxide gate layers, thus creating additional difficulties in the design phase. In the case of thin oxide gate layers, the stress conditions between the two pins are smaller than in the case of transistors with thick oxide gate layers. All thin gate oxide transistors are manufactured in such a way that the voltage difference between any two of their pins (gate–drain (V_{gd}), gate–source (V_{gs}), drain–source (V_{ds})) can reach a voltage equivalent to the low supply voltage (VDD) [76], and therefore their use in circuits

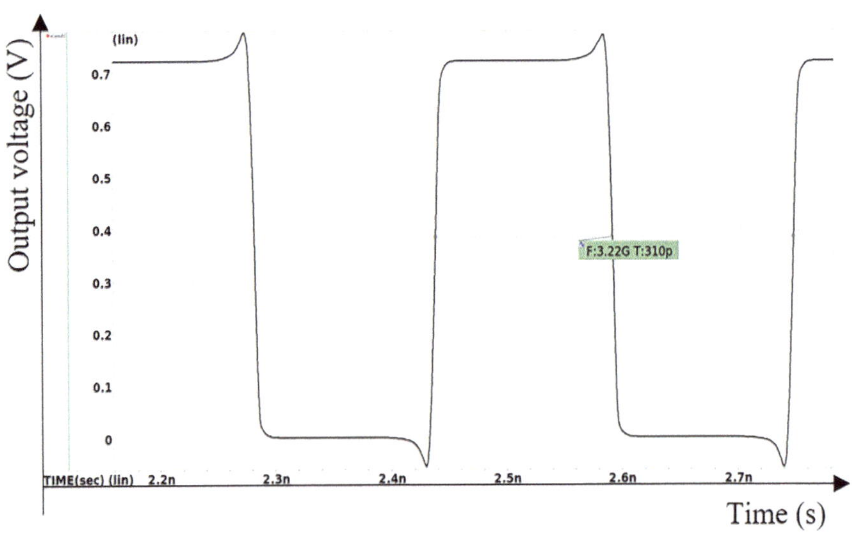

Fig. 2.9 View of VCO output signal

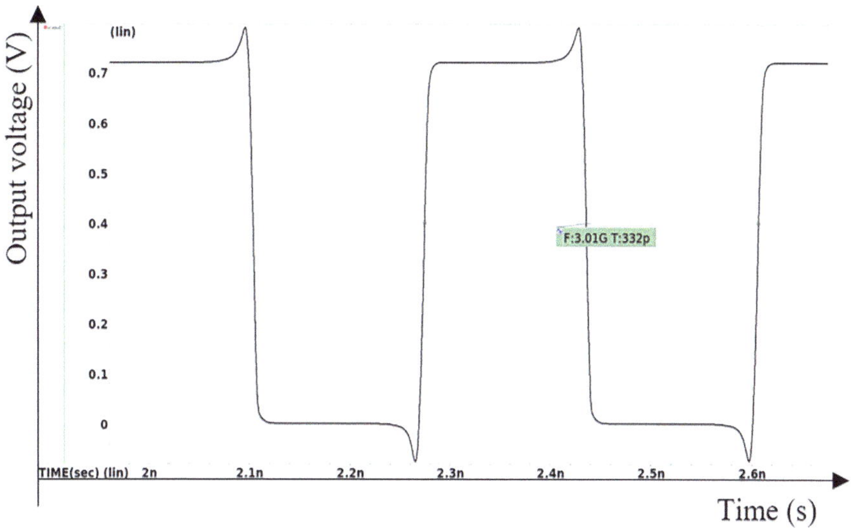

Fig. 2.10 The view of the VCO output signal after 10 years of aging

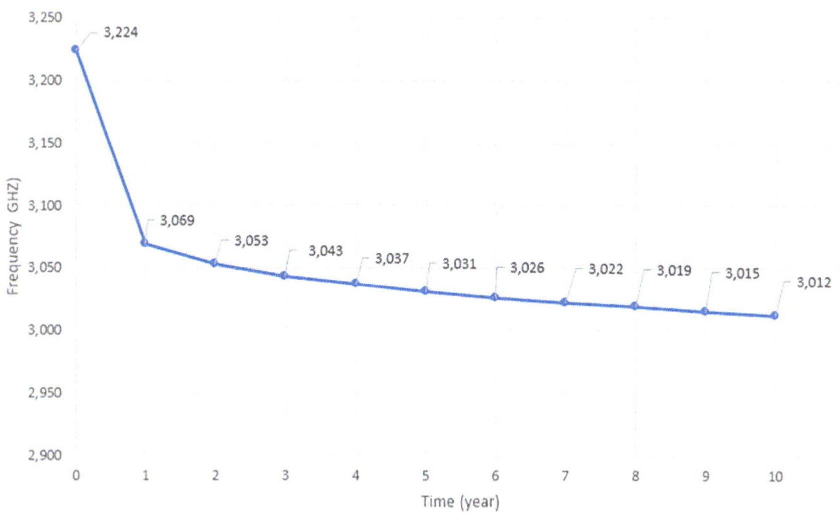

Fig. 2.11 Dependence of output frequency on years of VCO

powered by a high supply voltage (VDDH) can create stress conditions that will lead to aging.

Operational Amplifier (OpAmp). Many circuits in analog and mixed-signal systems, among which OpAmps are the most important and sensitive, are powered from VDDH to provide high gain and wide operating range. Until now, OpAmps have been designed with thick oxide gate transistors, and replacing them with thin oxide gate transistors can cause stress on some transistors, which in turn can lead to aging phenomena, changing the values of a number of important parameters of the OpAmp over time and disrupting its operation [76].

The important parameters of OpAmps are gain, bias voltage, noise immunity, operating range, speed, phase margin, gain margin, cut-off gain frequency, setup time, rise time, maximum, and worst case values of the supply noise suppression factor, etc. [5, 77].

There are different types of architectures for implementing OpAmps, depending on the problems being solved and the requirements. Three types are presented below.

Two-Stage OpAmp. The two-stage OpAmp is one of the most common OpAmps, which provides high gain and is used in biasing circuits [5, 76–80]. One of the well-known two-stage OpAmp circuits was designed using the SAED 14 nm FinFET (Fig. 2.12) [76], and was simulated in HSPICE (Fig. 2.13) [76]. It was possible to achieve a gain of 54.2 dB [76].

The transistors marked with dotted lines (Fig. 2.12) [76] are used to bring the circuit to the off state. P1, P2, and P3 transistors close all branches through which current can flow, and the P4 transistor connects the output to VDDH. Almost all modern analog circuits have the capability to switch to the off state to reduce power consumption; some circuits may even spend more time in the off state than in the on

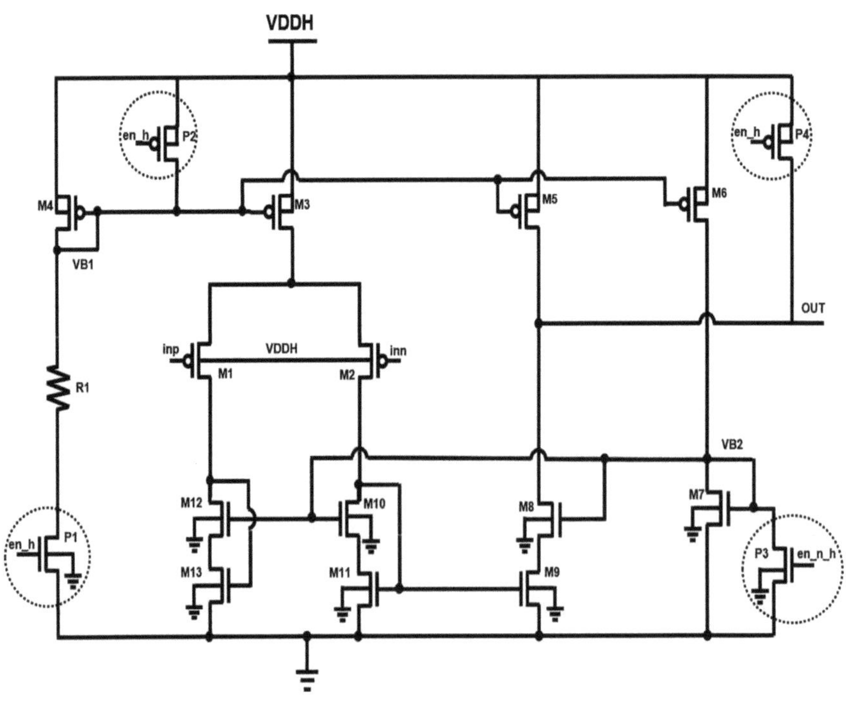

Fig. 2.12 Schematic view of the designed two-stage OpAmp

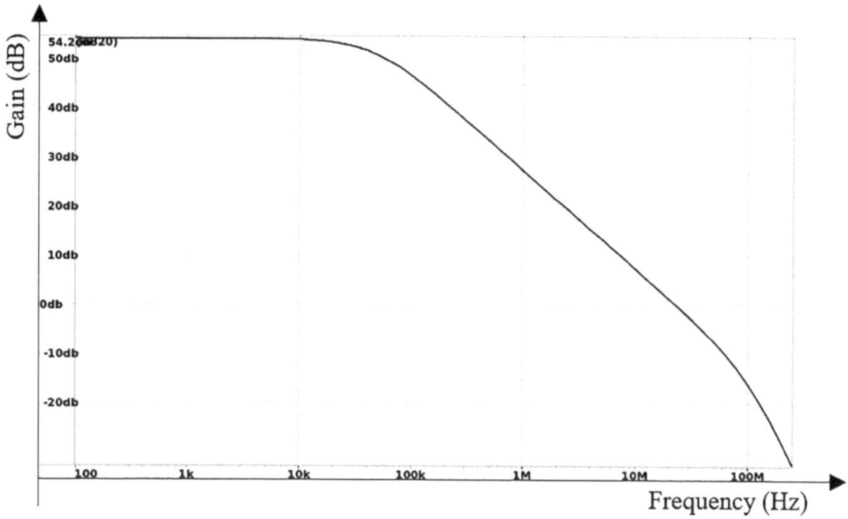

Fig. 2.13 Gain of the designed two-stage OpAmp

Table 2.1 Maximum deviations of I_d and V_{th} of transistors as a result of 10 years of aging of a two-stage OpAmp

	On state		Off state	
Transistor	ΔV_{th} (mV)	ΔI_d (%)	ΔV_{th} (mV)	ΔI_d (%)
M6	22	2.1	175	17.4
M8	19	1.8	172	16.9
P1	152	15.5	127	12.8
P2, P4	0.1	0.1	198	21.6
P3	0.1	0.1	198	21.6

state during their lifecycle [76]. Therefore, circuits must be designed in such a way that stress situations do not arise even in the off state in order to avoid aging.

To study the aging of the designed two-stage OpAmp, HSPICE simulations were performed, aging the circuit for 10 years in both the on and off states. In the on state, the key transistors of the OpAmp did not undergo significant aging, and the changes in the I_d and V_{th} parameters were within acceptable limits, except for the P1 transistor (Fig. 2.1) [76], as a result of which the key parameters of the OpAmp did not undergo significant changes. In contrast, in the off state, serious aging was observed in several key and turn-off transistors, as a result of which the gain of the OpAmp decreased by approximately 14 dB. The I_d and V_{th} parameters also underwent significant changes (Table 2.1) [76].

The reason for the aging of the P1 transistor in the on state is the voltage levels of the control signal coming to the input, which vary from 0 to VDDH. In the on state, the P1 transistor is in the open state, and the control signal has a level of VDDH, as a result of which V_{gs} is equal to VDDH, which is not permissible, since it causes stress.

P2 and P4 transistors in the off state should be open, and their control signal should be 0, in the result of which V_{gs} and V_{gd} of both are equal to VDDH. The gate of transistor P3 is given VDDH; that is, V_{gs} and V_{gd} are equal to VDDH. Transistor P1 is in the off state, and the gate is given 0, but transistor P2 connects the output of transistor P1 to VDDH. Thus, V_{gd} of P1 becomes equal to VDDH, which is also not permissible, and stress occurs, although it is in the off state. Transistor M6 is in the off state because transistor P2 connects the gate of transistor M6 to VDDH, and transistor P3 connects the drain of transistor M6 to 0. Thus, V_{gd} of M6 is equal to VDDH. Transistor M8 is also in the off state since transistor P3 connects the gate of transistor M8 to 0, and transistor P4 connects the drain of transistor M8 to VDDH. Thus, V_{gd} of M8 is equal to VDDH.

Thus, the aging simulation results of the two-stage OpAmp showed that there are no significant problems in the on state of the circuit. On the contrary, in the off state of the circuit, there are quite serious problems that can disrupt the operation of the OpAmp; therefore, it is necessary to develop methods to solve the problem.

Folded Cascode OpAmp. Folded cascode OpAmps are widely used in voltage regulators to provide high gain [5, 76, 81, 82]. One of the well-known folded cascode OpAmp circuits was designed using SAED 14 nm FinFET technology (Fig. 2.14) [76] and was simulated in HSPICE (Fig. 2.15) [76]. As a result, a gain of 85.2 dB was achieved [76].

The transistors marked with dotted lines (Fig. 2.14) [76] are used to bring the circuit to the off state. P1, P2, P3, P4, and P5 transistors close all branches through which current can flow, and transistor P6 connects the output to VDDH.

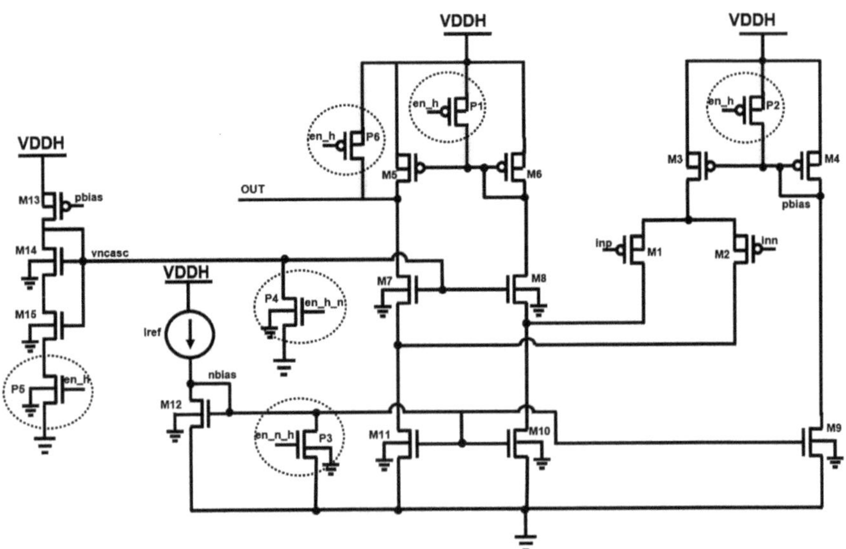

Fig. 2.14 Schematic view of the designed folded cascode OpAmp

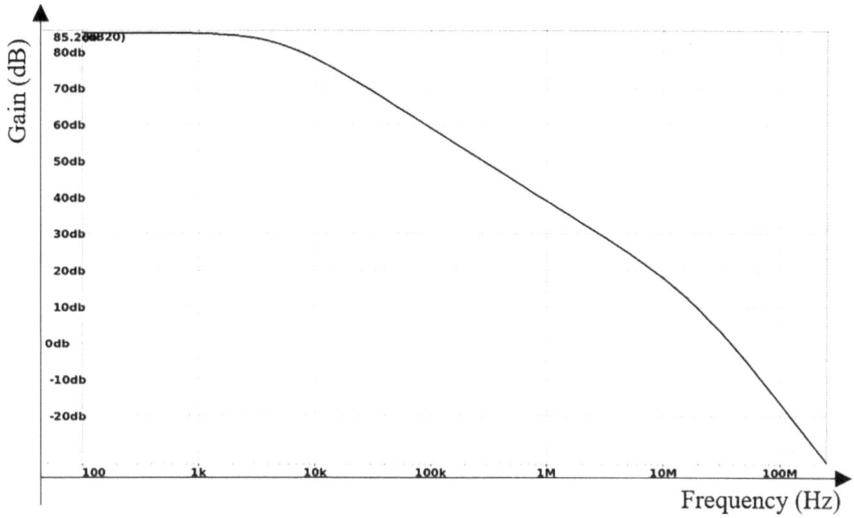

Fig. 2.15 Gain of the designed folded cascode OpAmp

Table 2.2 Maximum deviations of I_d and V_{th} of transistors as a result of 10 years of aging of a folded cascode OpAmp

	On state		Off state	
Transistor	ΔV_{th} (mV)	ΔI_d (%)	ΔV_{th} (mV)	ΔI_d (%)
M7, M8	24	3.1	192	19.6
M9	20	2.8	198	20.1
P1, P2, P6	0.1	0.1	252	24.2
P3, P4	0.1	0.1	254	24.3
P5	148	14.2	0.1	0.1

To study the aging of the designed folded cascode OpAmp, HSPICE simulations were performed, aging the circuit for 10 years in both the on and off states. In the on state, the key transistors of the OpAmp did not undergo significant aging, and the changes in the I_d and V_{th} parameters were within acceptable limits, with the exception of the P5 transistor (Fig. 2.2) [76], as a result of which the key parameters of the OpAmp did not undergo significant changes. In contrast, in the off state, serious aging was observed in several key and turn-off transistors, as a result of which the gain of the OpAmp decreased by approximately 19 dB. The I_d and V_{th} parameters also underwent significant changes (Table 2.2) [76].

The aging of the P5 transistor in the on state and the aging of the other transistors responsible for switching off in the off state are due to the signal controlling them, which was discussed when considering the two-stage OpAmp.

In the off state, transistor M9 is closed because P3 connects the gate of M6 to 0, and P2 connects the drain of M6 to VDDH. Thus, the V_{gd} of M9 is equal to VDDH. Transistors M7 and M8 are in the same closed state since P4 connects the gates of M7 and M8 to 0, P6 connects the gates of M7, and P1 connects the outputs of M8 to VDDH, and the sources are in a floating state.

Thus, the aging simulation results of the folded cascode OpAmp showed that there are no serious problems in the on state of the circuit, but in the off state, they are present and can disrupt the operation of the OpAmp; therefore, it is necessary to develop methods to solve the problem.

Rail-to-Rail OpAmp. Rail-to-rail OpAmps allow for a very wide operating range and are mainly used in digital delay lines, charge pumps, voltage and temperature compensation circuits [83–85]. One of the well-known rail-to-rail OpAmp circuits was designed using the SAED 14 nm FinFET technology (Fig. 2.16) [83] and was simulated in HSPICE (Figs. 2.17 and 2.18) [83]. A gain of 52.3 dB was achieved [83].

The transistors marked with dotted lines (Fig. 2.16) [83] are used to turn the circuit off. Transistors P1, P2, P3, P4, P5, and P6 close all branches through which current can flow.
Study the aging of the designed OpAmp from one cycle to another, HSPICE simulations were performed, aging the circuit for 10 years in both the on and off states. In the on state, it was found that some key transistors of the OpAmp underwent aging, but the main parameters of the OpAmp underwent minor changes (Fig. 2.3) [4, 8]. In the off state, serious aging was observed in all transistors responsible for switching, mainly the main and switching, as a result of which the gain of the OpAmp

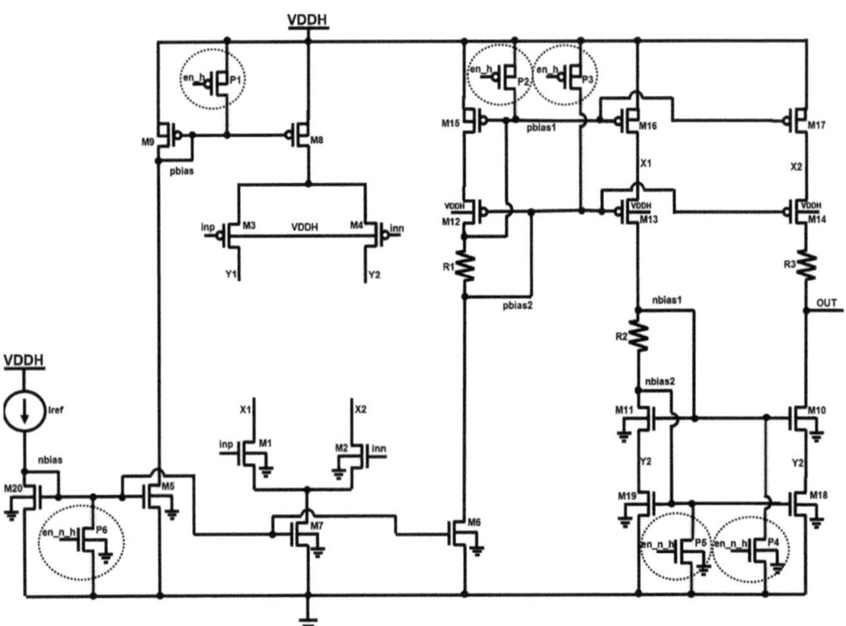

Fig. 2.16 Schematic view of the designed rail-to-rail OpAmp

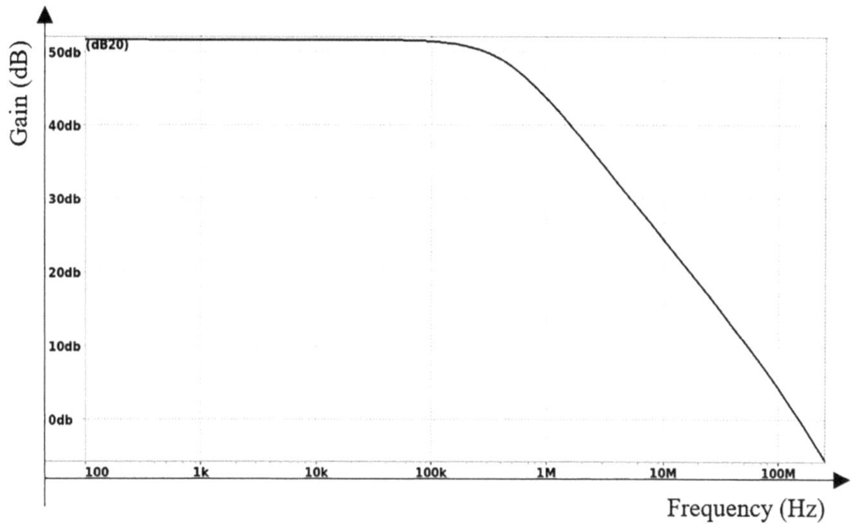

Fig. 2.17 Gain of the designed rail-to-rail OpAmp

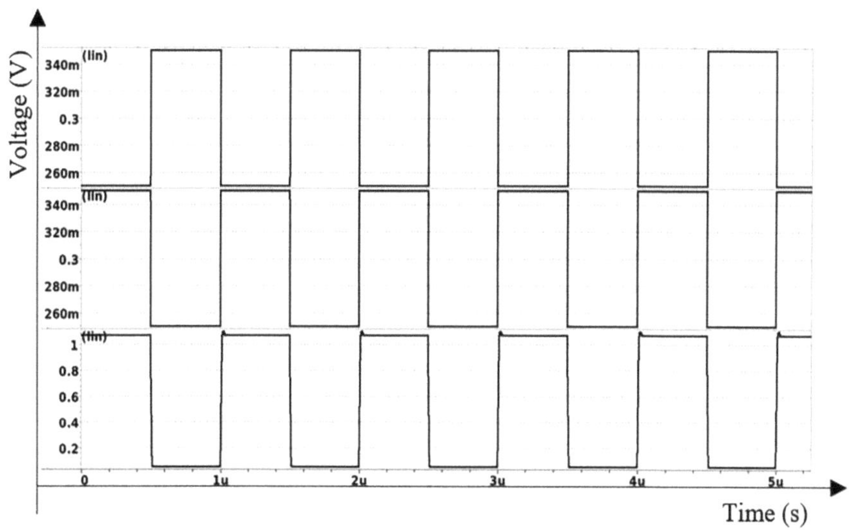

Fig. 2.18 Timing analysis results of the designed rail-to-rail OpAmp

decreased by approximately 12 dB. The I_d and V_{th} parameters also underwent significant changes (Table 2.3) [83].

The aging of M1 and M2 transistors in the on state is due to the fact that their drain voltage reaches VDDH while the gate voltage is approximately 300 mV. This difference causes stress. The drain of M10, which is the output of the OpAmp, can reach VDDH, while the gate voltage is approximately 200 mV, which causes stress.

Table 2.3 Maximum deviations of I_d and V_{th} of transistors as a result of 10 years of aging of a rail-to-rail OpAmp

	On state		Off state	
Transistor	ΔV_{th} (mV)	ΔI_d (%)	ΔV_{th} (mV)	ΔI_d (%)
M1, M2	119	15.2	26	4.1
M5, M6	25	3.8	178	18.5
M10	109	14.6	0.1	0.1
M13, M14	31	4.5	182	19.3
P1, P2, P3	0.1	0.1	252	22.4
P4, P5, P6	0.1	0.1	252	22.4

In the off state, the aging of all the transistors responsible for switching off is due to the signal controlling them, which was discussed when considering the two-stage OpAmp. Transistors M5 and M6 are in the same off state since P6 connects the gates of M5 and M6 to 0, and P1 connects the drain of M5, and P3 connects the drain of M6 to VDDH. Thus, the V_{gd} of both is equal to VDDH. Transistor M13 is in the off state because P3 connects the gate of M13 to VDDH, and P4 connects the source of M13 to 0; the drain is floating and can be equal to VDDH. The M14 transistor is also in the same state.

Thus, the results of the rail-to-rail OpAmp aging simulation showed that there are certain problems in the on and off states of the circuit and can disrupt the operation of the OpAmp; therefore, it is necessary to develop methods to solve the problems that arise.

The Impact of Aging Phenomena on Digital Integrated Circuits

In modern ICs, digital ICs also play a major role, constituting the majority of the total IC. Among them, standard cells have a special place [86–88], in which transistors are also affected by aging phenomena. Due to the influence of known phenomena, the V_{th} of the transistors changes which leads to an increase in the cell delay, which in turn can lead to failures in critical paths, or new critical paths can occur [61, 89]. In case of a CMOS buffer, HSPICE simulations were performed to observe the changes in V_{th} and delay over a period of 10 years (Figs. 2.19 and 2.20) [89].

To observe aging phenomena in digital ICs, standard cell libraries are designed that take aging phenomena into account. To observe the BTI phenomenon, it is necessary that constant stress conditions are applied to the cell inputs. Therefore, it should depend on the duty cycle of the input signals [89]. In case of a CMOS buffer, HSPICE simulations were also performed after 10 years of aging to observe the dependence of the delay change on the duty cycle (Fig. 2.21) [89].

Let us consider the effect of phase shift between input signals on the change in the delay of standard cells due to aging. The observation is made on one of the most complex standard cells, the CMOS XOR cell (Fig. 2.22) [89].

The effects of BTI and HCI phenomena are considered separately. To study the dependence of BTI phenomenon on phase deviation, two cases were considered (Fig. 2.23) [89].

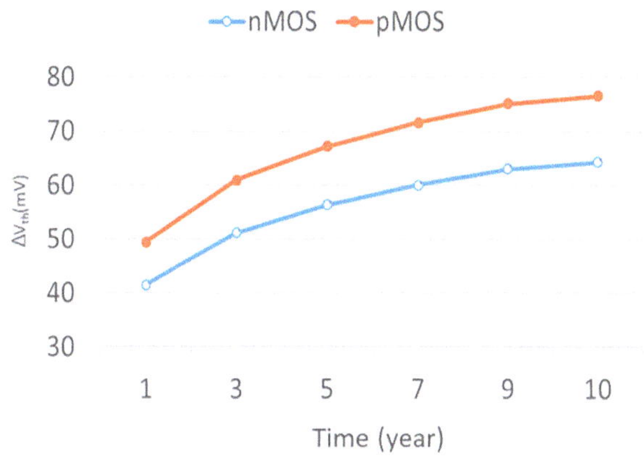

Fig. 2.19 Change in V_{th} of a CMOS buffer over years

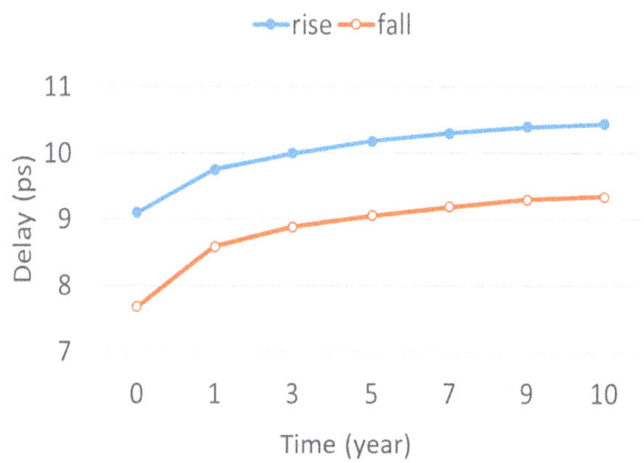

Fig. 2.20 Delay changes in CMOS buffer over years

For different phase shifts, each transistor will have a different BTI effect. For example, transistor P5 will be most stressed when input a is set to 0, but this will depend on the state of P4. When P4 is closed, the stress will be at a minimum (Fig. 2.23a) [89], and when P4 is open, it will be at a maximum (Fig. 2.23b) [89]. It follows that the BTI phenomenon will have the greatest effect in this case when the phase shift between the inputs is 180°.

To study the dependence of the HCI effect on phase deviation, two cases were also considered (Fig. 2.24) [89].

The effect of the HCI phenomenon is minimal when the output signal of the circuit is switched off (Fig. 2.24a) [89], and maximum when the number of switches is as high as possible (Fig. 2.24b) [89]. In this case, it can be seen that the HCI phenomenon will have the greatest effect when the phase shift between the inputs is 90° or 270°.

HSPICE simulations of the XOR cell were performed, varying the phase shift between the inputs at a 50% fill factor of the input signals, aging the circuit for 10 years, and observing the change in delay (Table 2.4; Fig. 2.25) [89].

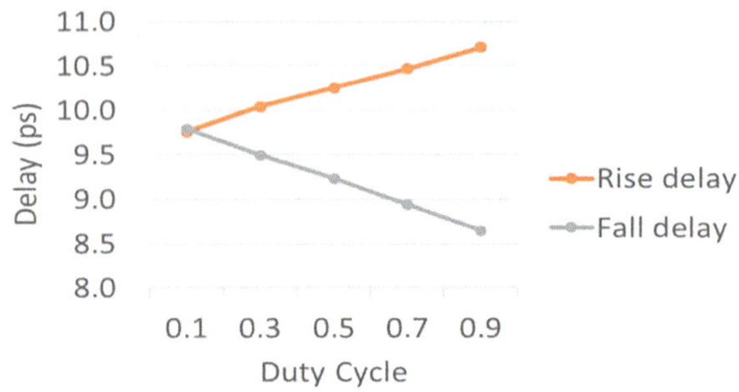

Fig. 2.21 Change in delay as a function of fill factor of a CMOS buffer after 10 years of aging

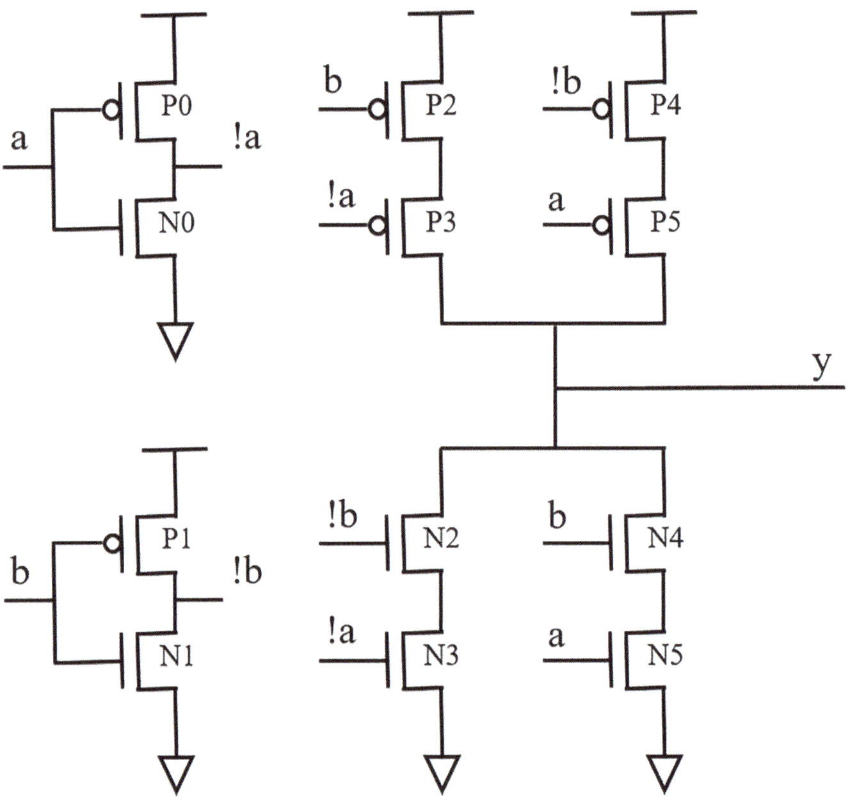

Fig. 2.22 CMOS XOR gate circuit

The results show that the change in delay due to aging can vary greatly depending on the phase shift of the input signals to a digital standard cell. Therefore, it is necessary to take into account possible phase shifts between the input signals of the cells in order to increase the accuracy of the prediction of standard cell libraries that take aging phenomena into account.

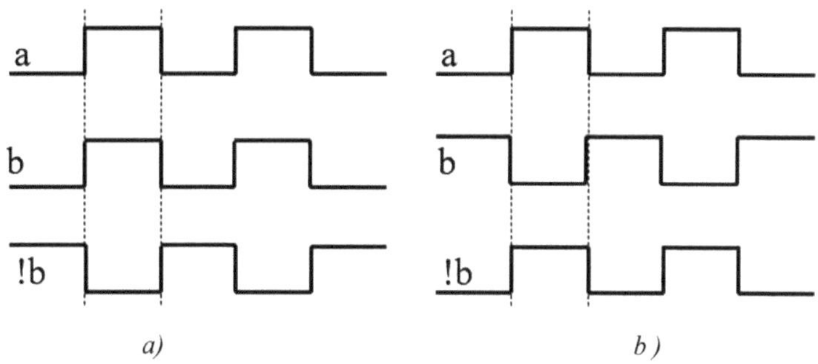

Fig. 2.23 Illustration of the dependence of the effect of the BTI phenomenon on the phase deviation of the input signals

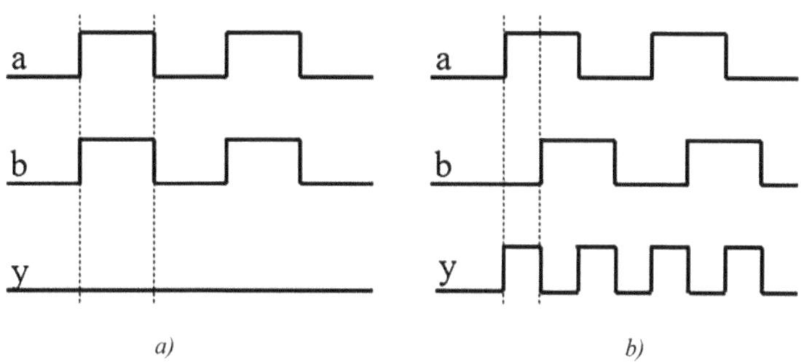

Fig. 2.24 Illustration of the dependence of the effect of the HCI phenomenon on the phase deviation of the input signals

2.1.3 Existing Solutions and Key Issues to Reduce the Impact of Aging Phenomena in Integrated Circuits

The above illustrated fundamental problems can cause failures in the operation of the mentioned circuits, which can cause serious damage to the overall IC. To avoid all this, it is necessary to develop new approaches that will allow reducing or compensating for the effects of aging phenomena. So far, a number of methods and techniques have been developed to solve these problems, but each of them has its advantages and disadvantages, and due to the challenges arising from the super-active scaling of technology, they may not meet current requirements. Below, several such methods and the problems and issues they present are considered.

Table 2.4 The change in delay due to 10 years of aging for different input signals of an XOR cell with a 50% duty cycle

Delay change (pv)	Phase shift 0°	90°	180°	270°	Minimum/maximum difference (%)
rise_rise_a	13.54	13.65	13.74	13.72	1.48
rise_fall_a	7269	6901	6006	6991	21.03
fall_rise_a	5049	5219	5547	5197	9.86
fall_fall_a	11.98	11.86	11.66	11.97	2.74
rise_rise_b	12.48	13	13.27	13.17	6.33
rise_fall_b	6664	6205	6027	6.15	10.57
fall_rise_b	4654	5842	6315	5914	35.69
fall_fall_b	12.59	12.6	12.52	12.61	0.72

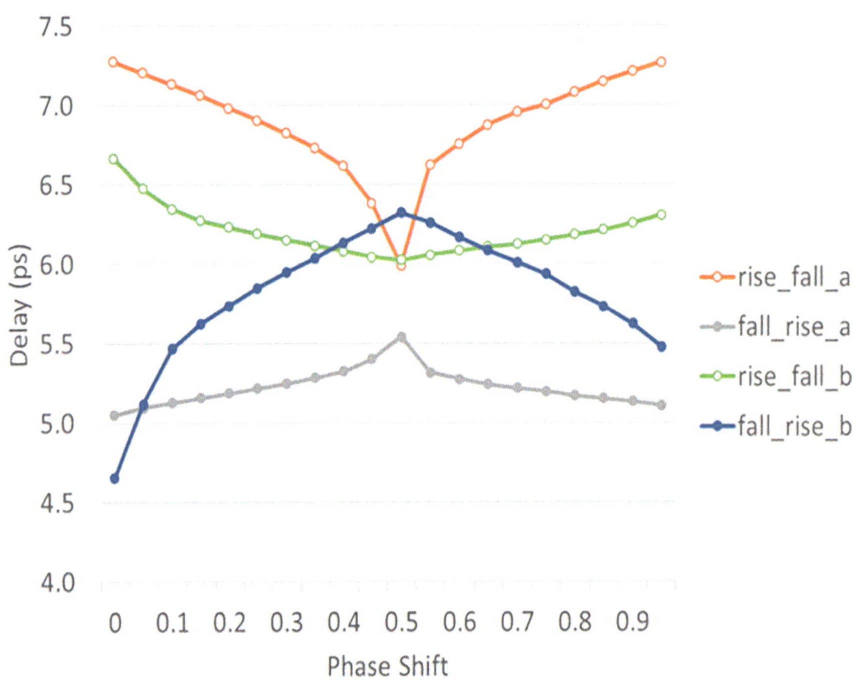

Fig. 2.25 Dependence of the change in delay due to 10 years of aging on the phase deviation of the input signals with a 50% fill factor of the XOR cell

Existing Measures to Reduce the Impact of Aging Phenomena in Voltage-Controlled Oscillators (VCO)

It was shown that aging phenomena cause a decrease in the output frequency of the VCO, and even a decrease of a few percent can have serious consequences in certain cases. There are many known solutions to this problem [49, 58, 90].

Let us consider the method of switching the supply voltages in the RO aimed at solving this problem [90]. The essence of the method is that the designers have changed the places of supply voltages (VDD and 0) of the inverters that are part of the RO (Fig. 2.26) [90].

As a result of this change, the effect of the BTI phenomenon is neutralized, since the V_{gs} voltages will already be equal to a maximum of <<VDD-V_{th} >> (Fig. 2.27) [90] which is not a sufficient condition for stress.

Since the proposed circuit already acts as a buffer, and the output voltage is either less or more than V_{th}, it was proposed to add a buffer after each cascade to restore the output signal level, and a NAND cell was added, one input of which is given an enabling signal, so that the signal is inverted each time, and the circuit acts as an RO (Fig. 2.28) [90].

The main disadvantages of this method are: it is necessary to add additional buffers, and the outputs of the modified inverters result in a voltage that is less than or more than V_{th}, which in modern technologies, when supply voltages can reach 0.6 V, will cause failures. Another disadvantage is that it neutralizes only the effect of the BTI phenomenon, while in VCOs, the HCI phenomenon has a greater effect.

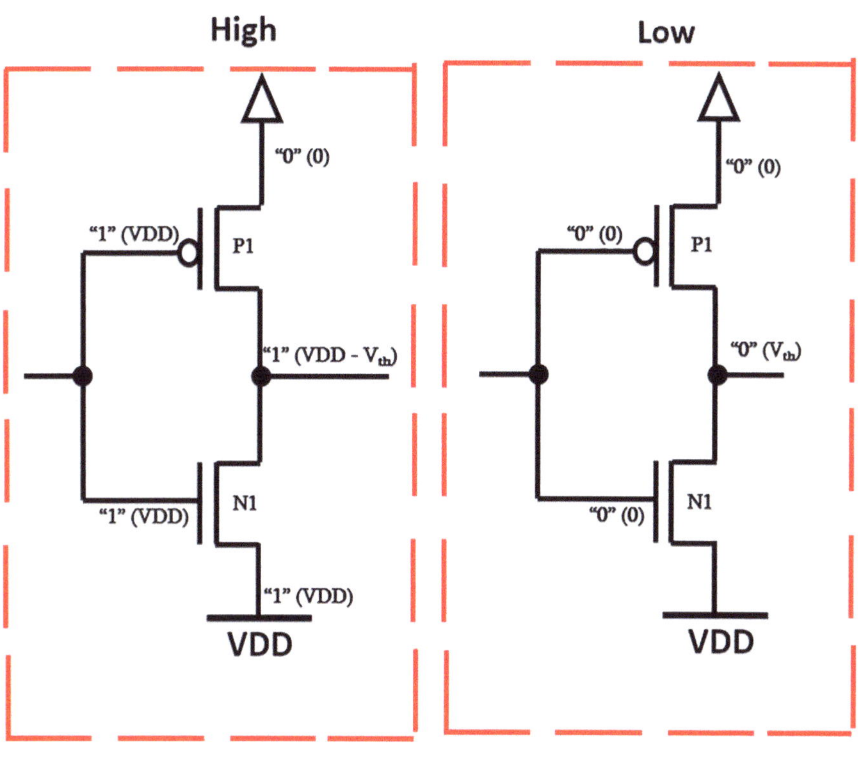

Fig. 2.26 Circuit diagram of an inverter for realization of neutralization method of BTI phenomenon of VCO

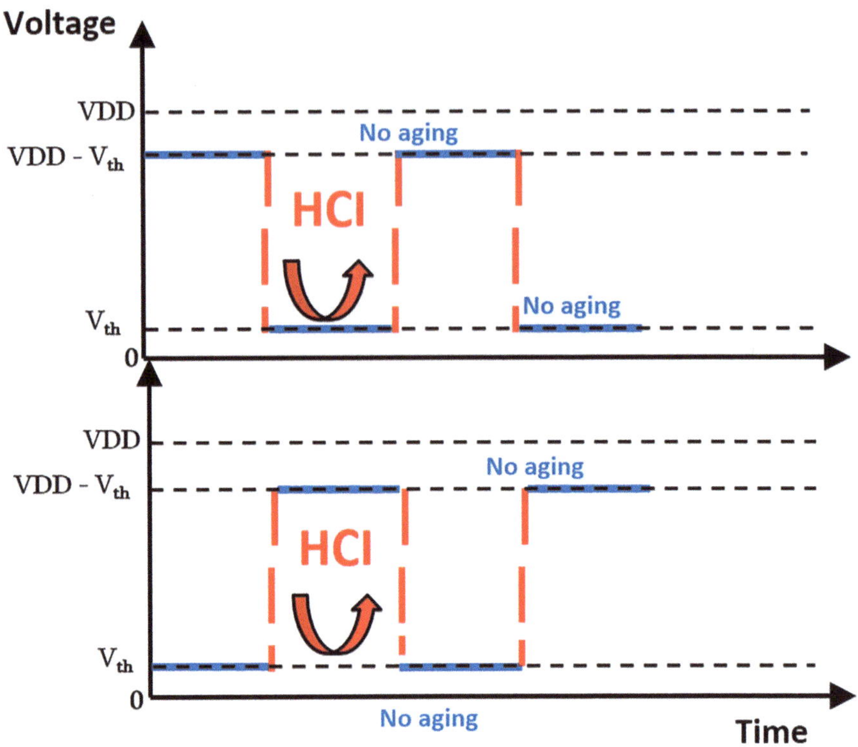

Fig. 2.27 Illustration of neutralization method of BTI phenomenon of VCO

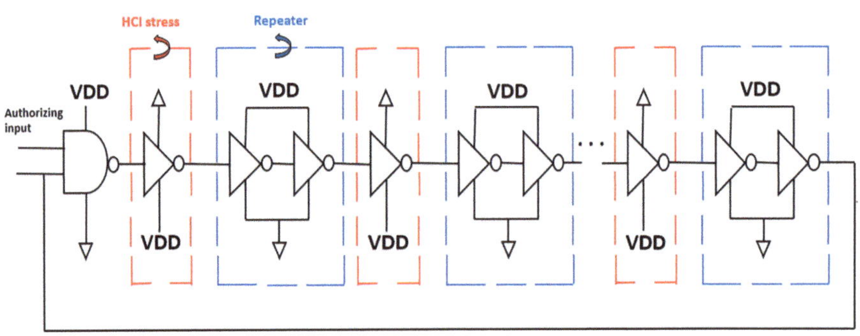

Fig. 2.28 Implementation of neutralization method of BTI phenomenon of VCO

Existing Methods for Reducing the Effects of Aging Phenomena in Operational Amplifiers with Thin Gate Oxide Transistors Only

When considering the problems of OpAmps above, it was noted that they arise due to the replacement of thick oxide gate transistors with thin oxide gate transistors.

Foundries are unable to produce thick oxide gate transistors, which leads to this problem. Thin oxide transistors are manufactured in such a way that the voltage difference between any two of their pins [gate–drain (V_{gd}), gate–source (V_{gs}), drain–source (V_{ds})] can reach a voltage equivalent to the low supply voltage (VDD), and therefore their use in circuits that are powered by a high supply voltage (VDDH) leads to the problems mentioned in the previous section.

This challenge has only recently emerged in 3 nm technologies, and there are still no known solutions to mitigate aging in OpAmps with only thin oxide gate transistors, which further increases the need to develop methods to address the problem. However, even thick oxide gate transistors can be stressed by voltage differences between their pins that are larger than the allowable voltage difference (Fig. 2.29) [5]. When V_{out} equals VDDH, V_{ds} of transistor M1 also equals VDDH and, depending on the value of VDDH, can cause stress on the transistor, which in turn will lead to aging phenomena.

One of the well-known methods for protecting transistors from stress is the cascoding method (Fig. 2.30) [5].

The essence of the method is as follows: by adding transistor M2, the output voltage of transistor M2 can reach a maximum of V_{bias}–V_{th3}. By choosing an appropriate

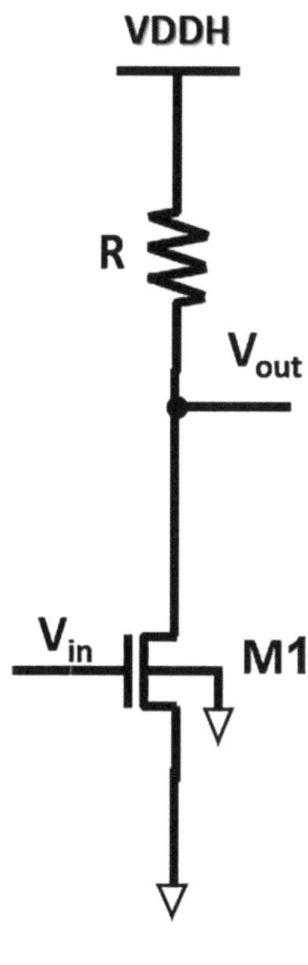

Fig. 2.29 Illustration of transistor stress conditions

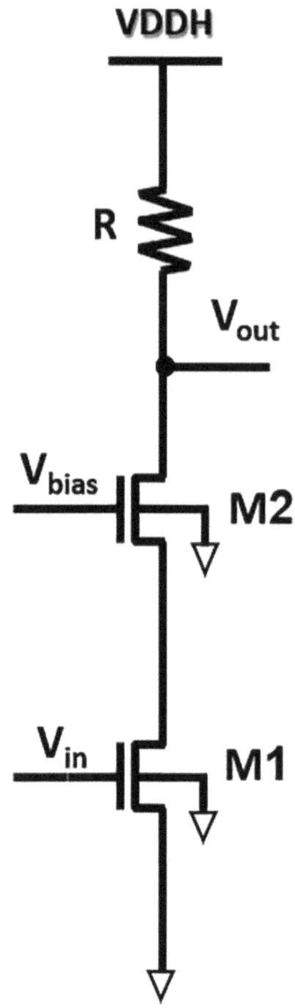

Fig. 2.30 Cascoding method

value of V_{bias}, it is possible to ensure that both transistors M1 and M2 are protected from stress, that is, from aging.

However, using this method alone will not be able to protect transistors from stress in all cases. Therefore, it is necessary to find local solutions as well.

Analysis of Existing Methods for Increasing the Accuracy of Design of Digital Standard Cell Libraries Taking Aging Phenomena into Account

The effect of aging phenomena on digital standard cells and its dependence on various parameters was presented. In order to increase the accuracy of digital standard cell libraries that take aging phenomena into account, it is necessary to perform preliminary measurements, which in turn must be accurate. There are various known approaches to increase accuracy [91–95].

In the first case, the aging of cells in a standard cell library was considered depending on the duty cycle of the input signals [92, 93]. In 1.2.3, this dependence was shown, as a result of which the effect of BTI phenomena on the cell was considered. The method developed four cases of discretization of the duty cycle sampling:

1. Uniform sampling at 5 points (U5): the sampling points are {0.1, 0.3, 0.5, 0.7, 0.9};
2. Non-uniform sampling at 3 points (NU3): the sampling points are {0.1, 0.5, 0.9};
3. Non-uniform sampling at 2 points (NU2): the sampling points are {0.1, 0.9};
4. Non-uniform worst case sampling at 2 points (NU2W): The sampling points are {0,1, 0.5}.

The disadvantage of this method is that the HCI phenomenon and possible phase deviations between the inputs of gѣџu are not taken into account.

In the other case, the aging of cells in standard cell library is considered, depending on both the input signal duty cycle and the input signal transition time [94]. This method also takes into account the effect of BTI phenomena on the cell. However, the disadvantage of this method is that the HCI phenomenon and possible phase deviations between the inputs of cells are also not taken into account.

Thus, this section showed that the existing known solutions considered have serious shortcomings, and the challenges of designing current ICs cannot be met, so it is necessary to propose new approaches to solve the emerging problems.

2.1.4 Proposed Principles to Reduce the Impact of Aging Phenomena in Integrated Circuits

Considering the shortcomings of the existing methods discussed above, the following principles have been proposed to develop new methods and solutions to reduce aging phenomena in VCOs and OpAmps with only thin oxide gate transistors and to increase accuracy in libraries of digital standard cells that take aging phenomena into account:

1. Adding a unitary code-controlled current compensation circuit to VCOs, which will make it possible to neutralize the decrease in the frequency of the output signal due to the effects of aging phenomena.

The results of simulations performed for one of the well-known VCO circuits showed that the output frequency of the VCO decreases by 6.5% after 10 years of aging, and what is worse, most of this decrease occurs during the first year. The other method neutralized only the effect of the BTI phenomenon and required addition of additional buffers, and also could lead to problems at a supply voltage of 0.6 V. It is proposed to add a unitary code-controlled leakage compensation circuit to VCOs, at the expense of a minimal increase in area, through which it will be possible to neutralize the decrease in the output signal frequency caused by the influence of aging phenomena.

2. Development of a method for continuously monitoring and recording aging phenomena and design of an intra-die circuit that will generate a digital code.

Based on the problems discussed, as well as the inadequacy of existing methods, it is proposed to develop a method for continuously monitoring and recording aging phenomena and to design an intra-die circuit that will generate a digital code, thanks to which it will be possible to neutralize the effects of aging phenomena with compensation circuits. The proposed method can be interconnected with method 1 as well as with many other existing or proposed compensation circuits in the future.

3. Reducing the effects of aging phenomena on the gain by neutralizing stress conditions in three types of OpAmps with only thin oxide gate transistors.

The results of simulations performed for the known circuits of three types of OpAmps designed only with thin oxide transistors showed that their gain decreases due to the influence of aging phenomena. Since this problem has arisen quite recently, there are no known approaches to solving the problem, and new approaches are needed. It is proposed to reduce the effects of aging phenomena in three types of OpAmps with only thin oxide transistors by neutralizing stress conditions with several types of solutions.

4. Increasing the accuracy of digital standard cell libraries that take aging phenomena into account, considering phase deviations between input signals of standard cells.

Thus, the effects of aging phenomena on standard cells were demonstrated through simulations, as well as the dependence of the effect of aging phenomena on the phase deviation between the input signals of the cell. None of the methods considered included the phase deviation between the input signals of the cell when designing digital standard cell libraries that take aging phenomena into account. Therefore, it is proposed to increase the accuracy of digital standard cell libraries that take into account the phase deviations between the input signals of standard cells.

Conclusions

1. With the continuous scaling of modern technologies, the increasing influence of aging phenomena has become a key issue in the design of integrated circuits. It can lead to a decrease in the reliability of integrated circuits, even to their failure. Therefore, the development of methods for the design of integrated circuits taking into account aging phenomena is an extremely urgent problem.
2. Taking into account aging phenomena, existing methods for designing integrated circuits are not sufficient to overcome current challenges, and in some cases they are even absent, which leads to the need to develop new methods.
3. Approaches to reduce the effects of aging phenomena in integrated circuits have been proposed, which ensure necessary operating conditions on the account of power consumption, occupied area and simulation time within permissible limits.

2.2 Proposed Solutions to Design Integrated Circuits Taking into Account Aging Phenomena

The problems described in Sect. 2.1 indicate the shortcomings of the methods for designing ICs with aging phenomena in mind. The increasing influence of aging phenomena in ICs has led to tightening requirements for IC reliability. It is necessary to solve the problems existing in the present methods or to develop new methods that should also take into account the challenges of designing modern ICs. The following are proposed methods for designing ICs with aging phenomena, which solve the shortcomings of the existing methods, taking into account the current challenges.

2.2.1 Method of Compensating for the Effects of Aging Phenomena in Voltage-Controlled Oscillators

As mentioned, the VCO circuit has a decrease in the frequency of the output signal, which occurs due to the influence of aging phenomena, due to which the V_{th} of transistors increases and the I_d decreases. The considered method made it possible to reduce the decrease in the frequency of the VCO output signal by changing the supply voltage locations of the RO cells and adding a buffer after them. As a result of applying the method, the area of the circuit increased due to the addition of buffers, and the voltage at the output of the RO cells was obtained by V_{th} less or more, which in modern technologies can lead to signal distortion due to low supply voltages. In addition, the method neutralized only the effect of the BTI phenomenon, when the main cause of aging in the VCO is the HCI phenomenon.

To overcome these shortcomings, a method has been proposed to compensate for the frequency decrease due to aging phenomena in VCOs [70].

In order to reduce the frequency drop, a modification was initially made to the voltage-to-current converter circuit (Fig. 2.31) [70].

In the modified circuit, the branches of the voltage-current converter have been separated, and instead of the control transistors, current mirrors have been connected [5, 77], which will make it possible to transmit a constant current that will not change due to the effects of aging phenomena. Such current can be transmitted from the support voltage sources [5, 77, 96, 97]. Thanks to this modification, it will be possible to keep the current flowing M5 and M6 transistors the same, since I_{ref} does not depend on aging, and as current mirrors age, their V_{gs} changes so that the current remains the same, i.e., they are self-compensating. The value of I_{ref} is chosen so that the same current flows in M5 and M6 transistors, and the n_ctrl and p_ctrl voltages are the same as before the change, resulting in the same frequency at the output. After this change, a time analysis was performed using HSPICE simulation, in the initial state, after and during 10 years of aging (Figs. 2.32 and 2.33) [70].

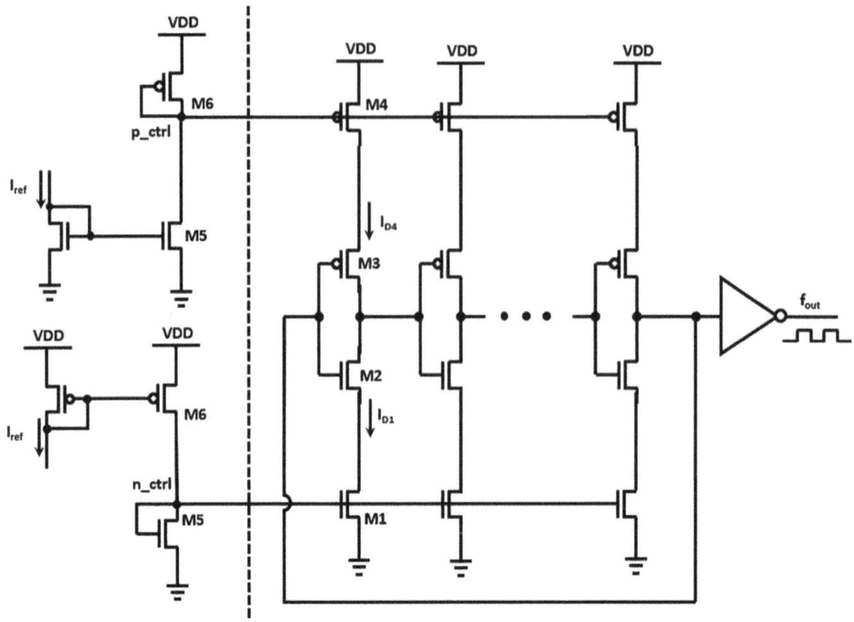

Fig. 2.31 VCO with modified voltage-current converter

The results show that the frequency was initially the same as before the change, and after 10 years of aging, it decreased less, and the decrease became 5%. This difference is still quite large. This indicates that the main reason for the decrease in frequency is the aging of RO cells, and it is necessary to develop a compensating method.

For this purpose, a method was developed and implemented, as a result of which 32 NMOS keys controlled by a digital unitary code and NMOS and PMOS transistors with the same current value performing the function of a current source were added to the circuit, the current of which is transferred from the support voltage sources through current mirrors (Fig. 2.34) [70].

The circuit works as follows: initially, when the aging phenomena have not appeared, all branches of the compensation circuit are closed, and the VCO works as in the previous cases. When, as a result of aging, the VCO frequency decreases by 10 MHz, a signal comes to the compensation circuit from the recording circuits, and one branch of the circuit opens, and the current of this branch is added to the total current, as a result of which the frequency increases and reaches its initial value. As a result of a further decrease in the frequency, the remaining branches open one by one, compensating for the decrease in frequency (Fig. 2.35) [70].

The results show that the largest frequency reduction is 0.31%. Comparing the frequency of the VCO output signal before and after applying the method (Fig. 2.36) (Table 2.5) [70], its effectiveness can be verified.

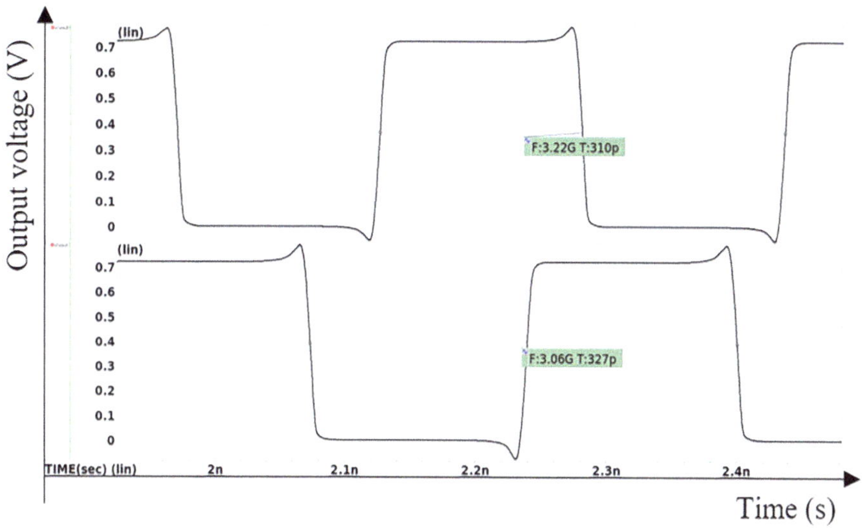

Fig. 2.32 View of VCO output signal with modified voltage–current converter, initially and after 10 years from aging

Thus, the proposed method, at the expense of only 13.6% of the area and 15.1% of the power consumption of the VCO circuit described above, provides a reduction in the output frequency decrease due to the aging effect of the circuit from 6.5% to 0.31%. The area of the circuit has increased due to the addition of compensating transistors. The method does not cause a decrease in the voltage level and can work without failure in modern technologies, and also compensates for the effects of both BTI and HCI phenomena.

2.2.2 Method for Aging Monitoring in Integrated Circuits

The proposed circuit in the method described in Sect. 2.2.1 requires an input code that must be generated by other, monitoring circuits. Most of the available means require additional pins on the IC, which is associated with high costs. For this purpose, a method for monitoring and recording aging phenomena in ICs has been proposed, and based on this, an intra-die circuit [98].

Initially, the effect of aging phenomena on a SAED 14 nm FinFET PMOS transistor (Fig. 2.37) was studied [98]. It should serve as a means to record aging. The source of the PMOS transistor is connected to VDD and the drain is connected to GND. Reference voltage is given to the gate which is equal to 0 V to ensure maximum V_{gs}. An HSPICE simulation was performed, which revealed that a current of 140 μA flows through the transistor [98].

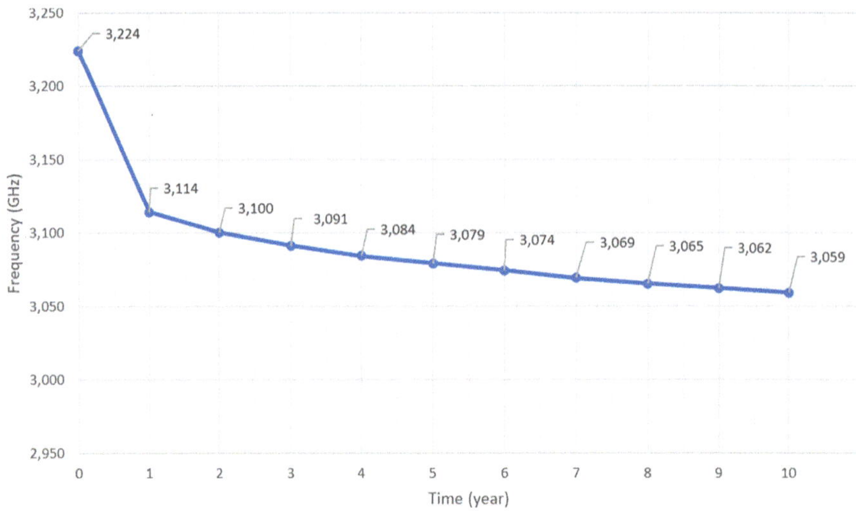

Fig. 2.33 Dependence of VCO output frequency with modified voltage–current converter on years

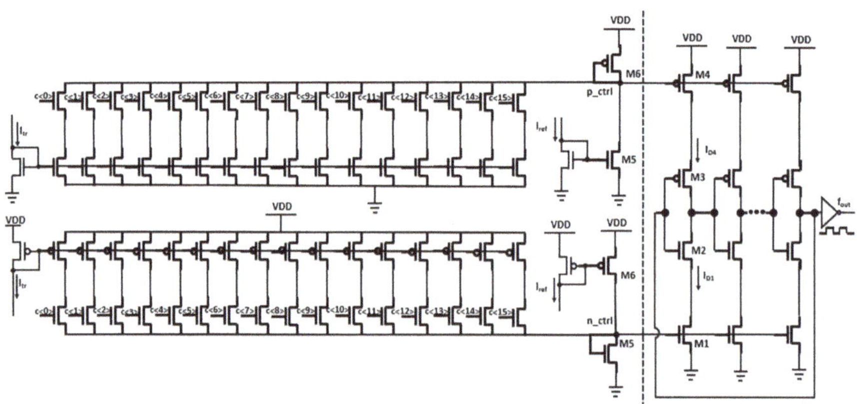

Fig. 2.34 View of VCO circuit after adding compensating circuit

HSPICE simulations were then performed, aging the circuit for 10 years. The results showed that I_{ds} decreased by 30 μA and is 110 μA, which is a fairly large change. HSPICE simulations were also performed from 1 to 10 years to understand how I_{ds} changes (Fig. 2.38) [98].

The results show that the largest decrease occurred during the first year, which further increases the need to design a circuit that captures aging and, starting from the first year, compensate for its effects in the necessary circuits.

Taking into account the above, a method has been developed, and based on it, a circuit has been designed that will record the aging phenomena and, depending on

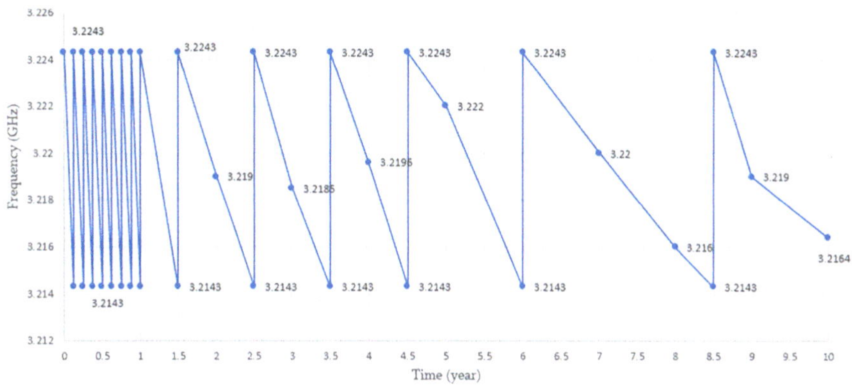

Fig. 2.35 Dependence of output frequency of VCO with compensating circuit on years

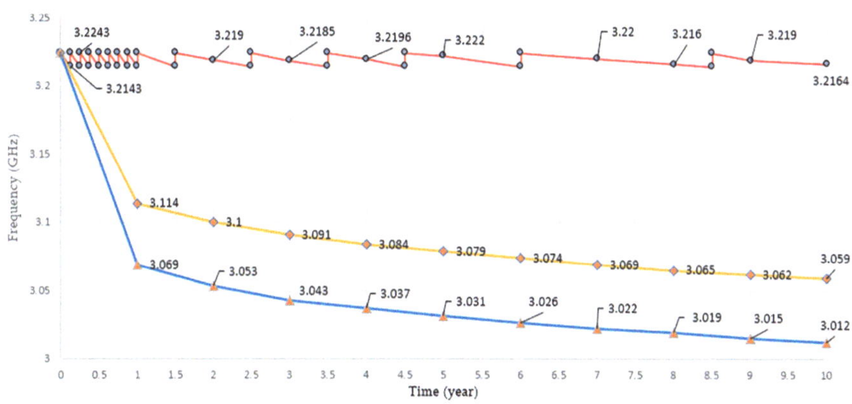

Fig. 2.36 Comparison of the output frequency dependences on years for the initial VCO circuit, modified with voltage–current converter circuit and a compensating circuit

Table 2.5 Comparison of the output frequency dependences on years for the original VCO circuit, the circuit with a modified voltage-to-current converter, and the compensator circuit

Circuit	Initial frequency (GHz)	Minimum frequency (GHz)	Difference (%)
Initial circuit	3.2243	3012	6.5
Circuit with modified voltage-to-current converter	3.2243	3059	5
Circuit with a compensating circuit	3.2243	3.2143	0.31

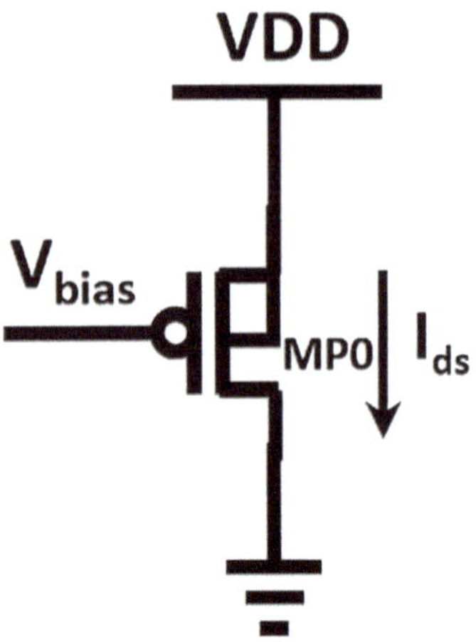

Fig. 2.37 PMOS transistor

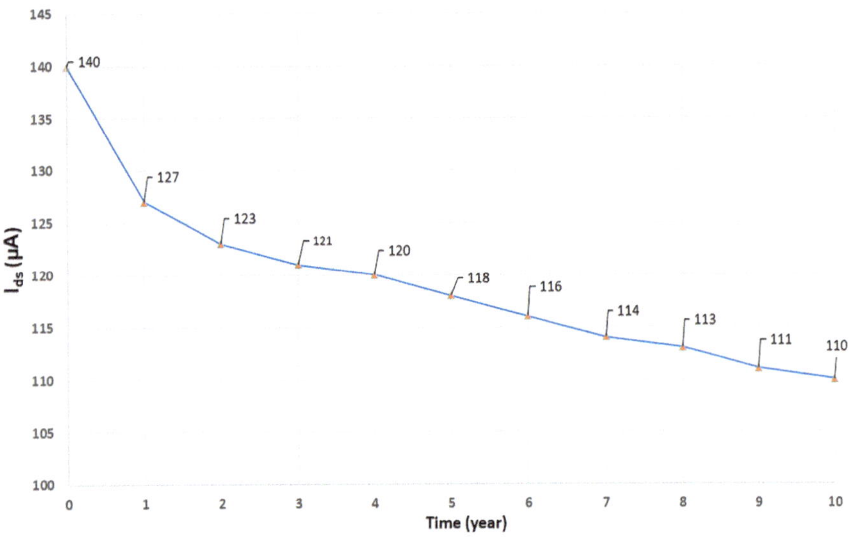

Fig. 2.38 Dependence of I_{ds} on years for a PMOS transistor

it, generate the corresponding digital code using the digital search algorithm (Fig. 2.39) [98, 99]. The generated code can be further used in existing compensating circuits as well as in the circuit proposed above [70].

To verify the accuracy of the method and to demonstrate the principle of operation, the circuit was initially designed with an ideal current source. The circuit consists of three cascades. In the first cascade, an ideal current source is connected, the current of which is reduced from 140 μA to 110 μA. A current mirror is sequentially connected to it which conducts a current of 125 μA, as an average value between 140 μA and 110 μA. After them, diode-connected transistors M1 (PMOS) and M2 (NMOS) are connected, which serve as a current subtractor. It is designed in such a

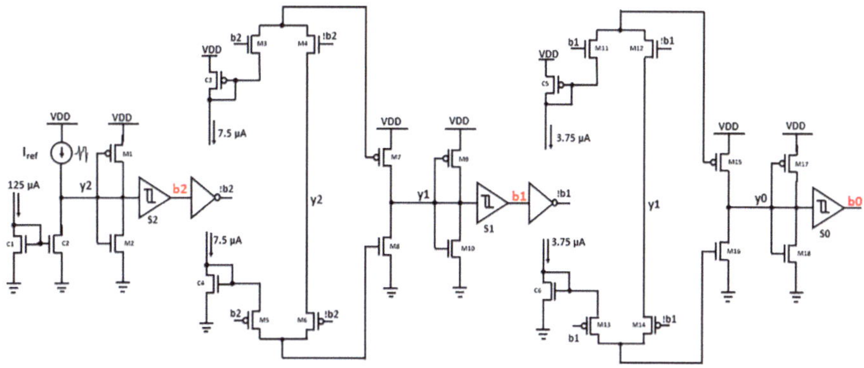

Fig. 2.39 Aging monitoring circuit designed with an ideal current source

Fig. 2.40 Schematic view of the designed Schmidt trigger

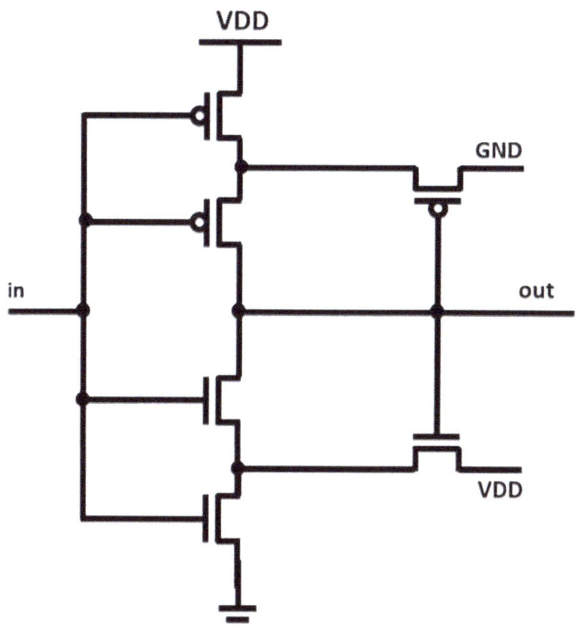

way that when the currents flowing through them are the same, the potential of point y2 is equal to VDD/2. y2 is also connected to the input of the S2 Schmidt trigger (Fig. 2.40) [77, 98, 100–103]. The designed Schmidt trigger has a threshold voltage of VDD/2, and uses positive feedback to add hysteresis, which provides a noise margin. The output of the S2 Schmidt trigger, b2, is the first bit of the generated digital code. b2 is inverted to drive the second cascade. When I_{ref} is >125 μA, their difference flows through transistor M2, which lowers the potential of y2, and the output of the S2 Schmidt trigger becomes 0. When I_{ref} is <125 μA, their difference flows through transistor M1, which raises the potential of y2, and the output of the S2 Schmidt trigger becomes 1. This produces the first bit of the code [98].

At the beginning of the second cascade, diode-connected transistors C3 and C4 are connected in two different branches, which conduct a current of 7.5 μA, which is half the maximum difference possible in the first cascade (15 μA). Transistors M3, M4, M5, and M6 serve as switches. When b2 is 0, transistors M3 and M6 are closed, and transistors M4 and M5 are open. Since M5 is open, M8 forms a current mirror with C4, and a current of 7.5 μA begins to flow through it, and M7 forms a current mirror with M1, and a current of the difference between I and 125 μA flows through it. If it is >7.5 μA, the difference will flow through transistor M10, which lowers the potential of y1, and the output of Schmidt trigger S1 becomes 0. If it is <7.5 μA, the difference will flow through transistor M9, which raises the potential of y1, and the output of Schmidt trigger S1 becomes 1. The output of Schmidt trigger S1, b1, is the second bit of the generated digital code. b1 is inverted to control the third cascade [98].

The third cascade has exactly the same appearance and function as the second one, with the only difference being that the current values are equal to 3.75 μA, which is already half the maximum difference possible in the second cascade (7.5 μA). There is no need for inversion there. The output of the third cascade is the signal b0, which is the third bit of the digital code [98].

To better understand the principle of operation of the circuit, two examples are considered. Let's assume that I_{ref} is equal to 135 μA. In this case, the difference between I_{ref} and 125 μA is 10 μA, which flows through transistor M2, and b2 becomes 0. The M5 switch opens, and a current of 7.5 μA begins to flow through M8. At the same time, the M4 switch also opens, and the difference in current in the first cascade, namely 10 μA, begins to flow through transistor M7. The difference between them (2.5 μA) flows through transistor M10, and b1 becomes 0. The M13 switch opens, and a current of 3.75 μA begins to flow through M16. At the same time, the M12 switch also opens, and the difference in current in the second cascade, namely 2.5 μA, begins to flow through the M15 transistor. The difference between them (1.25 μA) flows through the M17 transistor, and b0 becomes 1. Thus, the resulting code is 001 [98].

The next example considers the case when I_{ref} is <125 μA. Let's assume that I $_{ref}$ is equal to 115 μA. In this case, the difference between I_{ref} and 125 μA is −10 μA, which flows through transistor M1, and b2 becomes 1. Switch M3 opens, and a current of 7.5 μA begins to flow through M7. At the same time, switch M6 also opens, and the difference in current in the first cascade, namely 10 μA, begins to flow through transistor M8. The difference between them (−2.5 μA) flows through transistor M9, and b1 becomes 1. Switch M11 opens, and a current of 3.75 μA begins to flow through M15. At the same time, the M14 switch also opens, and the difference in current in the second cascade, namely 2.5 μA, begins to flow through the M16 transistor. The difference between them (1.25 μA) flows through the M18 transistor, and b0 becomes 0. Thus, in this case, the resulting code is 110.

To verify the correct operation of the designed circuit, HSPICE simulations were performed. The value of the ideal current source I_{ref} was changed from 140 μA to 110 μA (Fig. 2.41) [98].

The circuit generates a corresponding 3-bit digital code for each current change interval (Table 2.6) [98].

After verifying the correct operation of the circuit, the ideal current source was replaced with a PMOS transistor acting as a current source (Fig. 2.42) [98].

HSPICE simulations were performed for this circuit over a period of 0 to 10 years (Fig. 2.43) [98].

As a result of simulations, it becomes clear that the proposed aging monitoring circuit performs its function, namely, it generates a digital code in accordance with the current change due to the aging effects. This code can be converted to other types of codes and used in existing compensating circuits. In addition, if greater accuracy is required, it is possible to increase the number of bits of the digital code by increasing the number of cascades according to the same principle [98].

Thus, due to its simple structure, the proposed aging monitoring circuit is easily implemented in CMOS technology using only MOS transistors. It does not require any additional outputs on the IC for its operation, and there is no need for testing. The designed circuit leads to a reduction in area, testing time, and cost of ICs, and an increase in reliability, occupying an area of 10 μm^2 and consuming ~300 μA of current.

Fig. 2.41 View of circuit outputs, depending on I_{ref} change

Table 2.6 Generated digital codes for current change ranges

ΔI_{ref} (μA)	Numeric code
0–3.75	000
3.75–7.5	001
7.5–11.25	010
11.25–15	011
15–18.75	100
18.75–22.5	101
22.5–26.25	110
26.25–30	111

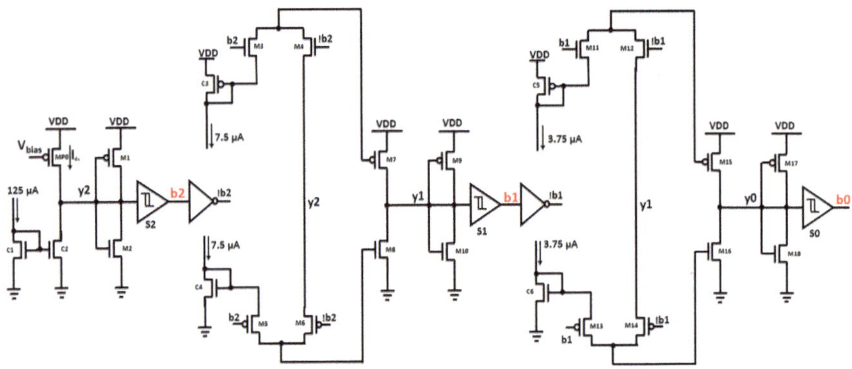

Fig. 2.42 Proposed aging monitoring and recording circuit

2.2.3 Method to Reduce the Effects of Aging Phenomena in Operational Amplifiers with Only Thin Oxide-Layer Transistors

When designing the known circuits of the three types of OpAmps considered above with only transistors with a thin oxide layer of the gate, deviations occurred due to the influence of aging phenomena, which led to a decrease in the gain of OpAmps. Aging phenomena appeared mainly in the off state of the circuits. Since then OpAmps have been designed with transistors with a thick oxide layer of the gate, and this problem has arisen quite recently, therefore, known methods for protecting OpAmps with only transistors with a thin oxide layer of the gate from stress conditions do not yet exist.

Taking these circumstances into account, methods for reducing the impact of aging phenomena in the three types of OpAmps with only thin oxide-layer transistors considered above have been proposed [76, 83]. These methods are considered below for each case separately.

Two-Stage OpAmp: To solve the problems described above, the following modifications were made to the two-stage OpAmp (Fig. 2.44) [76].

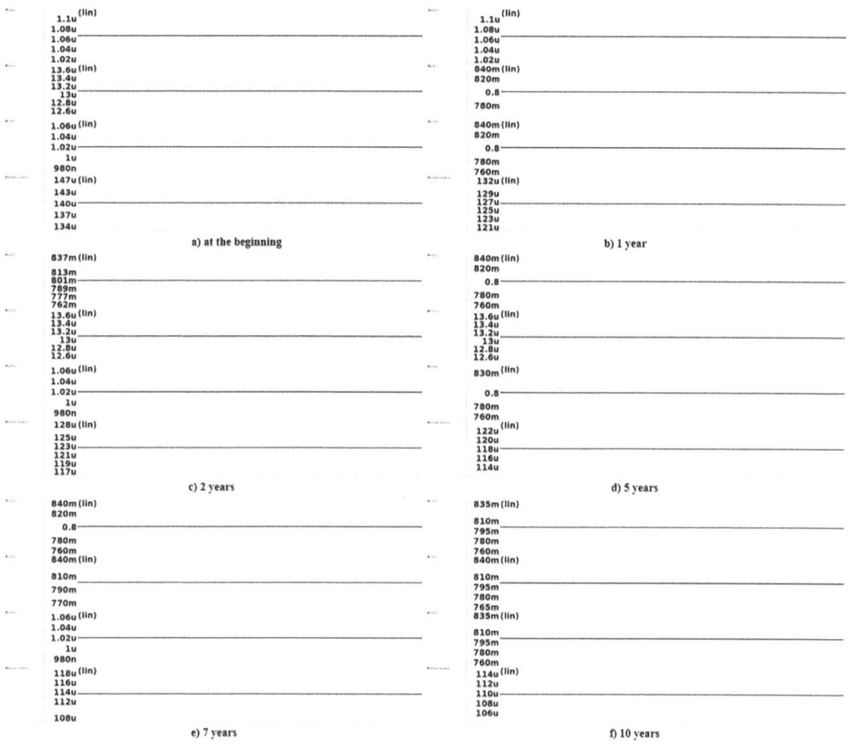

Fig. 2.43 View of the outputs of the designed circuit over the years

1. The switching signal was separated for NMOS and PMOS transistors. Their logical 0 and 1 levels were also changed using a multi-output voltage level converter [0 to vnbias (0.8 V) for NMOS and vpbias (0.4 V) for VDDH for PMOS] [104].
2. A cascoding method was used by adding transistors D1, D2, and D3 (marked with red dotted lines).

The vnbias and vpbias voltages are formed from voltage dividers or voltage-suppressing sources [5, 96, 97, 105]. Transistors D1, D2, and D3 are used for cascoding, protecting neighboring transistors from stress conditions, and do not change the functionality of the circuit. After the modifications, HSPICE simulations were performed, aging the circuit for 10 years. The results showed that there were no significant changes in the V_{th} and I_d parameters in either the on or off state (Table 2.7) [76], as a result of which the gain was not changed.

Folded Cascode OpAmp: To solve the problems described above, the following modifications were made to the folded cascode operational amplifier (OpAmp) (Fig. 2.45) [76].

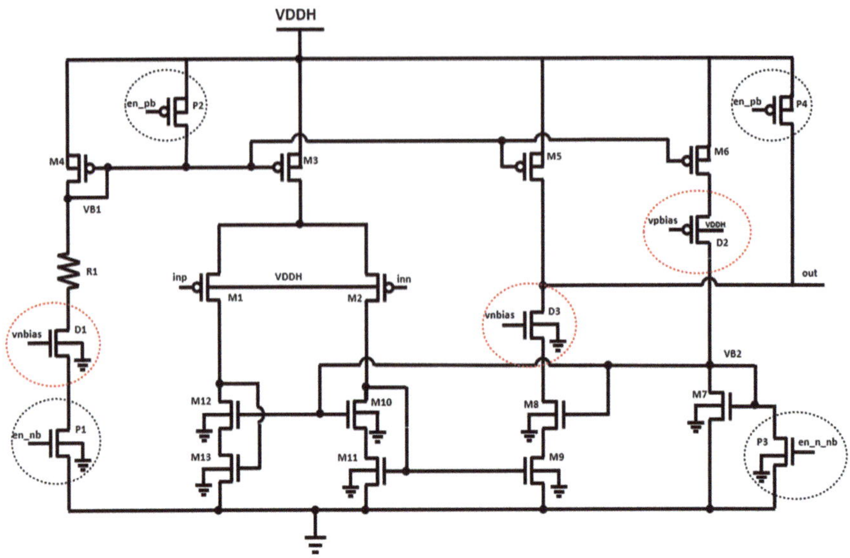

Fig. 2.44 Schematic view of a two-stage OpAmp after modifications

Table 2.7 After modifications, in the result of 10-year aging of the two-stage OpAmp, maximum deviations of I_d and V_{th} of transistors

	On state		Off state	
Transistor	ΔV_{th} (mV)	ΔI_d (%)	ΔV_{th} (mV)	ΔI_d (%)
M6	18	1.7	0.1	0.1
M8	15	1.2	0.1	0.1
P1	17	1.4	0.1	0.1
P2, P4	0.1	0.1	6	1.2
P3	0.1	0.1	6	1.2

1. The switching signal was separated for NMOS and PMOS transistors. Their logic 0 and 1 levels were also changed using a multi-output voltage level converter [0 to vnbias (0.8 V) for NMOS and vpbias (0.4 V) for VDDH for PMOS] [104].
2. The cascoding method was used by adding the D1 transistor (marked with red dotted lines).
3. Transistors D2, D3, and D4 were added to connect the floating pins to vnbias in the off state of the circuit (marked with red dotted lines).
4. The P4 NMOS transistor has been replaced with a PMOS transistor, and the source has been connected to vnbias instead of gnd (indicated by the red dotted lines).

Transistor D1 protects transistor M9 from stress conditions. Transistor P4 connects the gates of transistors M7 and M8 to vnbias in the off state of the circuit, while transistors D2 and D3 connect the sources of transistors M7 and M8 to vnbias,

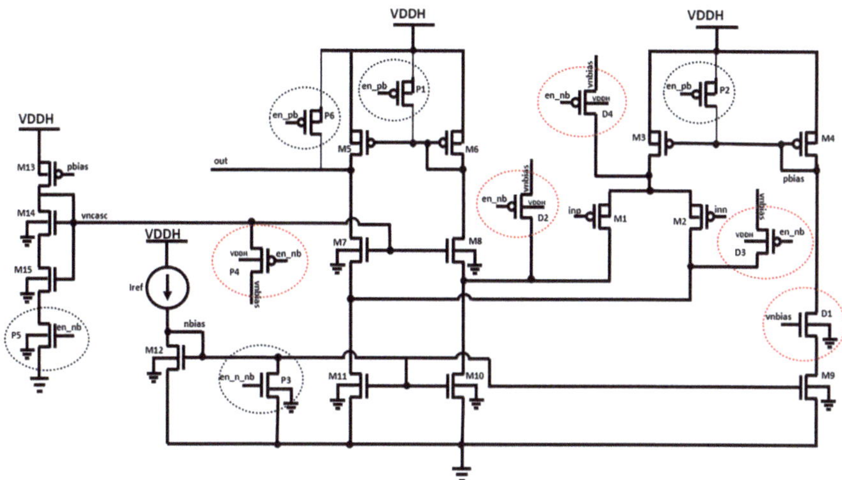

Fig. 2.45 Schematic view of the folded cascode OpAmp after modifications

Table 2.8 After modifications, in the result of 10-year aging of the folded cascode OpAmp, maximum deviations of I_d and V_{th} of transistors

	On state		Off state	
Transistor	ΔV_{th} (mV)	ΔI_d (%)	ΔV_{th} (mV)	ΔI_d (%)
M7, M8	22	2.9	0.1	0.1
M9	18	2.6	0.1	0.1
P1, P2, P6	0.1	0.1	20	2.8
P3, P4	0.1	0.1	20	2.8
P5	17	2.1	0.1	0.1

respectively. As a result, they will be closed in the off state, and there will be no stress conditions between their pins. After these modifications, HSPICE simulations were performed, aging the circuit for 10 years. The results showed that there are no significant changes in the V_{th} and I_d parameters in either the on or off state (Table 2.8) [76], as a result of which the gain did not change.

To solve the problems described above, the following modifications were made to the rail-to-rail OpAmp (Fig. 2.46) [83].

1. The switching signal was separated for NMOS and PMOS transistors. Their logic 0 and 1 levels were also changed using a multi-output voltage level converter (0 to vnbias (0.8 V) for NMOS and vpbias (0.4 V) for VDDH for PMOS) [104].
2. The cascoding method was used by adding transistors D1, D2, D3, D4, D5, and D6 (marked with red dotted lines).
3. A T1 transfer gate has been added to separate the pbias1 and pbias2 wires when the circuit is off (marked with red dotted lines).

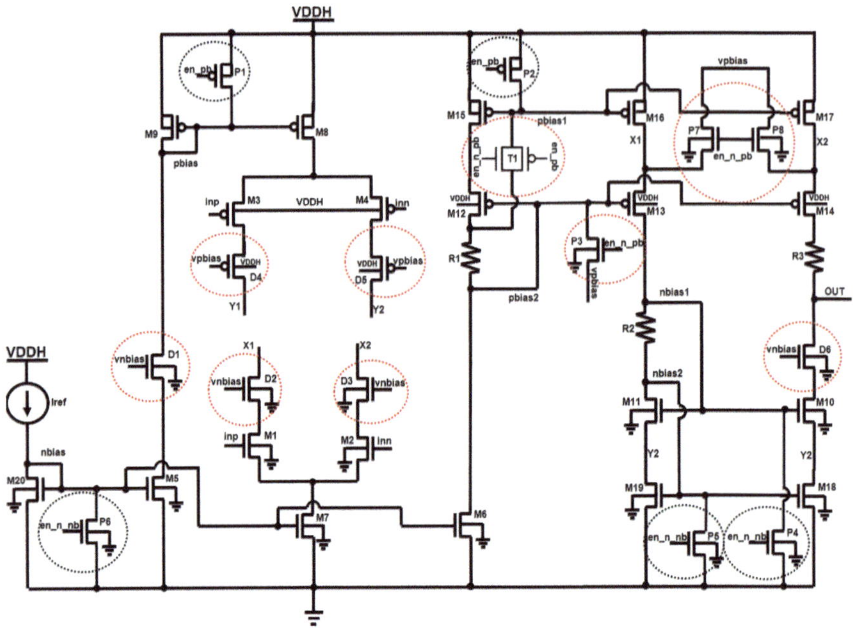

Fig. 2.46 Schematic view of a rail-to-rail OpAmp after modifications

4. The P3 PMOS transistor has been replaced with an NMOS transistor, and the source has been connected to vnbias instead of VDDH (indicated by the red dotted lines).
5. Transistors P7 and P8 have been added to connect the x1 and x2 wires to vpbias, respectively, when the circuit is off (marked with red dotted lines).

Transistor D1 protects transistor M5 from stress. Transistors D2 and D3 protect transistors M1 and M2, respectively. Transistors D4 and D5 will protect transistors M3 and M4 when the input signal voltage level is high. And transistor D6 protects transistor M10. Transistor P3 connects the gates of transistors M13 and M14 to vpbias, at the same time, transistor P7 connects the sources of M13 and transistor P8 connect the sources of M14 to vpbias, therefore, there will be no stress conditions between their pins. The output of transistor M6 is connected to vpbias, so it will not be stressed either. It is necessary to separate pbias1 from the R1 resistor pin using the T1 transfer gate, since otherwise one pin of the R1 resistor would be connected to VDDH and the other to vpbias, and a constant current would flow through the R1 resistor, which would cause undesirable power consumption. After the modifications were made, HSPICE simulations were performed, aging the circuit for 10 years. The results showed that there were no significant changes in the V_{th} and I_d parameters in either the on or off state (Table 2.9) [83], as a result of which the gain did not change.

Table 2.9 After modifications, in the result of 10-year aging of the rail-to-rail OpAmp, maximum deviations of I_d and V_{th} of transistors

	On state		Off state	
Transistor	ΔV_{th} (mV)	ΔI_d (%)	ΔV_{th} (mV)	ΔI_d (%)
M1, M2	21	3.5	8	1.4
M5, M6	22	3.6	0.1	0.1
M10	18	2.7	0.1	0.1
M13, M14	28	3.9	0.1	0.1
P1, P2, P3	0.1	0.1	15	2.1
P4, P5, P6	0.1	0.1	15	2.1

Table 2.10 Parameter comparison before and after modifications in the result of 10-year aging of OpAmps

	Two-stage OpAmp		Folded cascode OpAmp		Rail-to-Rail OpAmp	
Parameter	Before modification	After modification	Before modification	After modification	Before modification	After modification
Gain (dB)	40.2	54.2	66.2	85.2	40.1	52.3
ΔV_{th} (mV)	198	18	254	22	252	28
ΔI_d (%)	21.6	1.7	24.3	2.9	22.4	3.9

Generalizing the results of the three OpAmps considered, it becomes clear that after the modifications made, it was possible to reduce the effect of aging phenomena on the parameters of the OpAmps (Table 2.10) [76, 83].

Thus, after the modifications made as a result of applying the method, due to the increase in the area of the two-stage OpAmp by 3.2%, the folded cascode OpAmp by 4.8%, and the rail-to-rail OpAmp by 7.1%, it was possible to reduce the effect of aging phenomena on the gain of the OpAmps. It turns out that it does not change over 10 years. Before the modifications, the gain of the two-stage OpAmp decreased by 14 dB over 10 years, that of the folded cascode OpAmp by 19 dB, and that of the rail-to-rail OpAmp by 12 dB.

2.2.4 *Method for Increasing the Accuracy of Designing of Digital Standard Cell Libraries Taking into Account Aging Phenomena*

In the previous section it was shown that the effect of aging phenomena can have different consequences on the delay variation of standard cells, depending on the phase deviation between the signals. Therefore, taking the latter into account at the design stage of standard cell libraries will increase their accuracy. The existing methods considered did not take into account the possible phase deviations between the input signals of standard cells, as well as the effect of the HCI phenomenon on the delay variation. Failure to take them into account can lead to low accuracy, which can result in failures in critical paths.

To avoid the above problem, a method has been developed, the essence of which is that it is proposed to take into account possible phase deviations between the input signals of the cells when designing digital standard cell libraries in order to increase the accuracy of the effect of aging phenomena on standard cell libraries [89].

Suppose there is need to design a library of standard cells, the cells of which are elements of the set A $[A_1, A_2, \ldots, A_n]$, where n is the number of cells in the library, and the points of the duty cycle are elements of the set B $[B_1, B_2, \ldots, B_m]$, where m is the number of points of the duty cycle under consideration. For each cell from set A, m combinations from set B must be described for each entry. Therefore, the number of possible combinations of the duty cycle for one cell will be [93]:

$$P_{\mathrm{SC}} = m^k, \tag{2.6}$$

where k is the number of entries in a standard cell.

To improve the accuracy of standard cell libraries, it is proposed to describe cells by taking into account the set of phase deviations C $[C_1, C_2, \ldots, C_h]$ between the input signals, where h is the number of deviations of the neighboring signals. The number of combinations per cell, considering the possible phase deviations between the input signals, will be [89]:

$$P_{\mathrm{SC}-g} = h^{k-1} m^k, \tag{2.7}$$

$k-1$, because the first signal does not need to be deflected.

The number of descriptions for the entire set A will be [89]:

$$P_{G-g} = \sum_{i=1}^{n} P_{\mathrm{SC}-g_i} = \sum_{i=1}^{n} h^{k_i-1} m^{k_i} \tag{2.8}$$

Depending on the effects of possible phase deviations between input signals, it is proposed to describe digital standard cell libraries in three main cases: best, worst and average cases. The worst and best cases contain information that corresponds to the maximum and minimum effect of the phase deviation between input signals on the delay change. This will allow to check the most sensitive parts of the ICs. In contrast to the worst and best cases, the average case corresponds to the average effect of the phase deviation between input signals on the delay change. Since the phase deviations between input signals of standard cells during the life of ICs can vary and have a random nature, the use of the proposed average case will increase the accuracy of delay measurements after the effects of aging phenomena, in contrast to existing methods. In order to obtain appropriate digital standard cell libraries in the mentioned cases, it is necessary to consider different phase deviations between input signals for each combination of duty cycles and then select three of them: best, worst and average.

The accuracy of the description of a library of digital standard cells depends on the choice of sets B and C. The sets will describe as many delay changes as they have sampling points. For set B, four different duty cycle discretization cases are known from the available methods: uniform sampling at 5 points (U5), non-uniform sampling at 3 points (NU3), non-uniform sampling at 2 points (NU2), and non-uniform worst case sampling at 2 points (NU2W) [93]. The NU3 case was chosen as the golden mean between the description accuracy and the library size. NU3 chooses sampling at 3 points: B [0.1, 0.5, 0.9].

Since the number of elements in the set C of phase deviations between the input signals has no effect on the size of the standard cell library (it only affects the simulation time and the computer resources used), the set C can be chosen as large as necessary to ensure high accuracy of the library description.

Summarizing the above, the essence of the method can be summarized (Fig. 2.47) [89].

1. Standard cell selection in set A,
2. Output load selection,
3. Selection of input edges,
4. Selection of duty cycles for set B,
5. Selection of phase deviations of set C,
6. Summary of results,
7. Worst, best and average case selection.

For the XOR cell, HSPICE simulations were performed for NU3 combinations of input signal duty cycles, varying the phase shift between inputs, and aging the circuit for 10 years, considering the worst case, best case, and average effects on the delay change for three main process-voltage-temperature (PVT) cases (Fig. 2.48) [89].

A comparison was made between the developed method and existing methods in case of the XOR cell (Table 2.11) [89]. The number of simulations was determined based on formulas 2.1–2.3.

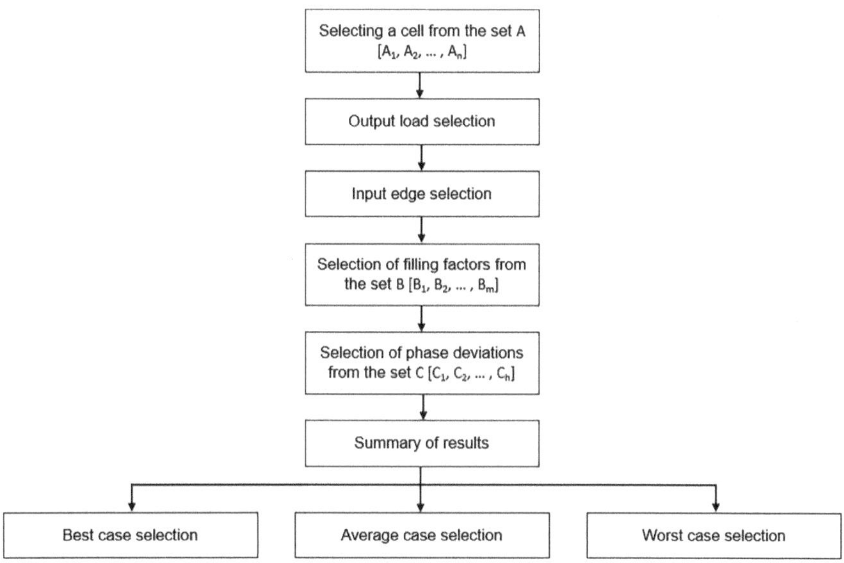

Fig. 2.47 The execution path of the proposed method

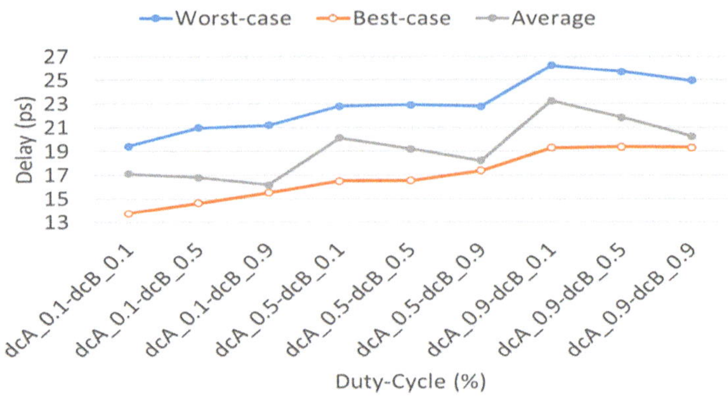

a) ff/125°C/1.1V

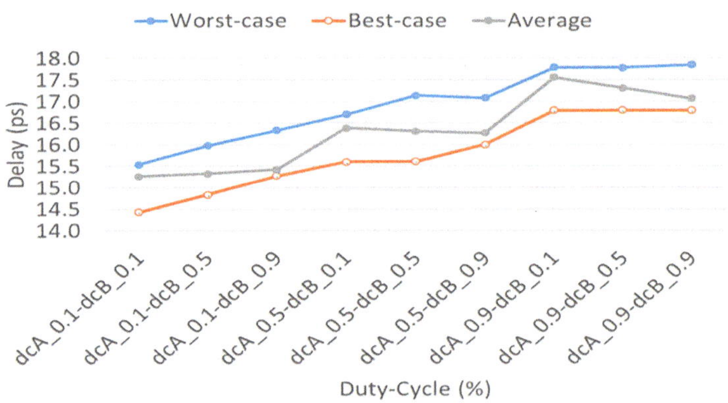

b) tt/25°C/1V

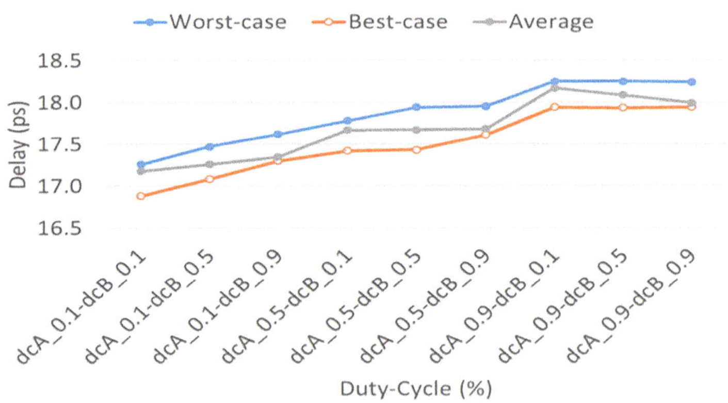

c) ss/-40°C/0.9V

Fig. 2.48 Best, worst, and average cases of the dependence of the delay change due to 10 years of aging on the phase deviation of the input signals in cases of NU3 combinations of the duty cycle of the input signals in three different PVT cases of the XOR cell. (**a**) ff/125 °C/1.1 V, (**b**) tt/25 °C/1 V, (**c**) ss/−40 °C/0.9 V

Table 2.11 Comparison of proposed and existing methods for three different PVT cases of the XOR cell

PVT	Change in growth delay (pv)				Error of the current method (%)		
		Recommended method					
	Available method	Best case scenario	Worst case scenario	Average case	Best case scenario	Worst case scenario	Average case
ff	18.4	16.5	22.9	19.2	11.5	24.5	4.4
tt	16.1	15.6	17.1	16.3	3.2	6.2	1.2
ss	17.6	17.4	17.9	17.7	1.2	1.7	0.6

Thus, thanks to the proposed method, taking into account the phase deviations between the input signals, it is possible to obtain up to 24.5% increase in design accuracy in the digital standard cell library, compared to existing methods, at the expense of a 2–3 times increase in simulation time.

Conclusions

1. A method for compensating for the effects of aging phenomena in voltage-controlled oscillators has been developed, as a result of which, thanks to the addition of a digital unitary code control circuit, it was possible to reduce the 6.5% decrease in the output signal frequency to 0.31%, at the expense of an increase in area of 13.6% and power consumption of 15.1%.
2. A method for monitoring and recording aging phenomena in integrated circuits has been developed, thanks to which it is possible to record aging phenomena by obtaining a digital code at the circuit output and using it in compensating circuits. The method was implemented only with MOS transistors, and as a result of not requiring unnecessary pins and testing on the IC, it leads to a reduction in area, testing time, cost of integrated circuits and an increase in reliability, occupying only 10 μm^2 of area and consuming 300 μA of current.
3. A method for reducing the effect of aging phenomena in three types of thin oxide-strip transistors in gate-only operational amplifiers has been designed: two-stage, folded cascode, and rail-to-rail. In the latter case, by adding cascode transistors and digital switches, as well as protecting the transistors from stress by changing the amplitude of the switching signal, it was possible to neutralize the decrease in gain of 25.8% for the two-stage OpAmp, 22.3% for the folded cascode OpAmp, and 23.3% for the rail-to-rail OpAmp, at the expense of 3.2%, 4.8%, and 7.1% increases in area, respectively.
4. A method for increasing the design accuracy of digital standard cell libraries that takes into account aging phenomena has been proposed, in which, taking into account possible phase deviations between the input signals of the cells, it is possible to obtain an increase in design accuracy of up to 24.5% at the expense of a 2–three-fold increase in the simulation duration.

References

1. A. Mocuta, P. Weckx, P. Demuynck, et al., Enabling CMOS scaling towards 3 nm and beyond, in *2018 IEEE Symposium on VLSI Technology. Honolulu, HI, USA*, (IEEE, 2018), pp. 147–148. https://doi.org/10.1109/VLSIT.2018.8510683
2. G. Yang, H. Jung, J. Lim, et al., Standard cell Design optimization with advanced MOL technology in 3 nm GAA process, in *2022 IEEE Symposium on VLSI Technology and Circuits (VLSI Technology and Circuits). Honolulu, HI, USA*, (IEEE, 2022), pp. 362–364. https://doi.org/10.1109/VLSITechnologyandCir46769.2022.9830450
3. S.C. Song, Design/system technology co-optimization for 3 nm node and beyond, in *2021 International Symposium on VLSI Technology, Systems and Applications (VLSI-TSA). Hsinchu, Taiwan*, (2021). https://doi.org/10.1109/VLSI-TSA51926.2021.9440089
4. K. Zhang, Circuit design in nano-scale CMOS technologies, in *2018 International Symposium on VLSI Design, Automation and Test (VLSI-DAT). Hsinchu, Taiwan*, (IEEE, 2018), p. 1. https://doi.org/10.1109/VLSI-DAT.2018.8373276
5. B. Razavi, *Design of Analog CMOS Integrated Circuits*, 2nd edn. (McGraw Hill India, 2017), 782p
6. J. Hoentschel, A. Wei, From the present to the future. Scaling of planar VLSI-CMOS devices towards 3D-FinFETs and beyond 10nm CMOS technologies; manufacturing challenges and future technology concepts, in *2015 China Semiconductor Technology international conference. Shanghai, China*, (IEEE, 2015), pp. 1–4. https://doi.org/10.1109/CSTIC.2015.7153333
7. A.A. Bait, N. Narkhede, S. More, et al., Performance enhancement and suppression of short channel effects of 14 nm double gate FETs by using gate stacked high-k dielectrics & work function variation, in *2016 International Conference on Inventive Computation Technologies (ICICT). Coimbatore, India*, (IEEE, 2016), pp. 1–5. https://doi.org/10.1109/INVENTIVE.2016.7824876
8. S. Ilin, D. Ryzhova, A. Korshunov, Comparative analysis of standard cell performance for 7 nm FinFET and 28 nm CMOS technologies with consideration for parasitic elements, in *2018 IEEE Conference of Russian Young Researchers in Electrical and Electronic Engineering (EIConRus). Moscow and St. Petersburg, Russia*, (IEEE, 2018), pp. 1360–1363. https://doi.org/10.1109/EIConRus.2018.8317349
9. N. Pande, C. Zhou, M. Lin, et al., Characterizing electromigration effects in a 16 nm FinFET process using a circuit based test vehicle, in *2019 IEEE International Electron Devices Meeting (IEDM)*, (IEEE, 2019), pp. 5.3.1–5.3.4. https://doi.org/10.1109/IEDM19573.2019.8993548
10. D. Kraak, I. Agbo, M. Taoui, et al., Degradation analysis of high performance 14 nm FinFET SRAM, in *2018 Design, Automation & Test in Europe Conference & Exhibition (DATE)*, (IEEE, 2018), pp. 201–206. https://doi.org/10.23919/DATE.2018.8342003
11. V.M. Santen, L. Schillinger, H. Amrouch, Self-heating effects from transistors to gates, in *2021 International Symposium on VLSI Design, Automation and Test (VLSI-DAT), Hsinchu, Taiwan*, (IEEE, 2021), pp. 1–4. https://doi.org/10.1109/VLSI-DAT52063.2021.9427356
12. O. Prakash, C.K. Dabhi, Y.S. Chauhan, H. Amrouch, Transistor self-heating. the rising challenge for semiconductor testing, in *2021 IEEE 39th VLSI Test Symposium (VTS). San Diego, CA, USA*, (IEEE, 2021), pp. 1–7. https://doi.org/10.1109/VTS50974.2021.9441002
13. P. Kushwaha, H. Agarwal, Y. Lin, et al., Characterization and modeling of flicker noise in FinFETs at advanced technology node. IEEE Electron Device Lett. **40**(6), 985–988 (2019). https://doi.org/10.1109/LED.2019.2911614
14. U. Mushtaq, V.K. Sharma, Design of 6T FinFET SRAM cell at 7nm, in *2019 International Conference on Communication and Electronics Systems (ICCES). Coimbatore, India*, (IEEE, 2019), pp. 104–108. https://doi.org/10.1109/ICCES45898.2019.9002138
15. S. Kiamehr, P. Weckx, M. Tahoori, et al., The impact of process variation and stochastic aging in nanoscale VLSI, in *2016 IEEE International Reliability Physics Symposium (IRPS). Pasadena, CA, USA*, (IEEE, 2016), pp. 1–6. https://doi.org/10.1109/IRPS.2016.7574590

16. H. Amrouch, V.M. van Santen, O. Prakash, et al., Reliability challenges with self-heating and aging in FinFET technology, in *2019 IEEE 25th International Symposium on On-Line Testing and Robust System Design (IOLTS). Rhodes, Greece*, (IEEE, 2019), pp. 68–71. https://doi.org/10.1109/IOLTS.2019.8854405
17. X. Yang, Y. Xue, Z. Dong, et al., Interconnection reliability on FinFET devices, in *2022 IEEE International Symposium on the Physical and Failure Analysis of Integrated Circuits (IPFA), Singapore, Singapore*, (IEEE, 2022), pp. 1–4. https://doi.org/10.1109/IPFA55383.2022.9915773
18. S. Kundu, Managing reliability of integrated circuits. Lifetime metering and design for healing, in *2016 IEEE 25th Asian Test Symposium (ATS)*, (IEEE, 2016), p. 227
19. A. Schaldenbrand, J. Xie, H. Elhak, Recent updates to transistor level reliability analysis, in *2019 IEEE International Reliability Physics Symposium (IRPS). Monterey, CA, USA*, (IEEE, 2019), pp. 1–8. https://doi.org/10.1109/IRPS.2019.8720540
20. K. Choi, H.C. Sagong, M. Jin, et al., Reliability on evolutionary FinFET CMOS technology and beyond, in *2020 IEEE International Electron Devices Meeting (IEDM)*, (IEEE, 2020), pp. 9.3.1–9.3.4. https://doi.org/10.1109/IEDM13553.2020.9371930
21. S. Panarello, C. Triolo, F. Garesci, et al., Improving ICs reliability with high speed thermal mapping, in *2017 14th International Conference on Synthesis, Modeling, Analysis and Simulation Methods and Applications to Circuit Design (SMACD)*, (IEEE, 2017), pp. 1–4. https://doi.org/10.1109/SMACD.2017.7981607
22. C. Liu, H. Sagong, H. Kim, et al., Systematic study of 14 nm FinFET reliability. From device level stress to product HTOL, in *2015 IEEE International Reliability Physics Symposium*, (IEEE, 2015), pp. 2F.3.1–2F.3.5. https://doi.org/10.1109/IRPS.2015.7112693
23. C. Prasad, Advanced CMOS reliability challenges, in *Proceedings of Technical Program - 2014 International Symposium on VLSI Technology, Systems and Application (VLSI-TSA). Hsinchu, Taiwan*, (IEEE, 2014), pp. 1–2. https://doi.org/10.1109/VLSI-TSA.2014.6839637
24. V.M. van Santen, P.R. Genssler, O. Prakash, et al., Impact of self-heating on performance, power and reliability in FinFET technology, in *2020 25th Asia and South Pacific Design Automation Conference (ASP-DAC)*, (IEEE, 2020), pp. 68–73. https://doi.org/10.1109/ASP-DAC47756.2020.9045582
25. G. Kamdi, T. Deotale, Effect of process variation on Power and delay in CMOS Circuits, in *2013 International Conference on Communication Systems and Network Technologies. Gwalior, India*, (IEEE, 2013), pp. 748–752. https://doi.org/10.1109/CSNT.2013.160
26. M. Avdalyan, A. Petrosyan, S. Ghukasyan, Z. Khojayan, Development of a method for reducing the power, area and delay time for a static random access memory cell, in *2022 International Conference on Electrical, Computer and Energy Technologies (ICECET). Prague, Czech Republic*, (IEEE, 2022), pp. 1–5. https://doi.org/10.1109/ICECET55527.2022.9872897
27. W. Ye, M.B. Alawieh, C.L. Hsu, et al., Dealing with aging and yield in scaled technologies, in *Embedded Systems*, (Springer, Cham, 2021), pp. 409–429. https://doi.org/10.1007/978-3-030-52017-5_17
28. P. Sharma, M. Rehal, P.K. Bangotra, VLSI defect analysis and yield management using interactive wafer map, in *2018 International Conference on Computing, Power and Communication Technologies (GUCON). Greater Noida, India*, (IEEE, 2018), pp. 1192–1196. https://doi.org/10.1109/GUCON.2018.8675028
29. S.S. Sapatnekar, What happens when circuits grow old? Aging issues in CMOS design, in *2013 International Symposium on VLSI Technology, Systems and Application (VLSI-TSA). Hsinchu, Taiwan*, (IEEE, 2013), pp. 1–2. https://doi.org/10.1109/VLSI-TSA.2013.6545621
30. D. Kraak, M. Taouli, S. Hamdiou, et al., Device ageing. A reliability and security concern, in *2018 IEEE 23rd European Test Symposium (ETS). Bremen, Germany*, (IEEE, 2018), pp. 1–10. https://doi.org/10.1109/ETS.2018.8400702
31. A. Ghosh, S. Satapathy, J.P. Kulkarni, P.D. Joshi, Aging effects on clock gated memory phase paths, in *2022 IEEE International Symposium on Defect and Fault Tolerance in VLSI*

and Nanotechnology Systems (DFT). Austin, TX, USA, (IEEE, 2022), pp. 1–5. https://doi.org/10.1109/DFT56152.2022.9962342
32. S. Morita, S. Bian, M. Shintani, T. Sato, Efficient analysis for mitigation of workload-dependent aging degradation. IEEE Trans. Computer-Aided Des. Integr. Circ. Syst. **41**(12), 5515–5525 (2022). https://doi.org/10.1109/TCAD.2022.3149856
33. A. Bhattacharjee, S.N. Pradhan, Impact of transistor aging on the reliability of the analog circuit, in *2020 International Conference on Computational Performance Evaluation (ComPE)*, (IEEE, 2020), pp. 212–216. https://doi.org/10.1109/ComPE49325.2020.9200055
34. S. Ramey, Y. Lu, I. Meric, et al., Aging model challenges in deeply scaled tri-gate technologies, in *2015 IEEE International Integrated Reliability Workshop (IIRW). South Lake Tahoe, CA, USA*, (IEEE, 2015), pp. 56–62. https://doi.org/10.1109/IIRW.2015.7437067
35. H. Amrouch, S.B. Ehsani, A. Gerstlauer, J. Henkel, On the efficiency of voltage Overscaling under temperature and aging effects. IEEE Trans. Comput. **68**(11), 1647–1662 (2019). https://doi.org/10.1109/TC.2019.2916869
36. Y. Sun, C. Zhan, Y. Fu, et al., Aging impacts on SOC timing in advanced FinFET technologies, in *2018 14th IEEE International Conference on Solid-State and Integrated Circuit Technology (ICSICT). Qingdao, China*, (IEEE, 2018), pp. 1–3. https://doi.org/10.1109/ICSICT.2018.8565644
37. *The Gold Standard for Accurate Circuit Simulation* (Synopsys, 2019), 4p
38. *PrimeSimTM HSPICE Reference Manual. Commands and Control Options* (Synopsys Inc., 2020), 823p
39. *FineSim® User Guide. Pro and SPICE Reference* (Synopsys Inc., 2020), 851p
40. http://www.ieca-inc.com/images/LTSPICE_Manual.pdf
41. https://picture.iczhiku.com/resource/eetop/WHIroZOqiUaWEmMX.pdf
42. https://www.seas.upenn.edu/~jan/spice/PSpice_UserguideOrCAD.pdf
43. https://www.eet.bme.hu/~szalai/Download/eldo_ur.pdf
44. B. Khapinhi, B. Omidi, H. Amrouch, et al., Estimating and mitigating aging effects in routing network of FPGAs. IEEE Trans. VLSI Syst. **27**(3), 651–664 (2019). https://doi.org/10.1109/TVLSI.2018.2886326
45. H.T. Kostanyan, H.V. Margaryan, V.A. Janpoladov, et al., The minimizaton method of transistor aging influence on modern voltage references, in *2020 IEEE 40th International Conference on Electronics and Nanotechnology (ELNANO)*, (IEEE, 2020), pp. 335–338. https://doi.org/10.1109/ELNANO50318.2020.9088844
46. Y.K. Law, I.C. Lin, C.C. Lin, Reducing aging on scratchpad memory using temporal- and FSM-based power management, in *2017 International Symposium on VLSI Design, Automation and Test (VLSI-DAT). Hsinchu, Taiwan*, (IEEE, 2017), pp. 1–4. https://doi.org/10.1109/VLSI-DAT.2017.7939675
47. H.S. Kalluru, P. Saha, A. Zahra, Z. Abbas, PVT and aging degradation invariant automated optimization approach for CMOS low-power high-performance VLSI circuits, in *2021 22nd International Symposium on Quality Electronic Design (ISQED). Santa Clara, CA, USA*, (IEEE, 2021), pp. 1–6. https://doi.org/10.1109/ISQED51717.2021.9424366
48. I. Sario, A.B. Channa, F. Ali Qureshi, et al., Study on early capture based VLSI aging monitoring techniques, in *2019 2nd International Conference on Computing, Mathematics and Engineering Technologies (iCoMET). Sukkur, Pakistan*, (IEEE, 2019), pp. 1–4. https://doi.org/10.1109/ICOMET.2019.8673402
49. M.A. Scarpato, Digital circuit performance estimation under PVT and aging effects. Ph.D., Université Grenoble Alpes, 2017, 157p
50. J. Pallavi, N. Mohan, Improved BIST design for detecting aging faults in automotive ICs, in *2021 Second International Conference on Electronics and Sustainable Communication Systems (ICESC). Coimbatore, India*, (IEEE, 2021), pp. 1701–1706. https://doi.org/10.1109/ICESC51422.2021.9532812

51. A. Khachatryan, A novel aging sensor with programmable resolution, in *2017 IEEE East-West Design & Test Symposium (EWDTS). Novi Sad, Serbia*, (IEEE, 2017), pp. 1–4. https://doi.org/10.1109/EWDTS.2017.8110097
52. D. Mirzoyan, A. Khachatryan, A new process variation monitoring circuit, in *2016 IEEE Computer Society Annual Symposium on VLSI (ISVLSI). Pittsburgh, PA, USA*, (IEEE, 2016), pp. 1–5. https://doi.org/10.1109/ISVLSI.2016.26
53. D. Mirzoyan, A. Khachatryan, A novel process corner detection circuit, in *2016 IEEE East-West Design & Test Symposium (EWDTS). Yerevan, Armenia*, (IEEE, 2016), pp. 1–4. https://doi.org/10.1109/EWDTS.2016.7807725
54. A. Khachatryan, D. Mirzoyan, A novel approach to detect temperature variation, in *2016 IEEE East-West Design & Test Symposium (EWDTS). Yerevan, Armenia*, (IEEE, 2016), pp. 1–4. https://doi.org/10.1109/EWDTS.2016.7807715
55. V. Melikyan, A. Khachatryan, D. Mirzoyan, On chip resistance value compensation technique for termination resistor, in *2016 IEEE 36th International Conference on Electronics and Nanotechnology (ELNANO). Kyiv, Ukraine*, (IEEE, 2016), pp. 378–381. https://doi.org/10.1109/ELNANO.2016.7493089
56. G. Aishwarya, B. Patil, P.V. Joshi, et al., A survey on use of FPGA in automotive system, in *2022 International Conference on Distributed Computing, VLSI, Electrical Circuits and Robotics (DISCOVER). Shivamogga, India*, (IEEE, 2022), pp. 51–56. https://doi.org/10.1109/DISCOVER55800.2022.9974941
57. S. Nagavarapu, S.S. Rout, A review on VLSI implementation in biomedical application, in *Innovations in Bio-Inspired Computing and Applications*, (Springer, 2021), pp. 130–138. https://doi.org/10.1007/978-3-030-49339-4_14
58. S. More, Aging degradation and countermeasures in deep-submicrometer analog and mixed signal integrated circuits. Ph.D. dissertation, Munich, 2011, 153p
59. N. Ruiz, Multilevel aging phenomena analysis in complex ultimate CMOS designs. Ph.D., Grenoble, 2012, 153p
60. N. Hellwege, Aging-aware design methods for reliable analog integrated circuits using operating point-dependent degradation. Ph.D., Bremen, 2015, 146 p
61. D. Lorenz, Aging analysis of digital integrated circuits. Ph.D., Technische University of Munich, Germany, 2012, 150p
62. B. Ullmann, M. Jech, K. Puschkarsky, et al., Impact of mixed negative bias temperature instability and hot carrier stress on MOSFET characteristics. Part I: experimental. IEEE Trans. Electron Devices **66**(1), 232–240 (2019). https://doi.org/10.1109/TED.2018.2873419
63. M. Jech, B. Ullmann, G. Rzepa, et al., Impact of mixed negative bias temperature instability and hot carrier stress on MOSFET characteristics. Part II. Theory. IEEE Trans. Electron Devices **66**(1), 241–248 (2019). https://doi.org/10.1109/TED.2018.2873421
64. B. Ullmann, M. Jech, S. Tyaginov, et al., The impact of mixed negative bias temperature instability and hot carrier stress on single oxide defects, in *2017 IEEE International Reliability Physics Symposium (IRPS). Monterey, CA, USA*, (IEEE, 2017), pp. XT-10.1–XT-10.6. https://doi.org/10.1109/IRPS.2017.7936424
65. D. Son, K. Hong, H. Shim, et al., New insight into negative bias temperature instability degradation during self-heating in nanoscale bulk FinFETs. IEEE Electron. Dev. Lett. **40**(9), 1354–1357 (2019). https://doi.org/10.1109/LED.2019.2930077
66. S. Peng, H. Zhou, T. Kim, et al., Physics-based compact TDDB models for Low-k BEOL copper interconnects with time-varying voltage stressing. IEEE Trans. VLSI Syst. **26**(2), 239–248 (2018). https://doi.org/10.1109/TVLSI.2017.2764880
67. https://www.synopsys.com/content/dam/synopsys/verification/white-papers/mosra-wp.pdf
68. https://mos-ak.org/beijing_2018/presentations/Tianlei_Guo_MOS-AK_Beijing_2018.pdf
69. https://www.cadence.com/content/dam/cadence-www/global/en_US/documents/tools/custom-ic-analog-rf-design/rethinking-reliability-wp.pdf
70. S. Ghukasyan, Aging compensation method for voltage controlled oscillator, in *2022 International Conference on Electrical, Computer and Energy Technologies*

(ICECET). Prague, Czech Republic, (IEEE, 2022), pp. 1–5. https://doi.org/10.1109/ICECET55527.2022.9872878
71. I. Lahbib, M.A. Doukkali, P. Martin, G. Imbert, Hot carrier injection and negative bias temperature instability induced NMOS and PMOS degradation on CMOS ring oscillator, in *2016 Annual Reliability and Maintainability Symposium (RAMS). Tucson, AZ, USA*, (IEEE, 2016), pp. 1–7. https://doi.org/10.1109/RAMS.2016.7448024
72. X. Gui, R. Tang, Y. Zhang, et al., A voltage-controlled ring oscillator with VCO-gain variation compensation. IEEE Microw. Wirel. Compon. Lett. **30**(3), 288–291 (2020). https://doi.org/10.1109/LMWC.2020.2967391
73. S. Suman, K.G. Sharma, P.K. Ghosh, Analysis and design of current starved ring VCO, in *2016 International Conference on Electrical, Electronics, and Optimization Techniques (ICEEOT)*, (IEEE, 2016), pp. 3222–3227
74. N.O. Adesina, M.A.U. Khan, J. Xu, Modeling the effects of SBD, HCI, and NBTI in CMOS voltage controlled oscillator design for PLL applications, in *2021 IEEE 12th Annual Information Technology, Electronics and Mobile Communication Conference (IEMCON)*, (IEEE, 2021), pp. 0720–0725
75. V. Melikyan, M. Martirosyan, G. Piliposyan, 14 nm Educational design kit. Capabilities, deployment and future, in *Proceedings of the 7th Small Systems Simulation Symposium 2018. Niš, Serbia, February 12–14*, (2018), pp. 37–41
76. V.S. Melikyan, S.A. Ghukasyan, S.S. Harutyunyan, et al., The aging influence minimization method for operational amplifiers with thin oxide transistors. Proc. Repub. Armen. Natl. Acad. Sci. Natl. Polytech. Univ. Armen. Ser. Tech. Sci. **75**(4), 527–538 (2022). https://doi.org/10.53297/0002306X-2022.v75.4-527
77. J.R. Baker, *CMOS circuit design, layout, and simulation*, 3rd edn. (John Wiley and & Sons, 2010), 1173p
78. A.A. Lyalinsky, Characterization of analog schemes based on cloud technologies. Problems of development of perspective micro- and nanoelectronic systems. **1**, 118–124 (2018). https://doi.org/10.31114/2078-7707-2018-1-118-124
79. B. Hesham, E.S. Hasaneen, H.H.F.A. Design, Procedure for two-stage CMOS Opamp using gm/ID design methodology in 16 nm FinFET technology, in *2019 31st International Conference on Microelectronics (ICM). Cairo, Egypt*, (IEEE, 2019), pp. 325–329. https://doi.org/10.1109/ICM48031.2019.9021511
80. S. Waykole, V.S. Bendre, Performance analysis of classical two stage Opamp using CMOS and CNFET at 32 nm technology, in *2018 Fourth International Conference on Computing Communication Control and Automation (ICCUBEA). Pune, India*, (IEEE, 2018), pp. 1–6. https://doi.org/10.1109/ICCUBEA.2018.8697461
81. R. Kumar, R. Nagulapalli, S.K. Vishvakarma, A novel gain enhanced folded Cascode OPAMP in 28 nm CMOS technology, in *2022 International Conference on Electrical, Computer and Energy Technologies (ICECET). Prague, Czech Republic*, (IEEE, 2022), pp. 1–4. https://doi.org/10.1109/ICECET55527.2022.9872766
82. R. Kammari, J. Gundla, S. Boyapati, V.S.R. Pasupureddi, Modeling and design of a compact low power folded Cascode OpAmp with high EMI immunity. IEEE Trans. Electromagn. Compat. **64**(2), 595–598 (2022). https://doi.org/10.1109/TEMC.2021.3119190
83. V.S. Melikyan, S.A. Ghukasyan, S.S. Harutyunyan, et al., A method for minimizing Aging influence on rail-to-rail operational amplifier with thin oxide transistors. Proc. Univ. Electron. **28**(3), 351–359 (2023). https://doi.org/10.24151/1561-5405-2023-28-3-351-359. EDN. UDMQWS
84. M.A.G. Lorenzo, A.A.S. Manzano, M.T.A. Gusad, et al., Design and implementation of CMOS rail-to-rail operational amplifiers, in *2007 International Symposium on Communications and Information Technologies*, (IEEE, 2007), pp. 61–66. https://doi.org/10.1109/ISCIT.2007.4391985
85. A.K. Dubey, R.K. Nagaria, P.K. Pal, R.K. Singh, Enhanced slew rate, constant-gm rail-to-rail OpAmp using 1:2 current mirror biasing technique, in *2016 International Conference*

on Computing, Communication and Automation (ICCCA), (IEEE, 2016), pp. 1612–1618. https://doi.org/10.1109/CCAA.2016.7813982
86. V.M. Movsisyan, *Digital Logical Design of Systems* (Armenian National University Publishing House, Yerevan, 2012), p. 284
87. A.K. Tumanyan, *Basics: Digital Design with Use Language Verlilog* (Изд-во GIUA, Yerevan, 2012), 504c
88. H. Chukhajyan, *Application of Verilog in the Design Process of Digital Systems*, 2nd edn. (Publishing House of the Armenian National University of Applied Sciences, Yerevan, 2014), 247p
89. V.S. Melikyan, O.H. Petrosyan, Z.M. Avetisyan, S.A. Ghukasyan, et al., Accuracy increasing approach in aging-aware standard cell libraries, in *2019 IEEE 39th International Conference on Electronics and Nanotechnology (ELNANO). Kyiv, Ukraine*, (IEEE, 2019), pp. 292–295. https://doi.org/10.1109/ELNANO.2019.8783524
90. Y. Kim, H. Shim, M. Jin, et al., Investigation of HCI effects in FinFET based ring oscillator circuits and IP blocks, in *2017 IEEE International Reliability Physics Symposium (IRPS). Monterey, CA, USA*, (IEEE, 2017), pp. 4C-2.1–4C-2.4. https://doi.org/10.1109/IRPS.2017.7936313
91. M. Katoozi, E.H. Cannon, T. Dao, et al., An age-aware library for reliability simulation of digital ICs, in *2013 IEEE International Reliability Physics Symposium (IRPS). Monterey, CA, USA*, (IEEE, 2013), pp. 3A.3.1–3A.3.5. https://doi.org/10.1109/IRPS.2013.6531973
92. S. Kiamehr, F. Firouzi, M. Ebrahimi, M.B. Tahoori, Aging-aware standard cell library design, in *Design, Automation & Test in Europe Conference & Exhibition (DATE)*, (IEEE, 2014), pp. 1–4
93. S. Kiamehr, M. Ebrahimi, F. Firouzi, M.B. Tahoori, Extending standard cell library for aging mitigation. IET Comput. Digit. Tech. **4**, 206–212 (2015)
94. H. Amrouch, B. Khapinhi, A. Gerstlauer, J. Henkel, Reliability-aware design to suppress aging, in *53rd ACM/EDAC/IEEE Design Automation Conference (DAC)*, (IEEE, 2016), pp. 1–6
95. G.A. Ivanova, D.I. Ryzhova, Methods of increasing the accuracy of modeling delays and peak currents of combinational CMOS schemes at the logical level of design. Izv. Univ. Electron. **23**(1), 40–51 (2018). https://doi.org/10.24151/1561-5405-2018-23-1-40-51
96. Q. Wang, X. Wang, J. Li, Y. Liu, A bandgap reference voltage source design for wide temperature range, in *2022 7th International Conference on Integrated Circuits and Microsystems (ICICM). Xi'an, China*, (IEEE, 2022), pp. 245–250. https://doi.org/10.1109/ICICM56102.2022.10011392
97. S.K. Wadhwa, N. Chaudhry, High accuracy, multi-output bandgap reference circuit in 16 nm FinFet, in *2017 30th International Conference on VLSI Design and 2017 16th International Conference on Embedded Systems (VLSID). Hyderabad, India*, (IEEE, 2017), pp. 259–262. https://doi.org/10.1109/VLSID.2017.52
98. S.A. Ghukasyan, A novel aging monitoring circuit. Proc. Repub. Armen. Natl. Acad. Sci. Natl. Polytech. Univ. Armen. Ser. Tech. Sci. **76**(1), 58–66 (2023). https://doi.org/10.53297/0002306X-2023.v76.1-58
99. https://www.khanacademy.org/computing/computer-science/algorithms/binary-search/a/binary-search
100. B. Khanna, R. Gupta, Design, implementation and analysis of different models of CMOS Schmitt trigger, in *2018 International Conference on Intelligent Circuits and Systems (ICICS). Phagwara, India*, (IEEE, 2018), pp. 63–68. https://doi.org/10.1109/ICICS.2018.00025
101. H.V. Vardanyan, S.A. Ghukasyan, VIX correlation improvement between spice and ibis using duty cycle. Physico-math. Nat. Sci. Yerevan **1**, 86–94 (2020)
102. A. Vijay, C. Duari, L. Garg, A.K. Singh, CMOS Schmitt trigger circuit and oscillator design. The impact of NBTI degradation, in *2023 7th International Conference on Computing Methodologies and Communication (ICCMC). Erode, India*, (IEEE, 2023), pp. 7–10. https://doi.org/10.1109/ICCMC56507.2023.10084272

103. M.B. Martin, F.R.G. Cruz, Design and simulation of low voltage dual supply schmitt trigger with variable hysteresis, in *2020 11th IEEE Control and System Graduate Research Colloquium (ICSGRC), Shah Alam, Malaysia*, (IEEE, 2020), pp. 123–127. https://doi.org/10.1109/ICSGRC49013.2020.9232581
104. H. Kostanyan, S. Harutyunyan, G. Voskanyan, S. Ghukasyan, Transistor aging minimization method for multi output level converter design. Physico-math. Nat. Sci. Yerevan **1**, 87–98 (2022)
105. https://www.khanacademy.org/science/electrical-engineering/ee-circuit-analysis-topic/ee-resistor-circuits/a/ee-voltage-divider

Chapter 3
Development of Means to Improve the Performance and Reliability of Power Supply and Control Systems Embedded in Integrated Memory Devices

3.1 General Issues of Developing Means to Improve the Performance and Reliability of Power Supply and Control Systems Embedded in Integrated Memory Devices

3.1.1 Need to Develop Means to Improve the Performance and Reliability of Power Supply and Control Systems Embedded in Integrated Memory Devices

The main goal of the development of technological processes for the manufacturing of semiconductor devices has been and remains to increase the number of semiconductor elements located per unit area. Increasing the density of transistors in ICs allows for providing the same functionality on a much smaller area. This creates an opportunity to add functions to ICs. All this not only leads to an increase in the speed and accuracy of ICs, but also allows obtaining comprehensive and widely applicable systems. The operation of such systems will require minimal intervention from the user [1].

In 1958, the first IC was produced, in which bipolar transistors were used as the main active component [2]. The dominance and scaling of technological processes for bipolar transistors continued up to a base width of 10 μm. During this time, the existing metal-oxide-semiconductor (MOS) transistors had high threshold voltages, low efficiency, and reliability, which is why they were not widely used [2]. In the 1970s, the use of complementary metal-oxide-semiconductor (CMOS) technology in digital circuits became widespread. Later, due to the development of manufacturing processes and a number of advantages of the MOS structure, field-effect transistors almost completely replaced bipolar ones. Technological developments and scaling up of MOS transistor manufacturing continue to this day. The minimum

V. Melikyan, K. Li, *Parametric Optimization Techniques of Integrated Circuits*,
https://doi.org/10.1007/978-3-032-20400-4_3

allowable channel length in the technological processes of the Taiwanese company TSMC, founded in 1987 and still a leader in the production of semiconductor integrated circuits, has decreased from 3 μm to 3 nm over the years (Fig. 3.1) [3].

ICs have wide application in almost all areas, including medical and mechanical engineering equipment, where IC reliability must comply most severe standards (Table 3.1) [4].

The ICs used in these devices often have multiple memory systems that provide various functions. They store security keys, perform post-manufacturing settings of the ICs, and perform a number of other important operations necessary for the smooth and safe operation of the device [4–6]. These ICs must be non-volatile (NV), i.e., they must retain the recorded information for years even in the absence of power (Fig. 3.2) [7]. NV memory cells have a complex and sensitive structure, and for data recording and reading they require precise input voltage values, which usually exceed the supply voltage value by several times [8]. It is advisable to use embedded power systems to supply these voltages to the memory devices present in the IC. This allows both not to occupy the limited number of IC pins and not to complicate the IC application from the user's point of view. To obtain and regulate the required high voltages, charge pumps and voltage stabilizers are used in embedded power systems.

In submicron technological processes, short channel secondary phenomena become prominent, which complicate the design process. ICs with MOS transistors with channel lengths of up to 3 nm are in mass production, while technologies with channel lengths of 2 nm and smaller are in limited production and experimental stages [9–13]. Compared to this thousand-fold scaling of transistor sizes over decades, the scaling of supply voltages is very slow (Fig. 3.3) [14]. The main reason for this is that the decrease in the threshold voltage of a MOS transistor is not directly proportional to the scaling of physical dimensions of the transistor. Thus, the scaling of supply voltages follows the rate of decrease in threshold voltages, not

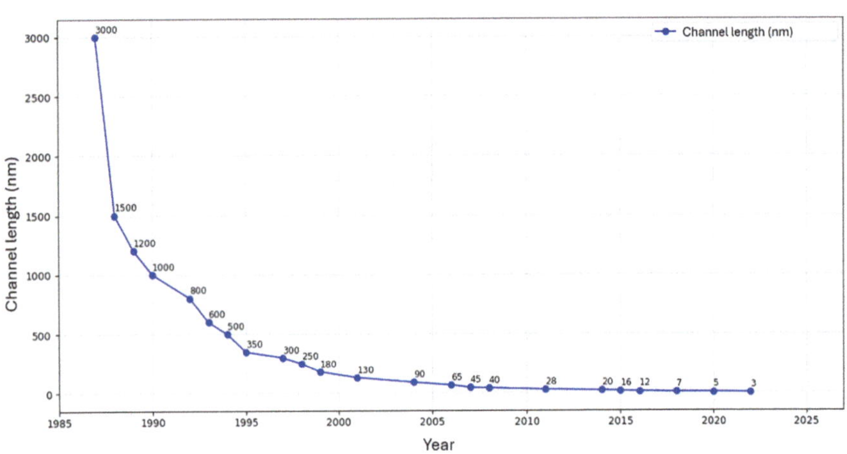

Fig. 3.1 TSMC technology processes development schedule

Table 3.1 IC reliability requirements for mechanical engineering quality systems

Operating temperature domain	−40 °C to +150 °C
Supply voltage domain	$V_{sup.nom}$ ±15%
High temperature lifecycle	1000 h at 150 °C
Defective parts per million	<1
Lifecycle	15 years

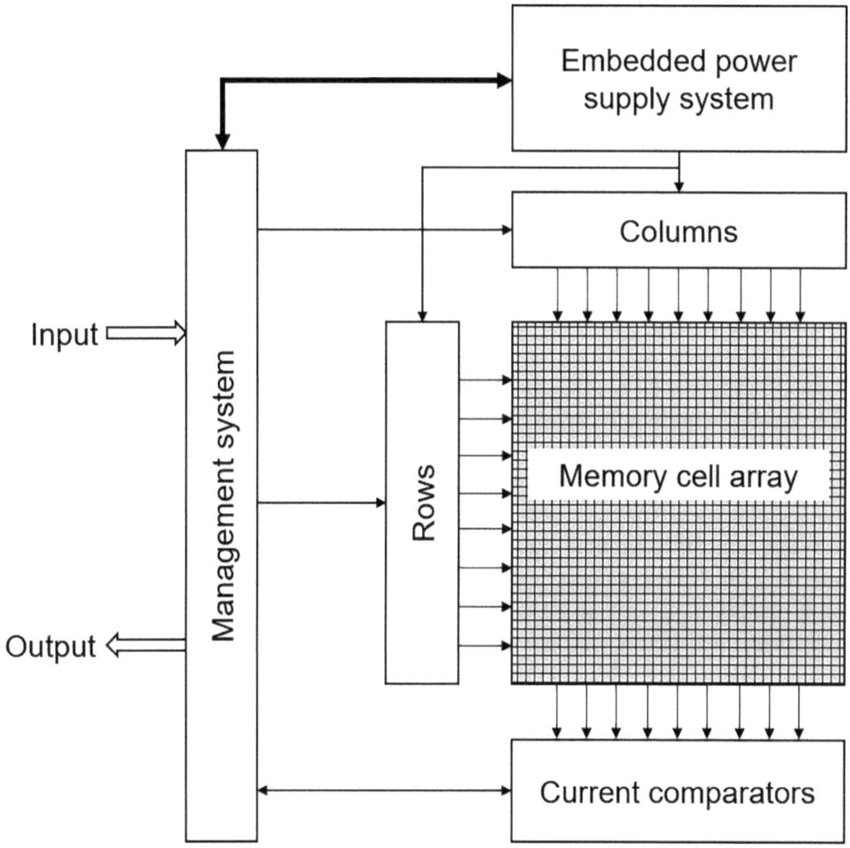

Fig. 3.2 The structure of an integrated memory device

in channel length. All this leads to an increase in the intensity of electric fields in the channel and causes serious reliability problems [15–19].

The control and power systems embedded in integrated memory devices are most vulnerable to these reliability issues, as their operating voltages can often significantly exceed the supply voltages. Under the influence of overvoltages and high current dissipation density, the physical characteristics of transistors, metal, and semiconductor interconnects in the IC deteriorate over time. One of the main factors

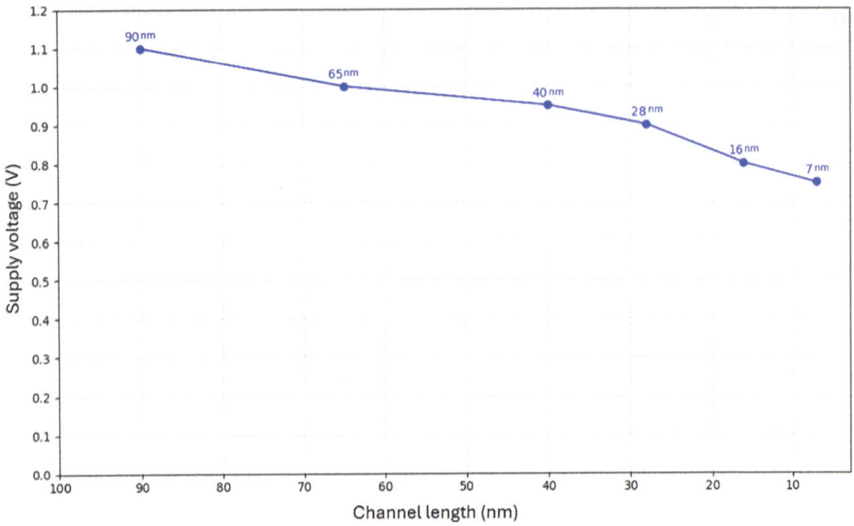

Fig. 3.3 Supply voltages in various technological processes

of deterioration is aging phenomena, which have different mechanisms of influence and operate in both high and low temperature conditions.

Aging phenomena are present in almost all nodes of the power supply and control systems embedded in memory ICs. In power supply systems, voltage multipliers and voltage stabilizers with charge pumps lose their maximum output load carrying capacity over time due to aging phenomena, which leads to incomplete data recording in the memory cell [20–25]. There is a need to repeat the programming operation, as a result of which the voltage multipliers and stabilizers waste additional time, power, and are exposed to overvoltage stress for a longer period than intended. Thus, the effects of aging phenomena have a cumulative nature. Because of them, the IC loses its functional efficiency, as a result of which it is sometimes necessary to repeat operations. This leads to additional aging and an increase in the raw bit error rate (RBER) (Fig. 3.4) [26].

The deterioration of the systems implemented in IMCs due to aging phenomena and other reliability issues directly affects the accuracy and long-term storage of the recorded information. Incomplete recording and loss of information can lead to a number of irreversible consequences. In the application of the memory IC as a security key to the device, in the event of loss of information, the user may gain access to and control the system without possessing the necessary passwords. An example of such vulnerability in the power and control system is one of the solutions presented in the 2024 competition for bypassing the security mechanisms of the RP2350 microcomputer proposed by the Raspberry Pi company [27]. A one-time programmable memory IC designed by the Sidense company was targeted as a security weak link in the RP2350. Through fluctuations in the supply voltage, it was possible to bypass the signals given to the memory IC from the control system and

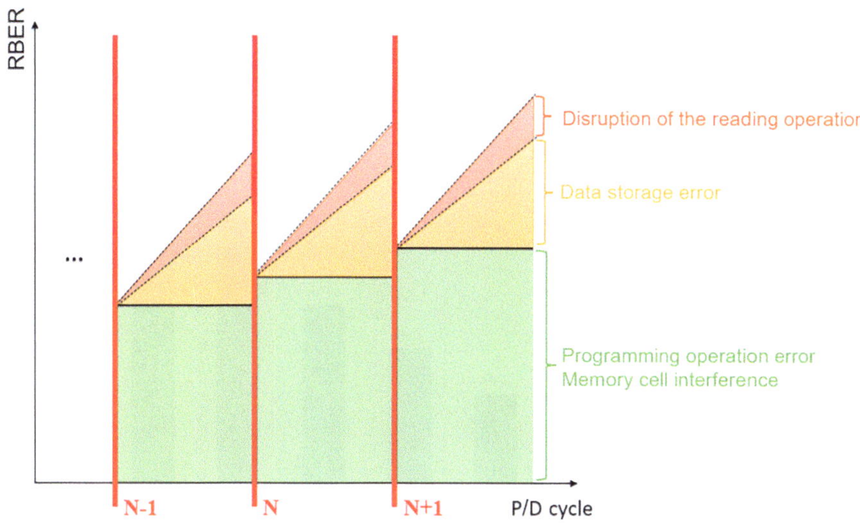

Fig. 3.4 The probability of a raw bit error depending on the number of program/erase cycles

bring the power system embedded in it to an unintended state. By repeating this operation periodically, it was possible to reprogram the security key stored in the IC and bypass the authentication stage.

Due to their low yield, designs with such reliability problems cannot meet the stringent requirements of the mechanical and medical sectors and would also be undesirable even in consumer-level production [28–32].

In parallel with the development of technological processes, the impact and weight of aging phenomena increase. Solving aging problems becomes a priority for obtaining competitive and reliable designs. Over time, the areas of application of ICs are expanding, from autonomous cars controlled by artificial intelligence to space satellites and medical devices installed in the human body [33–39]. Over time, the number of electronic elements in such devices and the associated costs have increased (Fig. 3.5) [40]. As a result, both reliability problems and their detection and prevention become more complicated.

Thanks to the cooperation of semiconductor device manufacturers and software developers, tools are being created and improved that allow for highly accurate simulation of the physical phenomena occurring in transistors at the design stage [41–43]. Due to this data, it becomes possible to assess the reliability of the IC at the design stage and identify the sources of problems affecting it. Various measures are taken at the design stage to solve the identified problems. In order not to occupy the limited footprints of the ICs and not to complicate the external control of the device, it is advisable to obtain solutions within the IC itself. In some nodes, it is possible to reduce or eliminate the effect of aging phenomena through circuit-technical solutions. It is also possible to introduce systems in the IC that constantly monitor the state of the circuit and operating conditions, which, if necessary, will compensate for deviations that appear due to aging phenomena [19, 44, 45].

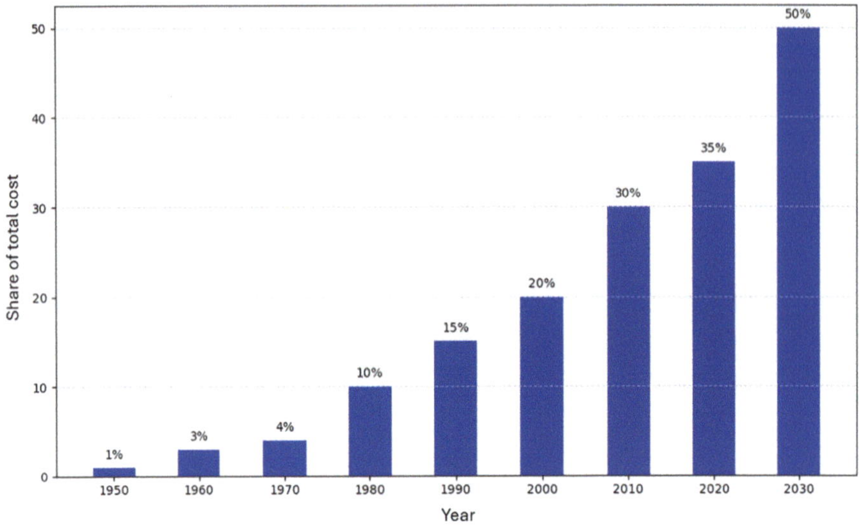

Fig. 3.5 The share of electronic devices in the production costs of a car

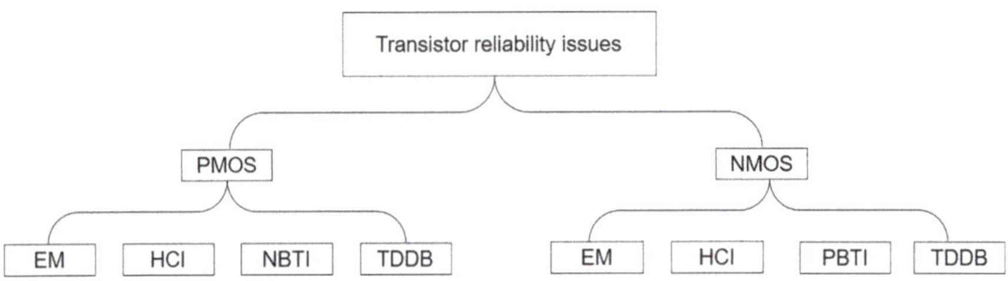

Fig. 3.6 Reliability issues in field-effect transistors

Thus, in order to ensure reliable operation of modern electronic equipment for 15 or more years under extreme conditions of technological deviations, temperature, and supply voltages, it is necessary to develop means to improve the reliability and operating parameters of the power supply and control systems embedded in their integrated memory devices.

3.1.2 *Main Problems of Reliability of Power Supply and Control Systems in Integrated Memory Devices*

In ICs, including non-volatile memories (NVM) and power and control systems, a number of mechanisms for transistor aging phenomena operate (Fig. 3.6) [46]. These include base temperature instability (BTI), hot charge carrier injection (HCI), time-dependent dielectric breakdown (TDDB), electromigration (EM), and self-heating [46–48]. These phenomena affect the physical structures and characteristics of semiconductor elements and interconnects, degrading their efficiency, performance, and reliability over time.

The BTI phenomenon occurs under the influence of a vertical electric field across the channel when $V_{gs} > |V_{th}|$ and is most expressed at high temperatures (100–150 °C). This phenomenon is divided into two groups: Negative BTI (NBTI) in p-type MOS transistors (PMOS), and Positive BTI (PBTI) in n-type MOS transistors (NMOS) [49]. Due to BTI, oxide traps are formed at the Si/SiO_2 surface as a result of the interaction of channel carriers and passivated hydrogen bonds present in the insulation. This leads to an increase in the absolute value of threshold voltage. In parallel with the scaling of gate oxide layer thickness, the effect of BTI and other aging mechanisms on the insulation permeability and gate leakage currents increases. To mitigate the effects of these phenomena, in 40 nm and smaller CMOS technologies, materials with a high dielectric constant (K) are used [50]. These materials provide smaller leakage currents compared to SiO_2. However, BTI phenomena do not cease to be one of the reliability problems in both analog and digital circuits.

After the stress conditions are removed, the BTI effect is partially restored. When the transistor is turned off ($V_{gs} = 0$), over time, some of the charge carriers in the oxide traps reappear from the insulator surface in the channel. This self-healing phenomenon cannot release all the charges that have appeared in the dielectric, since the stress and recovery states are not equivalent electrical conditions. Threshold degradation occurs under the influence of an electric field, while recovery occurs in the absence of it (Fig. 3.7) [51]. A long rest state compared to the operating time can contribute to a significant threshold recovery, but in modern devices, ICs do not usually have such small duty cycles.

The change in the threshold voltages of PMOS and NMOS transistors under the influence of NBTI and PBTI phenomena is modeled by this formula [52]:

$$\Delta V_{\text{th}} = A\left(\frac{V_g}{t_{\text{ox}}}\right)^m e^{\frac{\Delta E}{kT}} L^{\alpha} W^{\beta} t^{n} \tag{3.1}$$

where V_g is the gate–source voltage, ΔE is the thermal energy, k is the Boltzmann constant, T is the temperature, L is the channel length, W is the channel width, t is the stress time, n is the time index, and A, m, α, and β are experimentally determined parameters.

HCI phenomena occur when carriers with high kinetic energy are injected into the gate insulation [53–55]. Under the influence of hot carrier injection, the transistor's threshold voltage increases, and the drain current decreases. HCI effect mechanisms are divided into four types: Channel Hot Electron (CHE) injection, Secondary Generated Hot Electron (SGHE) injection, Substrate Hot Carrier (SHE) injection, and Drain Avalanche Hot Carrier (DAHC) injection.

1. CHE injection occurs when "lucky" electrons do not encounter other particles on their way from the source to the drain and continuously accelerate. These carriers

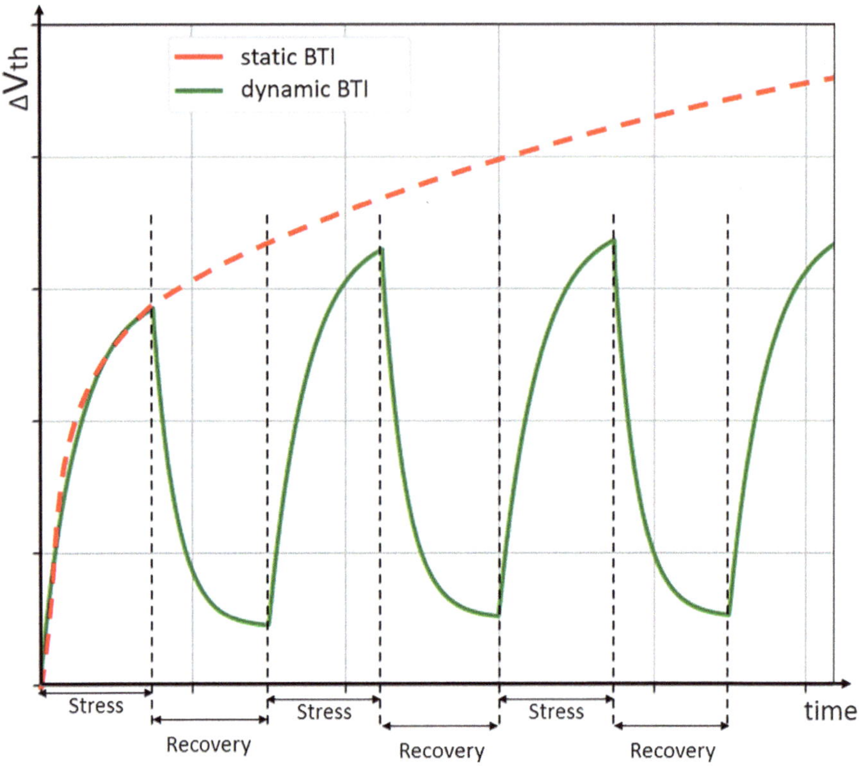

Fig. 3.7 Change in transistor threshold voltage over time as a result of BTI

eventually gain sufficient kinetic energy to overcome the potential barrier of the insulation under the influence of the gate-substrate electric field (Fig. 3.8) [53–55]. NMOS transistors are more vulnerable to this phenomenon because electrons have a smaller effective mass than holes.

2. DAHC injection occurs in the drain region of a MOS transistor operating in deep saturation mode. Electron-hole pairs are generated as a result of impact ionization on the channel (Fig. 3.9) [53–55]. The holes move toward the substrate, increasing the substrate current. Some of the “hot” electrons generated as a result of the collision gain sufficient energy and overcome the potential barrier of oxide.
3. Carriers with high kinetic energy, which are generated because of subsequent collisions of hot electrons formed during DAHC, are called secondary-generated hot electrons (SGHE). They can also gain enough energy to appear inside the gate insulator.
4. Injection of substrate hot electrons and holes occurs for a PMOS transistor when the substrate is positively biased, and for an NMOS transistor when it is negatively biased (Fig. 3.10) [53–55].

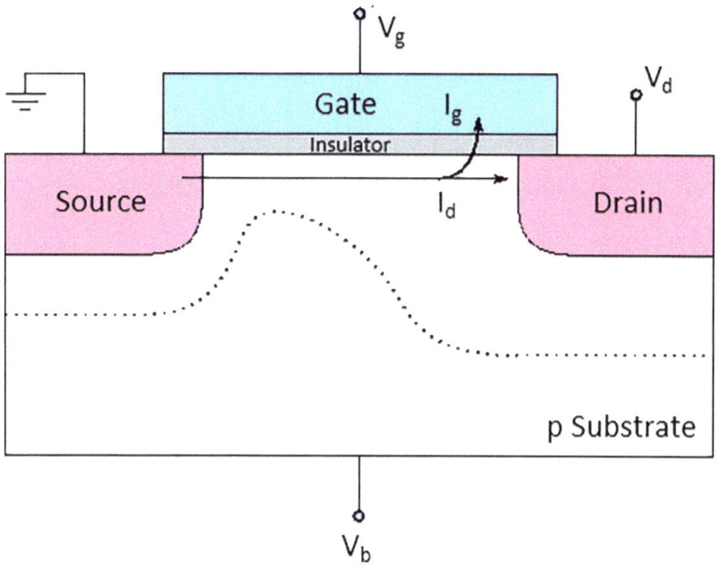

Fig. 3.8 Channel Hot Electron (CHE) injection phenomenon

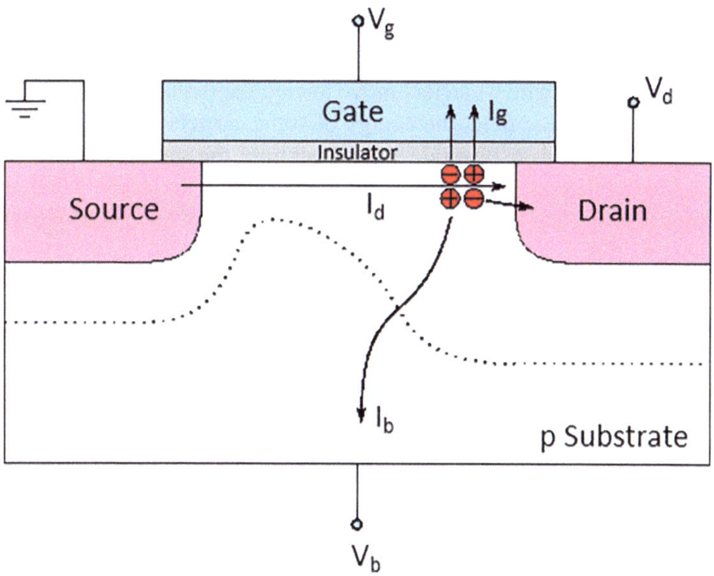

Fig. 3.9 Drain Avalanche Hot Carrier (DAHC) injection phenomenon

HCI phenomena damage the physical structure of the channel and gate insulation, and their effect, unlike BTI, is not recoverable. HCI phenomena are especially pronounced in transistors with short channel lengths because the probability of a "hot" carrier occurring there is higher [53–55].

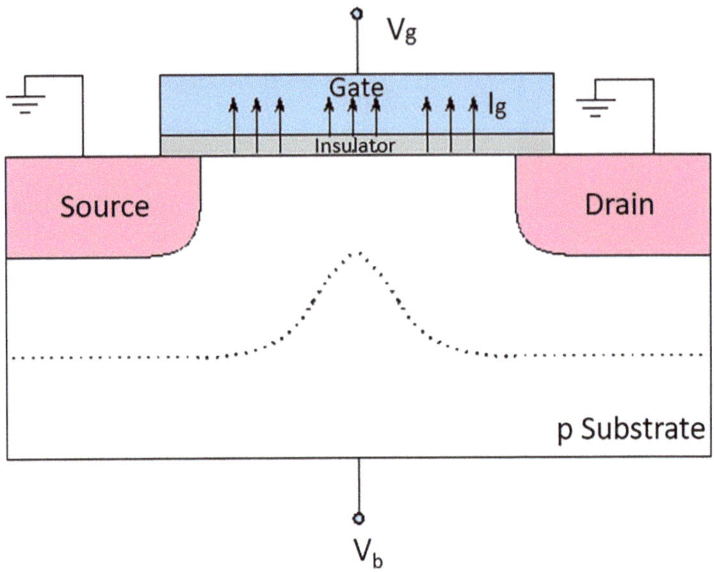

Fig. 3.10 Substrate Hot Carrier (SHE) injection phenomenon

TDDB phenomenon is the gradual degradation of the gate insulation layer because of the electric field. The gate oxide is subjected to high electric field stress, under the influence of which carriers move toward the oxide layer. During the prolonged application of the field, the energy of some carriers allows them to collide with the SiO_2 lattice and create oxide traps. Over time, the number of traps increases, which leads to an increase in the leakage current through the oxide. When the number of traps reaches a critical level, a breakdown path is formed along the oxide (Fig. 3.11) [56, 57].

Electromigration is the gradual displacement of metal atoms as a result of momentum transfer from electrons flowing with high density in interconnects and temperature and concentration gradients. This displacement is most pronounced at the edges and corners of the conductor. Over time, because of atom displacement, hillocks appear in some sections of the conductor and holes appear in the initial positions of the displaced atoms. As a result, both open circuits and short circuits between two adjacent wires can occur.

In interconnects deformed by the effect of electromigration, bridges created between two metal wires due to hillocks and breaks very close to the cut-off state can be seen under a microscope (Fig. 3.12) [58–61].

To more accurately model the effect of electromigration, it is necessary to consider the self-heating phenomenon [62–65]. Insufficient dissipation of thermal energy through the substrate or the environment leads to a local increase in temperature. A local increase in temperature in semiconductor elements and interconnects leads to an increase in the resistance of the conductive material in those areas. As a

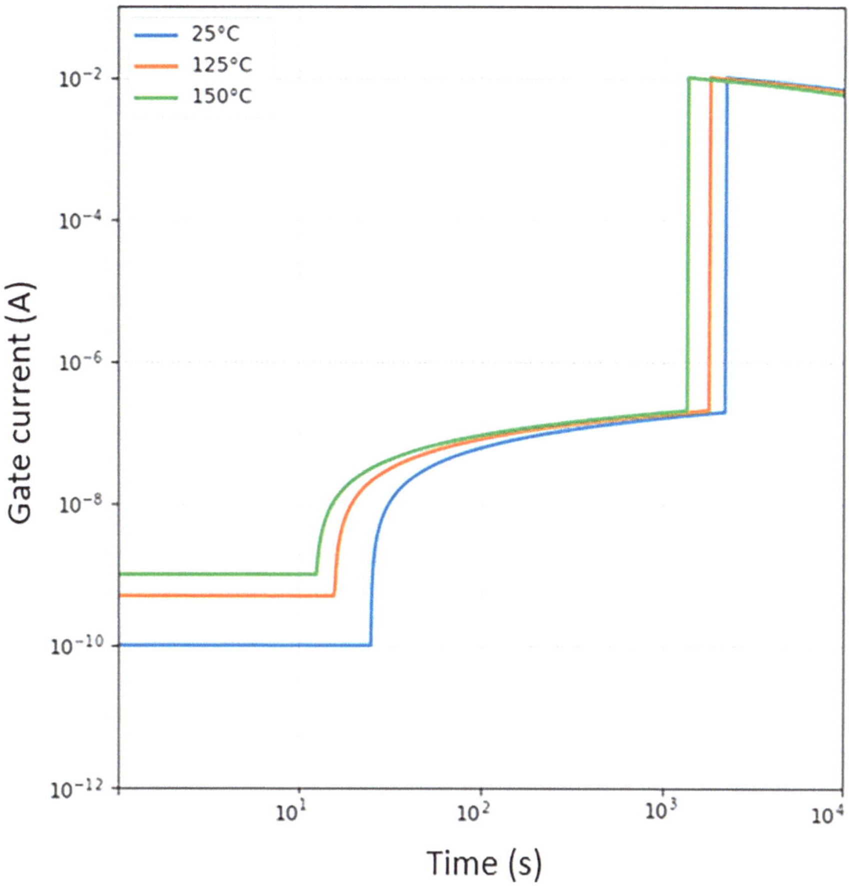

Fig. 3.11 Effect of TDDB phenomenon on gate current

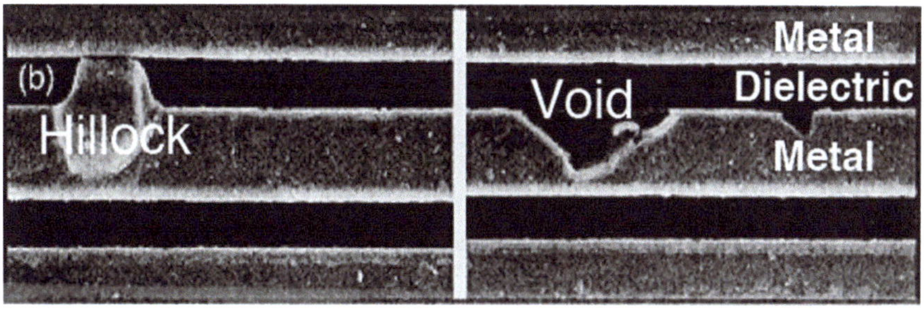

Fig. 3.12 Effect of electromigration on metal wires

result, the dissipation power increases, which in turn generates more thermal energy. This positive feedback is called self-heating.

In the simulation tools used in the design phase of modern ICs, it is necessary to introduce dynamic models of the physical characteristics of transistors [66–68]. The introduced models will allow obtaining an accurate timeline of the change in the IC's reliability and performance at the simulation stage.

In the power supply nodes embedded in NVMs, the degradation of transistor characteristics due to BTI, HCI, TDDB, and EM is most pronounced in high-voltage regulators (VR) and charge pumps (CP) generating high voltages [69–71]. Aging of MOS transistors in VRs leads to deterioration of the amplitude–frequency and phase–frequency characteristics of differential amplifiers and reduction of the maximum load capacity in power converters. The accuracy and speed conditions ensuring feedback are violated [72–75].

CPs often have adaptive control systems for the clock signal frequency [76–78]. The degradation of the threshold voltage and speed of transistors in the CP circuit leads to charge loss during transient processes, which is expressed in the deterioration of the output voltage value. To restore the output voltage value, the adaptive system controlling the clock signal increases the charging frequency. As a result of all this, the dissipation power increases, and the more frequent switching of transistors accelerates the effect of aging phenomena. The classical circuit of a voltage doubler with cross-coupled NMOS transistors has been considered (Fig. 3.13) [79].

The two symmetrical branches of the circuit are in anti-phase states during operation. During the active half-cycle of the V_{clk} signal, $V_1 = 2 \times V_{in}$, and during the passive half-cycle, $V_2 = 2 \times V_{in}$. Here, a number of factors contributing to aging phenomena are obvious. During the doubler's operation, the gate-substrate voltages of transistors M1 and M2 are twice the supply voltage. This creates a large vertical electric field along the transistors' gates, which provides favorable conditions for HCI, BTI, and TDDB phenomena.

These two transistors, in the active phase of their corresponding branches, must serve as a closed switch, and in the passive phase, charge the branch capacitances as quickly as possible. As a result of the listed aging phenomena, the speed of the transistors decreases, and the leakage currents increase. As a result of the deterioration of these characteristics, transistors lose the efficiency of performing the required functions. Due to the increased threshold voltage and leaks caused by aging, the charging rate of the capacitances decreases, and the charge loss through the transistor in the off state increases.

Although the M1–M2 pair is the most vulnerable link in the circuit, other components of the circuit also contribute to the loss of efficiency and aging. The cross-coupled connection of M3 and M4 PMOS transistors is a selector that outputs the maximum of the input voltages. In case of large output currents, it is advisable to design these transistors with the smallest possible channel resistance in order not to create a significant voltage drop across them. However, this creates a risk of accelerating the HCI phenomenon. The short channel length makes the presence of "hot" carriers already in a large current more likely. In case of reducing the effect of HCI phenomena and the voltage drop, both the channel length and width of M3 and M4

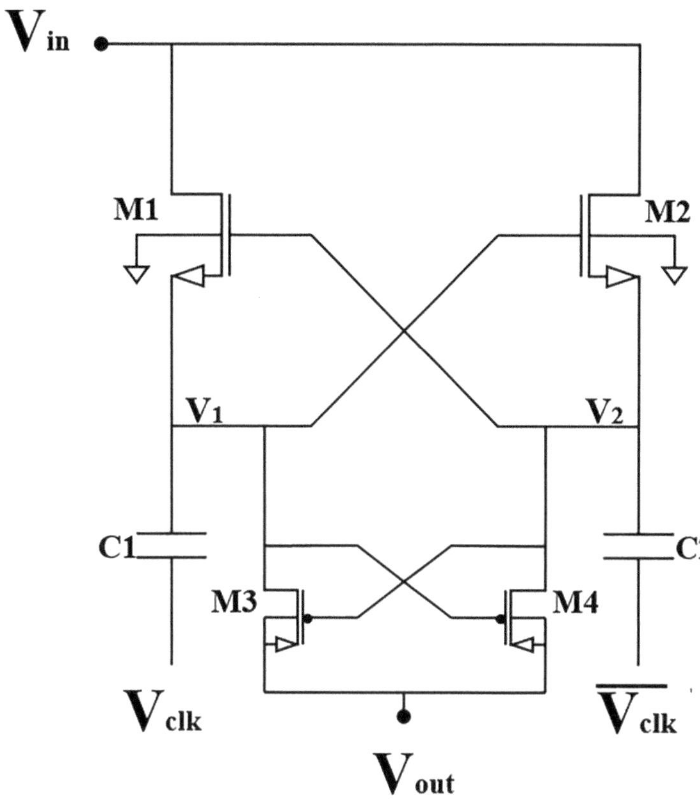

Fig. 3.13 Voltage doubler circuit with cross-coupled nMOS switches

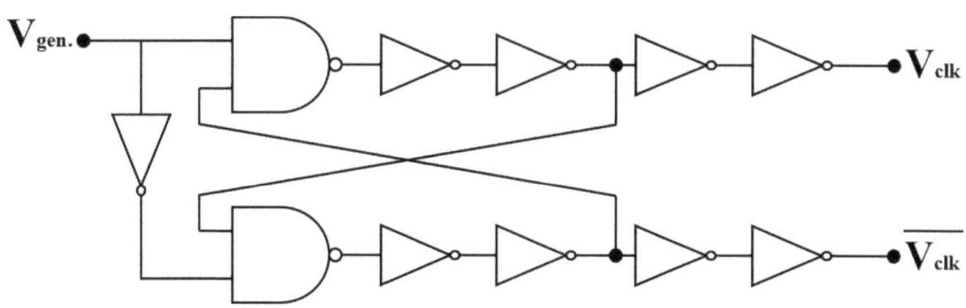

Fig. 3.14 Anti-phase clock signal generator

will increase. That is, a partial prevention of the problem will be obtained at the cost of a significant increase in the circuit area.

V_{clk} and $\overline{V_{clk}}$ anti-phase signals must be obtained so that their passive parts do not overlap (Fig. 3.14) [80–82]. In that case, during the operation of the voltage doubler, the value of at least one of the selector's inputs will always be $2 \times V_{in}$.

From the first-order output load formula (3.2) [80–82] it can be seen that it is directly proportional to the clock signal frequency f, C_{pump} capacitance and ΔV maximum allowable output voltage wave:

$$I_{out} = f \cdot C_{pump} \cdot \Delta V \tag{3.2}$$

This equation does not take into account a number of secondary and parasitic phenomena that become apparent at high frequencies. The switching loss power increases according to the following relation (3.3) [80]:

$$P_{loss} \propto f \cdot C_{par} \cdot V^2 \tag{3.3}$$

where C_{par} is the parasitic capacitances formed in the circuit.

At high frequencies, the switching times of transistors affect the charge transfer efficiency, and the phase overlap coefficient increases (Fig. 3.15) [80–82].

As a result of all this, the maximum output load capacitance obtains a non-linear dependence on the clock signal frequency. After a certain deterioration of the CP, it becomes impossible to provide the required output power even under conditions of sharply increasing input power.

The described CP circuit was designed for the SAED 14 nm FinFET [83] technological process. Using HSPICE simulation, the dependence of the output current capacitance on the clock signal frequency was obtained for an output voltage of $1.8 \times V_{in}$ (Fig. 3.16) [70]. In high-temperature conditions, due to large leakage currents, the output capacitance saturates and begins to decrease at lower frequencies than at room conditions. Under these conditions, a higher input frequency is required to provide the same output power.

At high temperatures, the effects of aging phenomena on the CP's efficiency also increase. To this, the deterioration of characteristics due to more frequent switching of transistors is added, and a sharp drop in operational reliability under high temperature conditions is obtained.

Using HSPICE MOSRA models, a multi-step simulation of the CP circuit's aging timeline was performed. The results obtained through this show that the main part of the deterioration of the circuit's multiplication factor due to aging phenomena occurs during the first year of operation (Fig. 3.17) [70]. Using exponential decay and Arrhenius models [84], the dependence of the input voltage multiplication factor at the output of this circuit on temperature and operating duration is obtained (3.4) [84]:

$$f(t,T) = (0.096 + 0.00176 \cdot T) \cdot e^{-(0.9788+0.00201 \cdot T) \cdot t} + (1.6625 - 0.0021 \cdot T) \tag{3.4}$$

where t is time expressed in years, and T is the temperature in Celsius.

This non-linear view of the timeline of the circuit's output power deterioration indicates that the absence of means to prevent high-voltage stress in the design has accelerated the effect of aging phenomena. Due to overvoltage stress on the

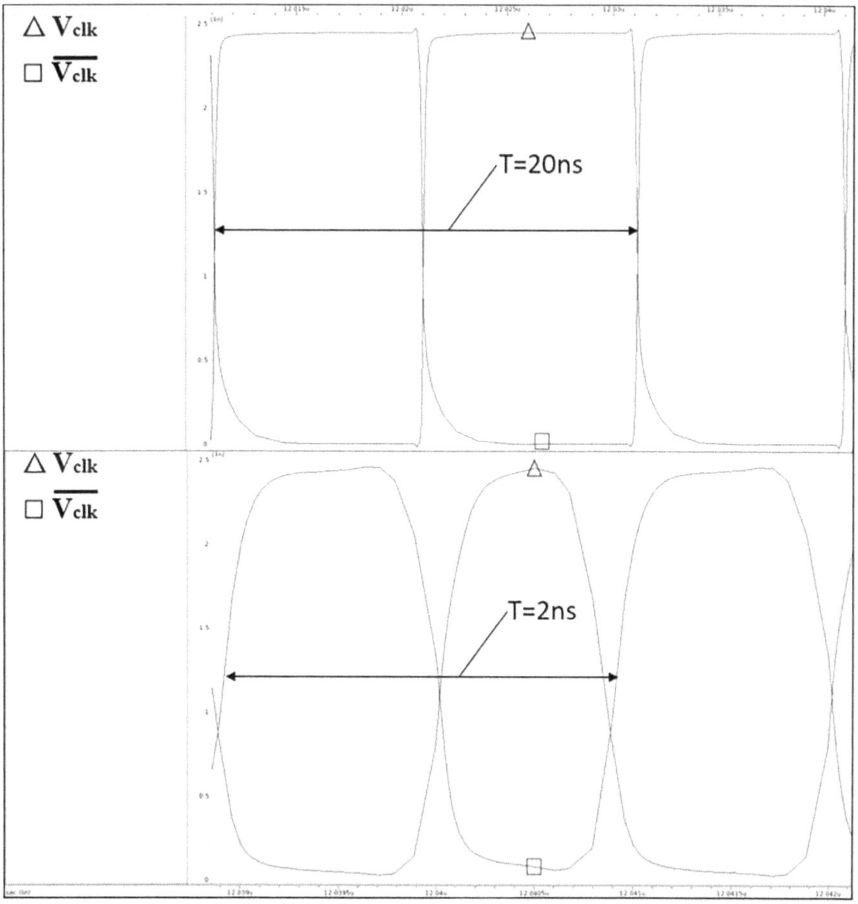

Fig. 3.15 Anti-phase clock signals at 50 MHz and 500 MHz

transistors, the output voltage value dropped sharply during the very first year of the circuit's operation. Due to this drop, the overvoltage stress on the transistors decreased, and further aging slowed down.

Such a circuit can be reliably used only in one-time programmable IMCs, where the only purpose of such a voltage multiplier is to provide the voltage necessary for the programming operation. However, its reliability will not satisfy either multiple programmable IMCs or one-time programmable IMCs, where a voltage greater than the supply voltage value is also required for the read operation.

In those IMCs where obtaining values higher than the supply voltages is also necessary to perform the read operation, VRs are also used in addition to voltage multipliers. Low dropout voltage regulators are widely used, which provide at the output a signal not much smaller than the input value, but with significantly suppressed noise. In the case of a large input voltage margin, it is possible to use a VR with an NMOS output transistor (Fig. 3.18) [85–88]:

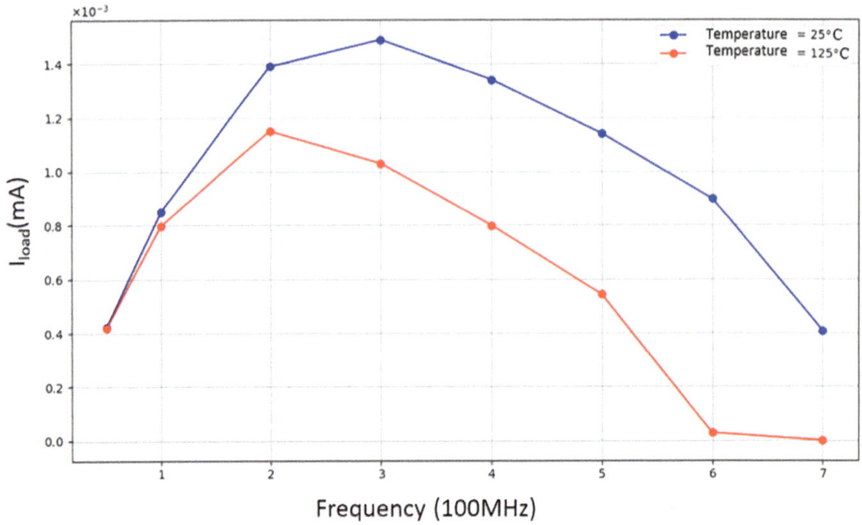

Fig. 3.16 Dependence of maximum output current value on frequency

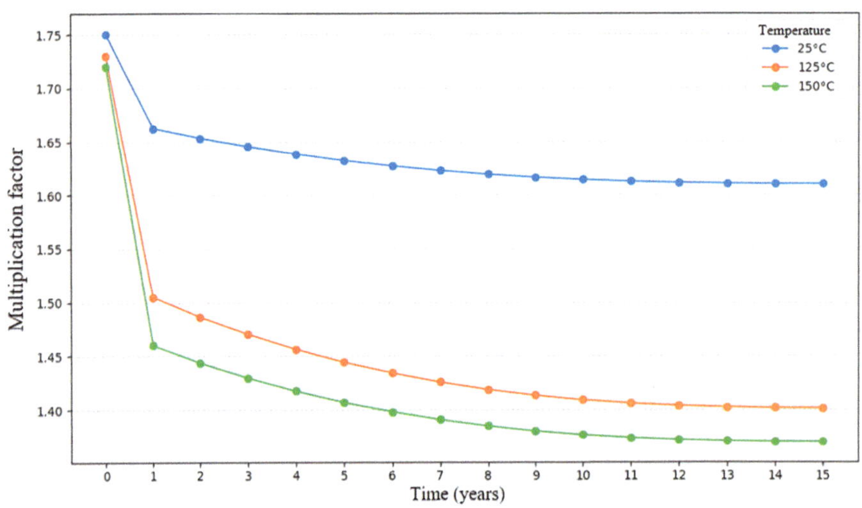

Fig. 3.17 Deterioration of the output voltage multiplication factor value over time

In modern technological processes, the applicability of such a design is decreasing both from the point of view of a small input signal margin and long-term reliability. The output signal cannot exceed the difference between the gate potential and the threshold voltage of an NMOS transistor. The gate potential, in turn, is limited by the input voltage, and the transistor's threshold voltage is significantly higher than the nominal value due to the substrate-source reverse junction. The following formula is obtained for the output voltage value (3.5) [49]:

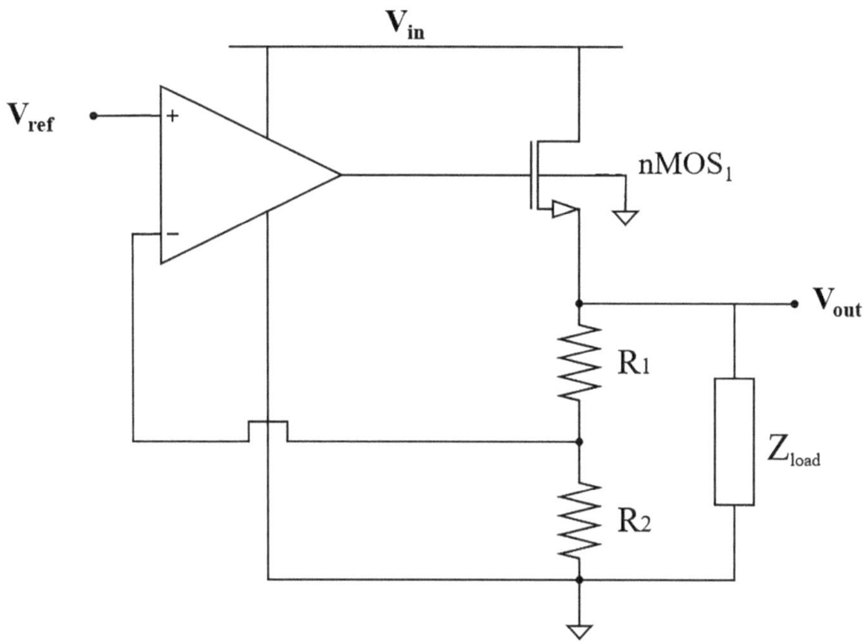

Fig. 3.18 Voltage regulator with NMOS output transistor

$$V_{out} \leq V_{in} - V_{sd} - \left(V_{th.nom} + \Delta V_{th.be}\right) \tag{3.5}$$

where V_{sd} is the source–drain voltage drop of the operational amplifier's (OpAmp) output transistor, $V_{th.nom}$. is the nominal threshold voltage, and $\Delta V_{th.be}$ is the threshold voltage deviation due to the body effect.

In a VR with a PMOS output transistor, the input voltage margin problem is solved, but reliability problems due to aging phenomena are also present in this circuit (Fig. 3.19) [89–91].

As a result of HSPICE simulation, the dependence of the deterioration of the PMOS transistor's drain current on the source–drain and gate–source voltages over 1000 h of operation was obtained (Fig. 3.20) [49]. The VR circuit is designed so that under operating conditions the voltages between the terminals of the output load-carrier transistor are in the low degradation zone shown in the figure. However, under the influence of many factors, the circuit can go beyond the limits of its intended normal operating conditions.

Often, in a system consisting of multiple IMCs, a single power supply node is embedded, which sequentially supplies the programming and read operations of memory cells. Local self-heating or an error in signals controlling IMD can lead to a sharp drop in load resistance. Through VR's negative feedback, the gate–source voltage of a PMOS transistor will increase, and after saturation of this voltage, the source–drain voltage will begin to increase. As a result, a transistor will find itself

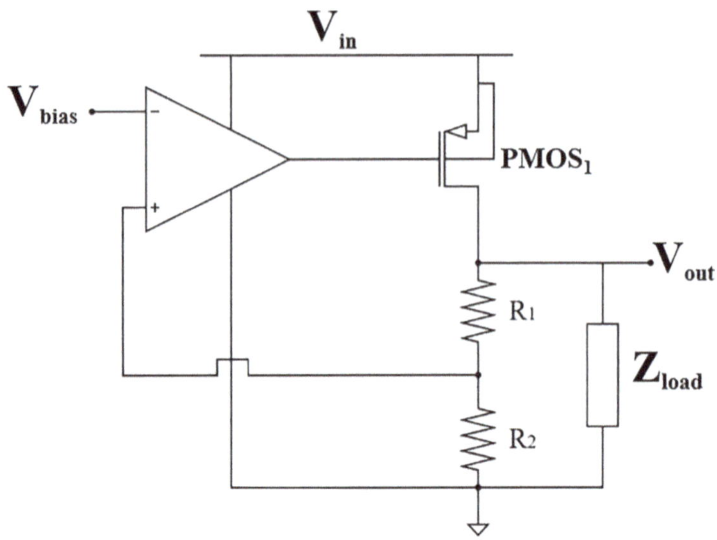

Fig. 3.19 Voltage regulator with PMOS output transistor

V_{gs} \ V_{sd}	0.1	0.2	0.3	0.4	0.5	0.6	0.7	0.8	0.9	1	1.1	1.2	1.3	1.4	1.5	1.6	1.7	1.8	1.9	2	2.1	2.2	2.3	2.4	2.5
2.5	100%	100%	100%	100%	100%	100%	100%	100%	100%	100%	100%	100%	100%	100%	100%	100%	100%	100%	100%	100%	100%	100%	100%	100%	100%
2.4	100%	100%	100%	100%	100%	100%	100%	100%	100%	100%	100%	100%	100%	100%	100%	100%	100%	100%	100%	100%	100%	99%	99%	98%	98%
2.3	100%	100%	100%	100%	100%	100%	100%	100%	100%	100%	100%	100%	100%	100%	100%	100%	99%	99%	99%	98%	98%	97%	96%	96%	95%
2.2	100%	100%	100%	100%	100%	100%	100%	100%	100%	100%	100%	100%	99%	95%	91%	87%	87%	97%	97%	96%	95%	95%	94%	94%	93%
2.1	100%	100%	100%	100%	100%	100%	100%	100%	99%	96%	92%	87%	81%	75%	69%	65%	71%	95%	94%	94%	93%	93%	93%	92%	92%
2	100%	100%	100%	100%	100%	97%	93%	87%	81%	75%	69%	62%	57%	51%	47%	45%	57%	93%	93%	92%	92%	92%	92%	91%	91%
1.9	100%	97%	92%	85%	79%	73%	67%	61%	56%	50%	46%	41%	36%	33%	30%	31%	46%	92%	92%	91%	91%	91%	91%	91%	91%
1.8	74%	68%	62%	57%	52%	47%	43%	39%	35%	32%	29%	26%	22%	20%	19%	21%	39%	91%	91%	91%	91%	91%	91%	90%	90%
1.7	48%	44%	39%	36%	33%	30%	27%	25%	22%	20%	18%	16%	14%	12%	12%	15%	33%	91%	91%	91%	90%	90%	90%	90%	90%
1.6	30%	27%	24%	22%	20%	18%	16%	15%	14%	12%	11%	9%	8%	8%	8%	11%	30%	90%	90%	90%	90%	90%	90%	90%	90%
1.5	18%	16%	15%	13%	12%	11%	10%	9%	8%	8%	7%	6%	5%	5%	5%	9%	27%	90%	90%	90%	90%	90%	90%	90%	90%
1.4	11%	10%	9%	9%	8%	7%	6%	6%	5%	4%	4%	3%	3%	3%	4%	8%	25%	90%	90%	90%	90%	90%	90%	90%	90%
1.3	7%	6%	6%	5%	5%	4%	3%	3%	3%	2%	2%	2%	2%	2%	2%	7%	23%	90%	90%	90%	90%	90%	90%	90%	90%
1.2	4%	4%	4%	3%	3%	3%	2%	2%	2%	1%	2%	1%	2%	2%	2%	6%	22%	88%	90%	90%	90%	90%	90%	90%	90%
1.1	2%	2%	2%	2%	2%	2%	2%	1%	1%	1%	1%	1%	1%	1%	2%	5%	21%	84%	90%	90%	90%	90%	90%	90%	90%
1	1%	1%	2%	1%	1%	1%	1%	1%	1%	1%	1%	0%	1%	1%	1%	5%	20%	80%	90%	90%	90%	90%	90%	90%	90%
0.9	1%	1%	1%	1%	1%	1%	1%	1%	0%	0%	0%	0%	0%	0%	1%	5%	19%	76%	90%	90%	90%	90%	90%	90%	90%
0.8	1%	1%	1%	1%	0%	0%	0%	0%	0%	0%	0%	0%	0%	0%	1%	5%	18%	72%	90%	90%	90%	90%	90%	90%	90%
0.7	1%	1%	1%	1%	1%	1%	0%	0%	0%	0%	0%	0%	0%	0%	1%	4%	17%	69%	90%	90%	90%	90%	90%	90%	90%
0.6	1%	0%	1%	0%	0%	0%	0%	0%	0%	0%	0%	0%	0%	1%	1%	4%	16%	65%	90%	90%	90%	90%	90%	90%	90%
0.5	1%	0%	0%	0%	0%	0%	0%	0%	0%	0%	0%	0%	0%	0%	1%	4%	16%	62%	90%	90%	90%	90%	90%	90%	90%
0.4	0%	0%	1%	0%	0%	0%	0%	0%	0%	0%	0%	0%	0%	0%	0%	0%	0%	0%	0%	0%	0%	0%	0%	0%	0%
0.3	0%	0%	0%	0%	0%	0%	0%	0%	0%	0%	0%	0%	0%	0%	0%	0%	0%	0%	0%	0%	0%	0%	0%	0%	0%
0.2	0%	0%	0%	0%	0%	1%	0%	0%	0%	0%	0%	0%	0%	0%	0%	0%	0%	0%	0%	0%	0%	0%	0%	0%	0%
0.1	0%	0%	0%	0%	0%	0%	0%	0%	0%	0%	0%	0%	0%	0%	0%	0%	0%	0%	0%	0%	0%	0%	0%	0%	0%

Fig. 3.20 Deterioration of PMOS transistor drain current

in conditions that accelerate degradation. After prolonged operation under such conditions, irreversible changes in physical characteristics can occur in a circuit, which will affect the system's future performance.

Voltage level shifters (LS) are used to control the above high-voltage circuits (Fig. 3.21) [92–94]. Thanks to them, the level of input low-voltage signals is increased and transmitted to analog circuits in power supply systems.

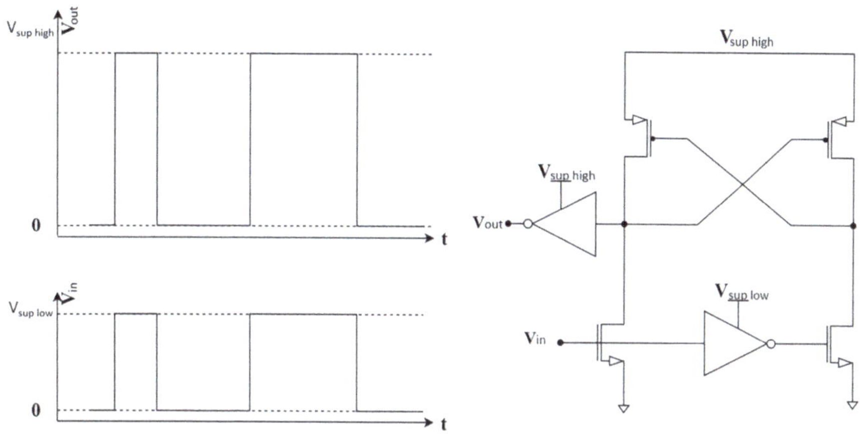

Fig. 3.21 Structure and timing characteristics of a voltage level shifter

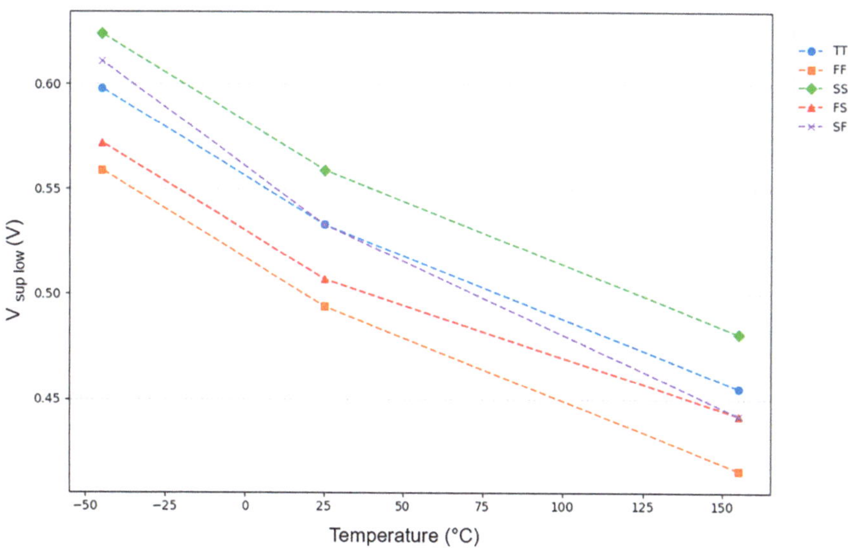

Fig. 3.22 Low supply voltage threshold, excluding output error

These LSs are very sensitive to process, voltage, and temperature (PVT), which is manifested by large deviations in operational parameters. Under extreme operating conditions, the probability of an error at the LS's output, depending on the previous state, increases. Using HSPICE simulation, the threshold that excludes the output error at low supply voltage values was determined, depending on PVT, for a high supply voltage value of 1.8 V (Fig. 3.22) [94].

Thus, simulations showed that classical circuits of high voltage generating and stabilizing systems embedded in IMCs, as well as their control circuits, have efficiency and reliability problems. The causes of these problems are both aging phenomena and circuit design solutions that have lost their relevance as a result of the development of modern technological processes. For these reasons, a strong need has arisen for the development of means to solve these problems.

3.1.3 Existing Solutions and Key Issues for Improving the Performance and Reliability of Power Supply and Control Systems Embedded in Integrated Memory Devices

The issues discussed above can cause failures in the operation of the mentioned circuits, which will reduce the reliability of power supply and control systems embedded in IMCs. This will lead to an increase in data recording failures and loss during storage in IMD, which will disrupt the accurate and reliable operation of an electronic system regulated or controlled through the information stored in the IMD. To avoid all this, it is necessary to develop new approaches through which it will be possible to improve the performance of ICs and reduce the effects of aging phenomena in them.

There are many existing means and methods developed so far to solve these problems, each of which has its advantages and disadvantages. In the conditions of hyperactive scaling of technological processes, even solutions that are several years old may no longer meet modern requirements. Several such solutions and their problems and key issues are discussed below.

Existing Measures to Improve Reliability in Charge Pumps

As shown above, in the classical circuits of high-voltage generators embedded in NVMs, significant efficiency losses occur as a result of the influence of aging phenomena, and reliability problems arise. There are many known means aimed at solving these problems [95–97].

From the point of view of not complicating the provision of clock signals controlling the circuit, the most effective existing solution to this problem is to physically isolate the substrates of NMOS transistors from the common grounded substrate of an IC [97]. This allows controlling the potential of the transistor's substrate and having a maximum voltage of V_{sup} between any terminals (Fig. 3.23) [97].

To isolate and control the substrate connection, it is proposed to place NMOS transistors in a region with a deep n-type substrate (Fig. 3.24) [97].

Through the M5–M6 cross-coupled transistors, the minimum of the source and drain potentials of the M1 transistor is dynamically selected and connected to their common substrate. The same operation is performed for the M2 transistor through the M7–M8 pair. As a result, the gate–substrate voltage in the on state is V_{sup}, and in the off state, it is 0.

As a result of this change, the effects of BTI and TDDB phenomena are reduced because the maximum static stress between any two terminals of the transistors is no more than V_{sup}. From the point of view of modern technological processes, this

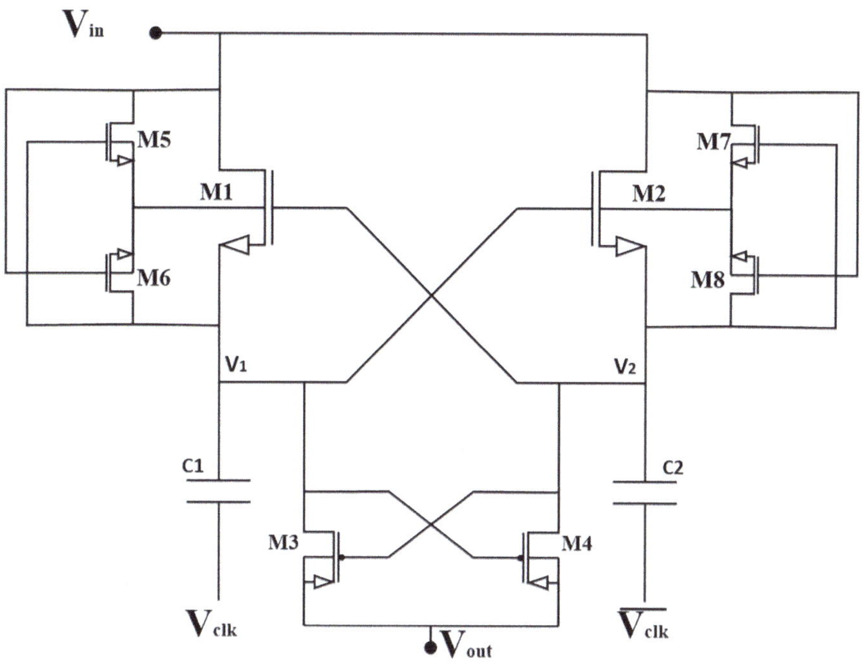

Fig. 3.23 Voltage doubler with controlled substrate

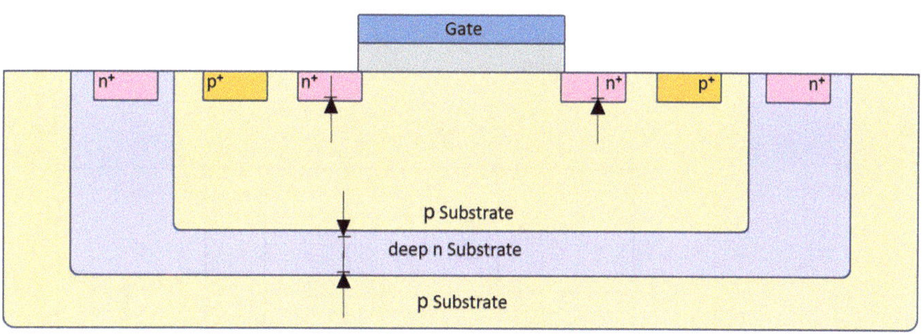

Fig. 3.24 Structure of an NMOS transistor with isolated substrate

method has a number of disadvantages. First, the number of transistors in the circuit increases, and due to the isolated substrate, their occupied area increases significantly. Due to the substrate isolation, the surface for dissipating thermal energy is very small, and self-heating occurs. The number of parasitic bipolar transistors and resistors and the probability of undesirable phenomena caused by them increases. In addition to the listed disadvantages affecting the IC area and operation, there are also a number of problems related to the manufacturing processes themselves. Thus, transistors with deep n implantation, which were widely used in previous generation ICs, have become an undesirable risk and are not even available in some low-voltage technological processes. It is necessary to develop new approaches to reduce aging phenomena in CPs that will satisfy the limitations of modern technological processes.

Existing Means of Improving Reliability in Voltage Stabilizers

There are a number of solutions to protect VRs from overloaded operating conditions [98–101]. To protect both VR and the circuits supplying it in case of short circuits and sharp load changes, constant and dynamic output current limiting behaviors are proposed.

The constant current limiting circuit monitors the gate voltage of the M1 output transistor and, through the M2 transistor, prevents its value from falling below the allowable threshold (Fig. 3.25) [100]:

In case of resistances smaller than the set threshold, the current value remains constant, and the output voltage value is determined by Ohm's law (3.6) [100]:

$$V_{out} = I_{in} \times R_{out} \tag{3.6}$$

The dynamic current limiting circuit to protect the VR from short circuits applies not only an upper current limit but also reduces the VR's output current in case of a further drop in load resistance (Fig. 3.26) [100].

The main feedback operates in the left part of the circuit, the short circuit protection circuit is in the middle, and the maximum allowable current value is set in the right part. As a result of the proposed solutions, the following curves of current dependence on output load are obtained (Fig. 3.27) [100].

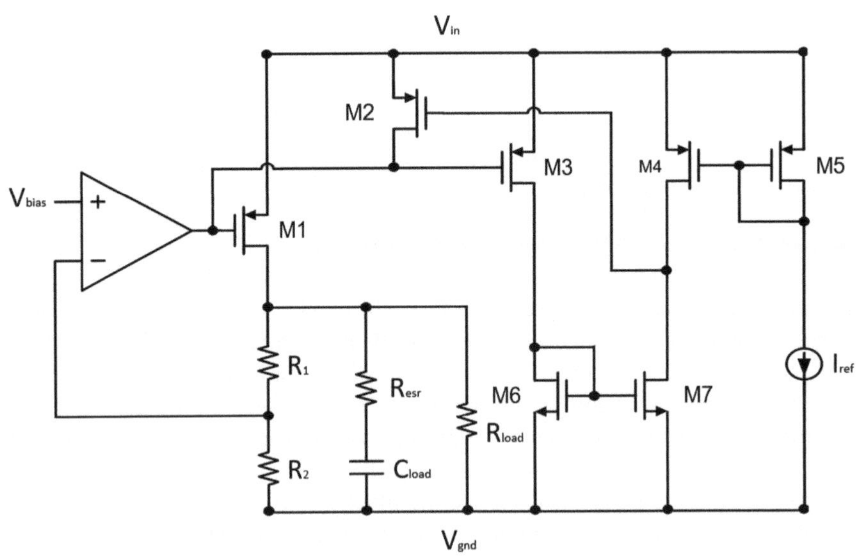

Fig. 3.25 Structure of the constant current limiting circuit

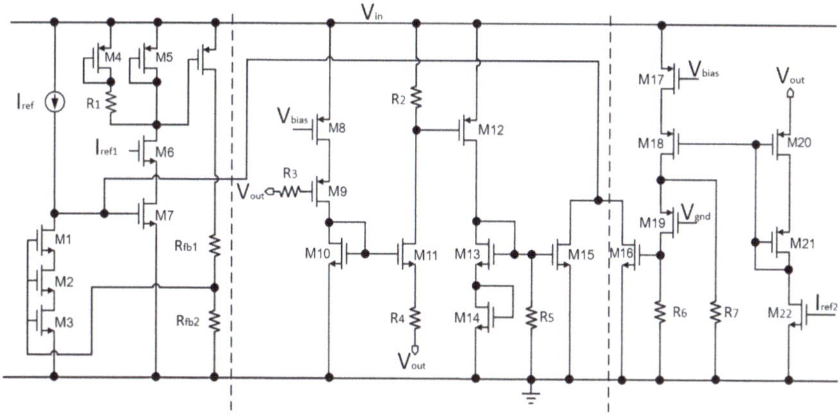

Fig. 3.26 Structure of the dynamic current limiting circuit

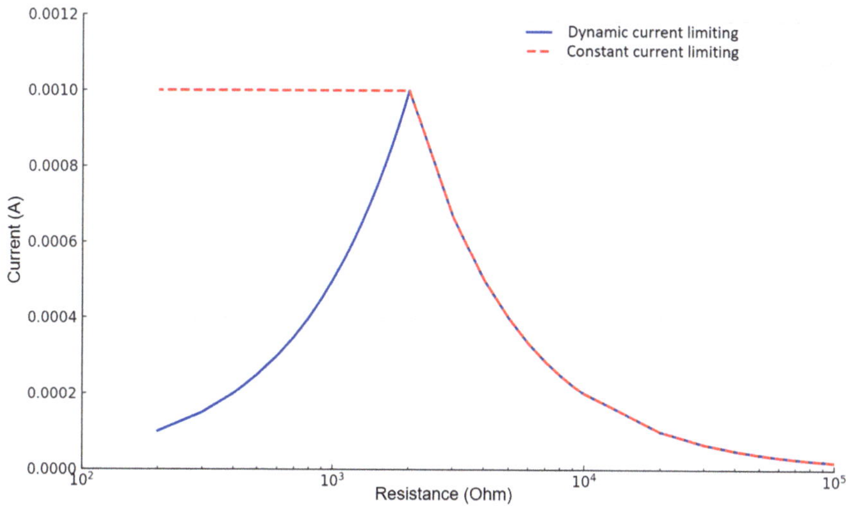

Fig. 3.27 Current dependence on load resistance

These solutions completely protect the output PMOS transistor from vertical overvoltage stress, but the protection from the large horizontal field along the channel is partial (Fig. 3.28) [49].

Thus, to reduce the deterioration of the VR over time, it is necessary to take into account the effects of aging phenomena and improve the existing approaches to protecting VRs from overload.

Existing Solutions to Reliability Problems in Power Supply Voltage Detection Circuits

For the reliability of systems operating with two or more supply voltage levels, circuits are needed that will allow the system to operate only when all supply

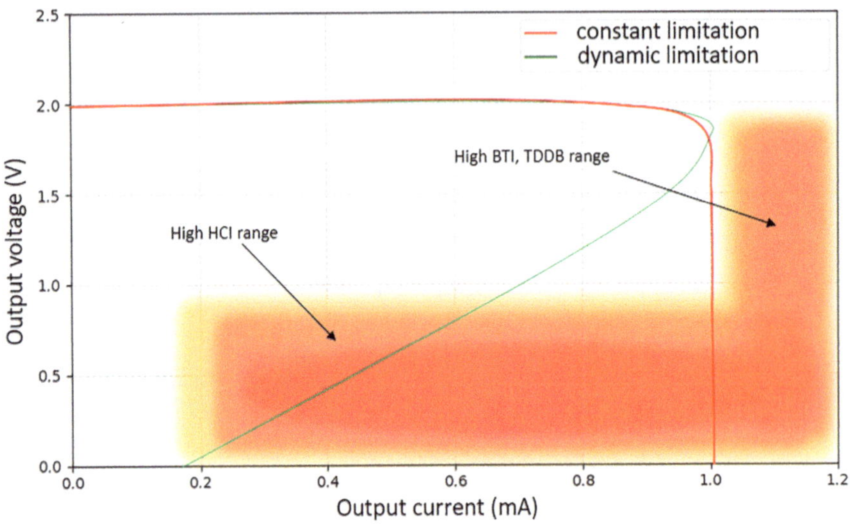

Fig. 3.28 Output current-limiting characteristics of the voltage regulator

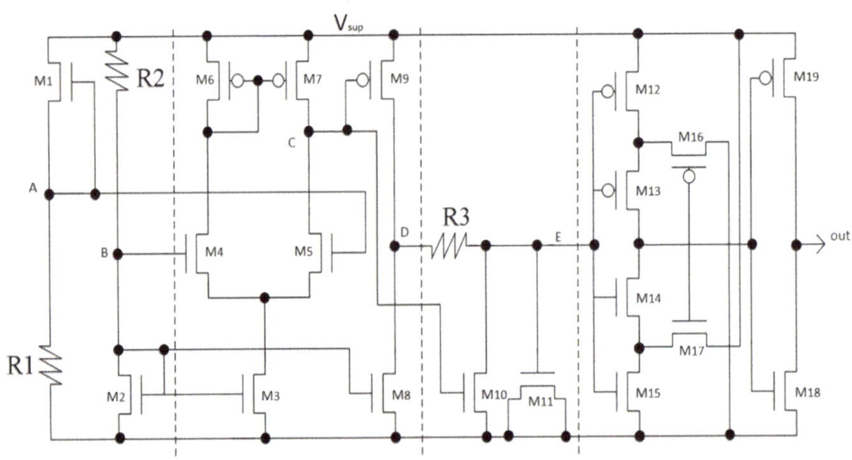

Fig. 3.29 Structure of the supply voltage detection circuit

voltages are within the intended operating range. There are known solutions that make it possible to perform this function without using reference voltage sources (Fig. 3.29) [102–104].

In the proposed circuit, the chain consisting of transistors M3–M9 compares the voltages of points A and B, whose values depend on the sizes of transistors M1 and M2, the resistances R1 and R2, and the supply voltage value. Transistors M12–M17 form a Schmitt trigger [102], which provides the circuit's hysteresis. As a result of designing and simulation this circuit with the SAED 14 nm technological process, the following inversely proportional dependence of the supply voltage detection threshold on temperature is obtained (Fig. 3.30) [102].

In control systems embedded in IMCs, such circuits are used to ensure the reliable operation of LSs. The minimum operating voltage of an LS modeled with the same technological process also shows a similar dependence on temperature (Fig. 3.31) [102].

The main disadvantage of this solution is that only the temperature dependence of LS's operating voltage is taken into account. As a result, the process dependence is ignored, and the threshold value is set for the worst case. There is a need to develop solutions through which the detection threshold will follow both the temperature and process deviations of the LS. This will allow aligning the margin held between the minimum operating and detection voltage thresholds in all PVT cases.

Thus, the existing known solutions considered have such disadvantages that these methods lose their relevance in developing technological processes, so it is necessary to develop and propose new approaches to solve the arising key issues.

3.1.4 *Proposed Principles to Improve the Performance and Reliability of Power Supply and Control Systems Embedded in Integrated Memory Devices*

Taking into account the disadvantages of the existing methods considered above, the following principles have been proposed for the purpose of developing new methods and solutions to reduce aging phenomena in CPs and VRs and to increase the accuracy of the PVT behavior of the supply voltage detection in the nodes controlling them.

1. Replacement of NMOS transistors with PMOS in CPs and the two-phase clock signal with a four-phase one, whereby the deterioration of transistor characteristics due to the influence of aging phenomena is reduced without loss of output power.

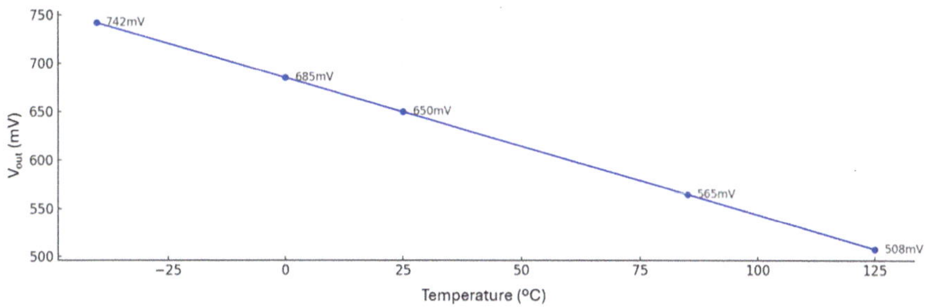

Fig. 3.30 Supply voltage detection threshold

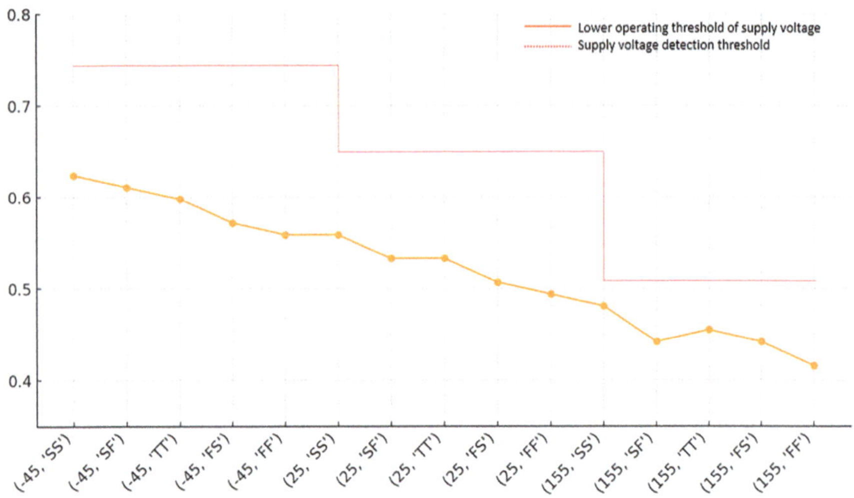

Fig. 3.31 Lower operating threshold of LS's low supply voltage

The results of the simulation performed for one of the known CP circuits above showed that its output voltage at room temperature decreases by 8% after 15 years of aging, and at 150 °C temperature by 20.4%. Most of the deterioration occurs during the first year of IC operation. The considered method allowed eliminating the overvoltage stress between the terminals of the NMOS transistors by isolating their bodies in deep n-type substrates. This complex structure led to a significant increase in area, complication of thermal energy dissipation, and undesirable phenomena due to the resulting additional parasitic elements. It is proposed to replace NMOS transistors in CPs with PMOS, and their control, instead of a two-phase clock signal, to be implemented with a four-phase clock signal, at the cost of a minimum increase in area.

2. Alignment of volt-ampere behavior of a VR with dynamic current limiting to the safe operating area of the driving PMOS transistor.

As a result of simulating the known VR circuits, reliability problems become apparent, which are most pronounced in regulators of high voltages generated by CPs. The proposed static and dynamic limiting methods only partially protect the VR's output driver transistor from the effects of aging phenomena. To protect the VR from overvoltage stress, it is proposed to apply a dynamic current limiting behavior whose volt-ampere characteristic will be entirely within the safe operating area of the output PMOS transistor.

3. Alignment of the threshold value of the supply voltage detection signal controlling LSs in multi-voltage control systems with PVT behavior of LS.

The simulation results performed for the LS circuit showed a large spread of its minimum operating supply voltage thresholds, depending on PVT. As a result, the supply voltage detection threshold, which was set according to the slowest PVT case, had an excessively large margin for other PVT cases. The method considered above compensated only for the temperature dependence. It is proposed to obtain an LS with a predictable failure state in the used LS circuit by adding asymmetries and weak conduction paths to the supply voltage, which will form the supply voltage detection signal.

Conclusions

1. With the ongoing scaling of modern technologies, the increasing impact of aging phenomena has become a critical issue in the design of power supply and control systems embedded in integrated memory devices. Aging phenomena can lead to a decrease in the reliability of integrated memory circuits, up to failure. Therefore, taking into account aging phenomena, the development of means to improve the performance and reliability of power supply and control systems embedded in integrated memory devices is a relevant task.
2. The existing methods for designing power supply and control systems embedded in integrated memory circuits, taking into account aging phenomena, are not sufficient to overcome current challenges, and in some cases, they are even absent, which leads to the need to develop new methods.
3. Approaches have been proposed to improve the accuracy of power supply and control systems embedded in integrated memory circuits and reduce the effects of aging phenomena, which, at the cost of a permissible increase in power consumption, occupied area, and simulation duration, provide the necessary operating conditions.

3.2 Proposed Solutions to Improve the Performance and Reliability of Power Supply and Control Systems Embedded in Integrated Memory Devices

The problems described in Sect. 3.1 show the existing shortcomings in the methods aimed at solving the performance and key reliability issues of power supply and control systems embedded in IMCs. Below, means are proposed that solve the reliability and performance problems in these systems, filling the gaps existing in the solutions so far.

3.2.1 Method to Reduce Aging Phenomena in Charge Pumps Through Multi-Phase Clocking and pMOS Only Structure

As noted in Sect. 3.1, the large gate–drain voltages in the NMOS transistors present in the CP circuit contributed to the effect of HCI, BTI, and TDDB aging phenomena, accelerating the deterioration of the CP's output power. With the proposed method, it was possible to reduce the effect of aging phenomena by isolating the substrates of vulnerable transistors in a deep n-type implanted layer. By applying this method, the dimensions of NMOS transistors increase by 3–4 times, as a result of which the total circuit area increases by 12–15%. Such an increase in area to reduce the effect of aging phenomena is common, and the main complications of applying this method are related to the manufacturing process. Obtaining a deep n-type implanted layer implies the use of additional masks in the manufacturing process, which leads to additional financial and timing costs. In processes with channel lengths of 16 nm and smaller, transistors have various three-dimensional structures, which complicate obtaining a substrate with deep n-type implantation. In many modern technological processes, this possibility is not available at all.

Taking these facts into account, it is proposed to use only PMOS transistors in the CP circuit, based on the simple control properties of their substrates. In the replaced PMOS transistors, it is also necessary to ensure unidirectional conductivity, so that in the passive half-cycle of the clock signal it charges the capacitance, and in the active half-cycle it serves as an open circuit. A diode connection solves this problem, but the ideal maximum output voltage value becomes $2 \times V_{sup} - V_{th}$ from $2 \times V_{sup}$ (Fig. 3.32) [105].

It is necessary to find a solution through which it will be possible to disconnect the PMOS's gate from the diode connection during the charging phase and allow the capacitance to charge up to V_{sup}. To obtain such behavior, it is proposed to control the diode connections through low-frequency RC filters, whose capacitances will be controlled by additional anti-phase clock signals (Fig. 3.33) [70].

In the steady state, the gate and drain potentials of the M1 transistor are equal to $V_{sup} - V_{th}$. Through V_{clk3} and the C_3 capacitance, it becomes possible to temporarily make the gate potential of the M1 transistor 0 V. Over time, with the current flowing through the R_1 resistance, it will equalize again with the drain potential, but this time delay will allow the drain to charge up to V_{sup}. Then, during the active edge of V_{clk3} and V_{clk1}, both the drain and gate potentials will become $2 \times V_{sup}$. The same operations take place in the opposite phase in the right branch of the circuit. To ensure this process, the following behavior of the four-phase clock signals is necessary (Fig. 3.34) [70].

The time constants of the R_1C_3 and R_2C_4 low-frequency filters must be chosen so that during the indicated segments in the diagram, the gates of the M1 and M2 transistors manage to charge to at least 90% of the drain voltage. The selection is made for the case of the clock signal generator's maximum operating frequency. Simulation performed under equal conditions shows that the proposed CP obtains a

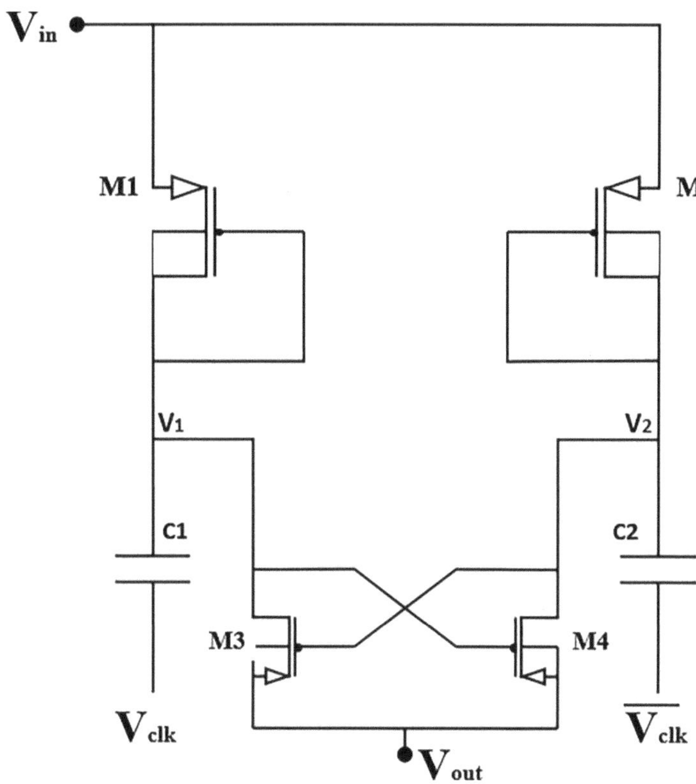

Fig. 3.32 CP structure with PMOS transistors in diode connection

larger output voltage than the CP with diode connection proposed in [105], but is inferior to the CP with classical NMOS structure [97] in case of low supply voltages (Fig. 3.35) [70].

The lower output voltage value is due to the incomplete charging of the capacitance during the passive half-cycle of the clock signal. To increase the efficiency of the charging phase, it is necessary to replace the resistances present in the circuit with elements with controlled conductivity. This will allow controlling the time constants of filters during different phases of CP's operation. By providing very small conductivity in the charging phase, the gate will be completely disconnected from the drain, and in the opposite phase they will be connected with greater conductivity. In this way, the reverse leakage of charge through the M1 and M2 transistors will be prevented, and the maximum drain voltage at the end of the charging phase will increase. A six-phase clock signal CP circuit consisting only of PMOS transistors is proposed, which performs the described function (Figs. 3.36 and 3.37) [70].

In the steady state, the gate voltage of the M2 transistor is equal to V_{sup}, but through V_{clk2} it can become 0 V or $2 \times V_{sup}$. In the first case, the transistor's channel will serve as a high conductivity path between points V_1 and V_2, and in the second,

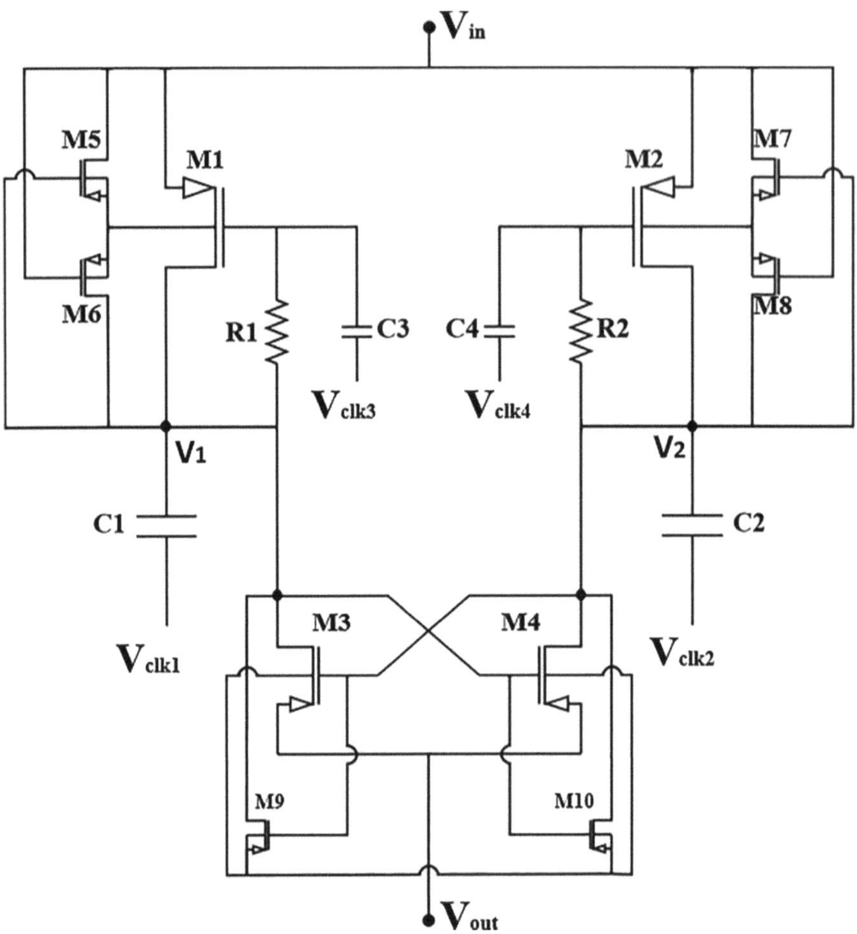

Fig. 3.33 Structure of the proposed CP with only PMOS transistors and four-phase clock signal

an open circuit. The complete isolation of these points allows controlling the gate of the M1 transistor through V_{clk3} without leaks, which leads to the maximum charging of the C1 capacitance. The highest potential in the branch is selected by the M3 and M4 transistors, and the substrates of all PMOS transistors are connected to that point. Due to their being in one substrate, the area increase is minimized as much as possible. To ensure the described process, the following behavior of the six-phase clock signals is necessary (Fig. 3.38) [70].

The simulation results show that the second proposed circuit, complementing the gaps of the first, provides a multiplication factor closer to the theoretical maximum (Fig. 3.39) [70] and more efficient power consumption (Fig. 3.40) [70].

These results are obtained at the cost of an increase in area and design complexity. The design complexity is due to the provision of the proposed four and six-phase clock signals and the incompatibility of the filters' time constants with high

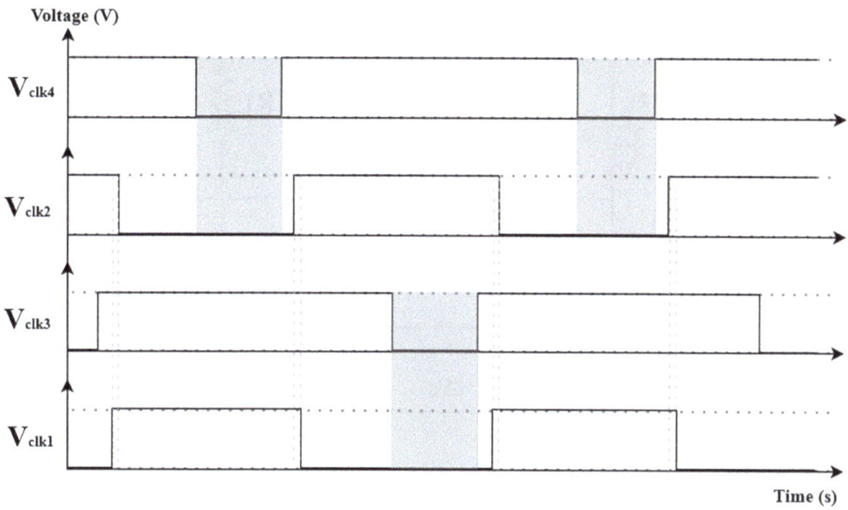

Fig. 3.34 Timing diagram of the proposed CP's clock signals

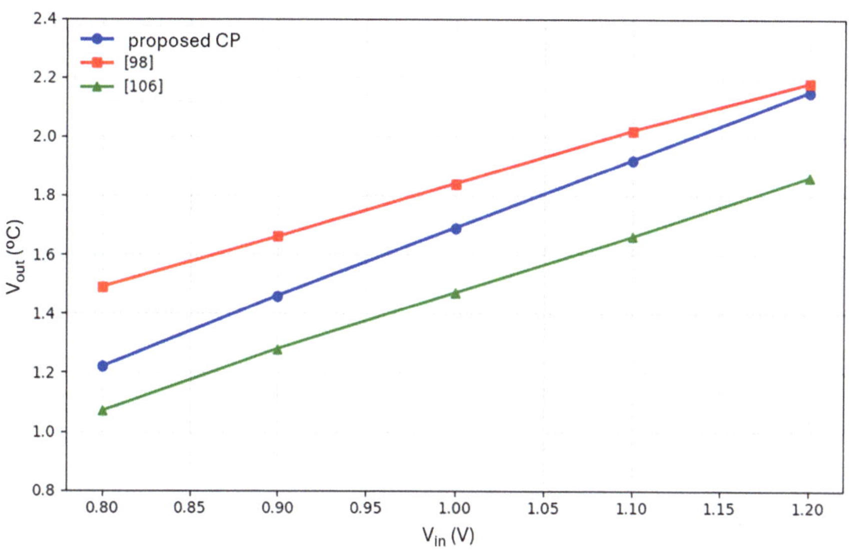

Fig. 3.35 Dependence of output voltage on supply voltage

frequencies. In a classical CP, in case of non-overlapping clock signal low levels, it was possible to increase the maximum output load value in a directly proportional manner to the frequency increase (Fig. 3.16) [70]. In the case of the first proposed solution, even in the case of ideal clock signals, the time constants of the low-frequency filters will not allow obtaining such behavior. Consequently, ensuring output load increase through frequency is more limited, and a need to increase the

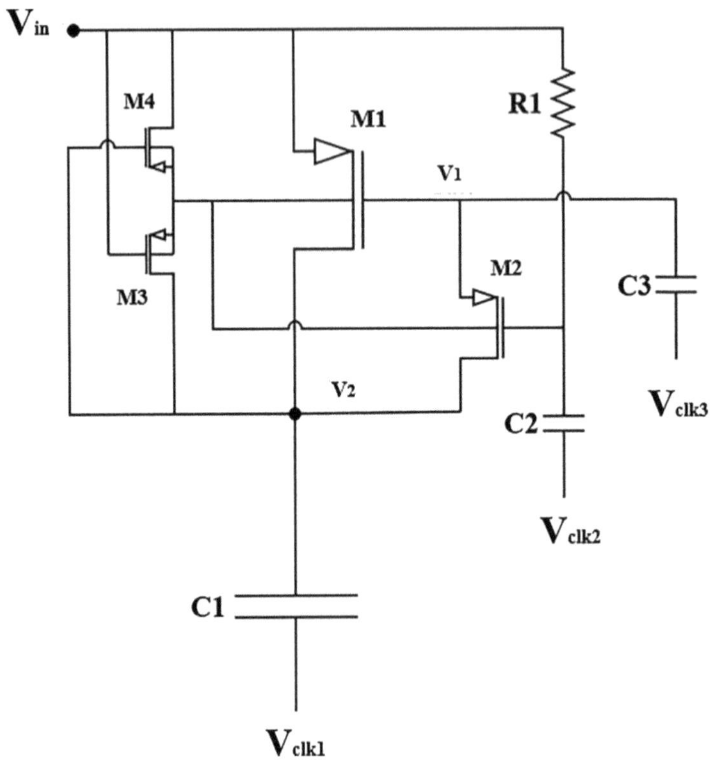

Fig. 3.36 Branch of the proposed CP circuit controlled with six-phase clock signal

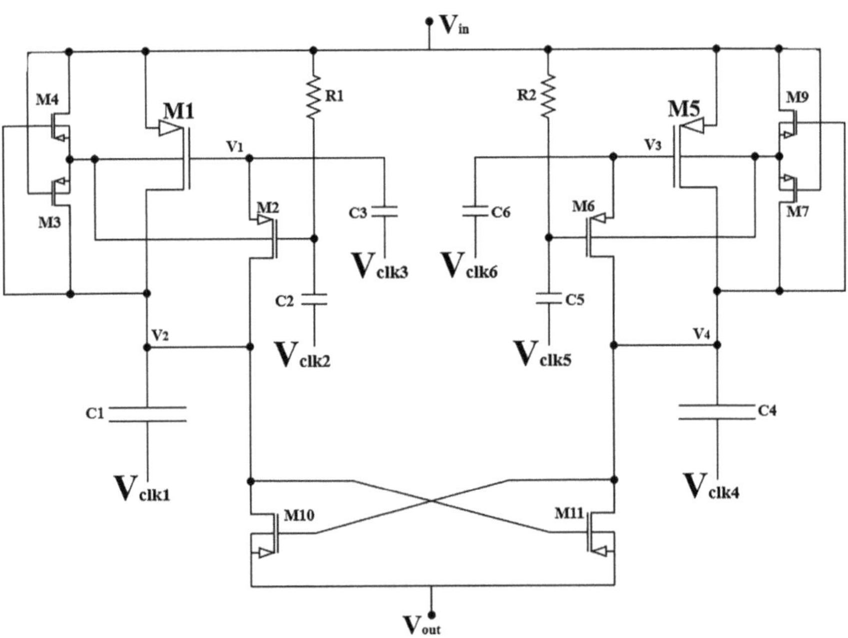

Fig. 3.37 Complete schematic view of the proposed CP controlled with six-phase clock signal

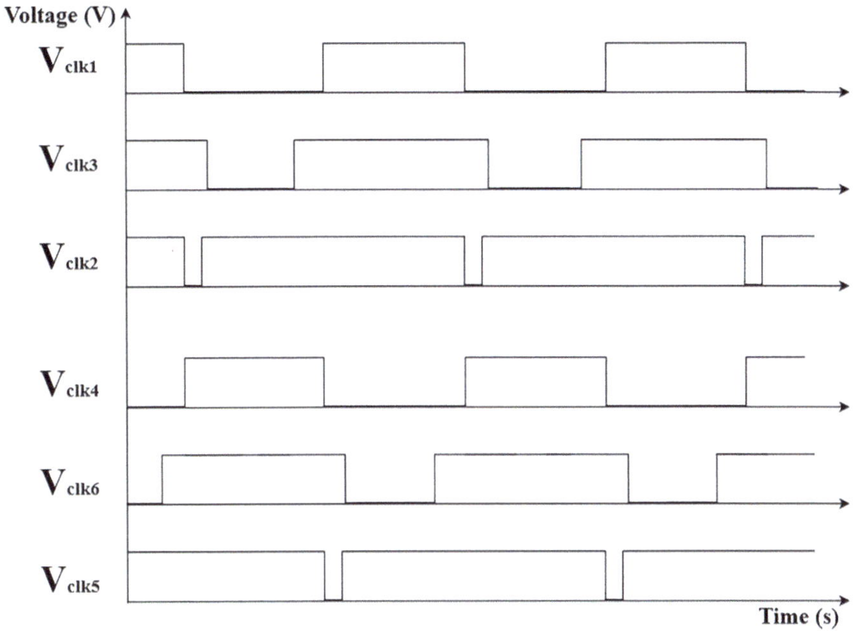

Fig. 3.38 Timing diagram of the proposed CP's clock signals

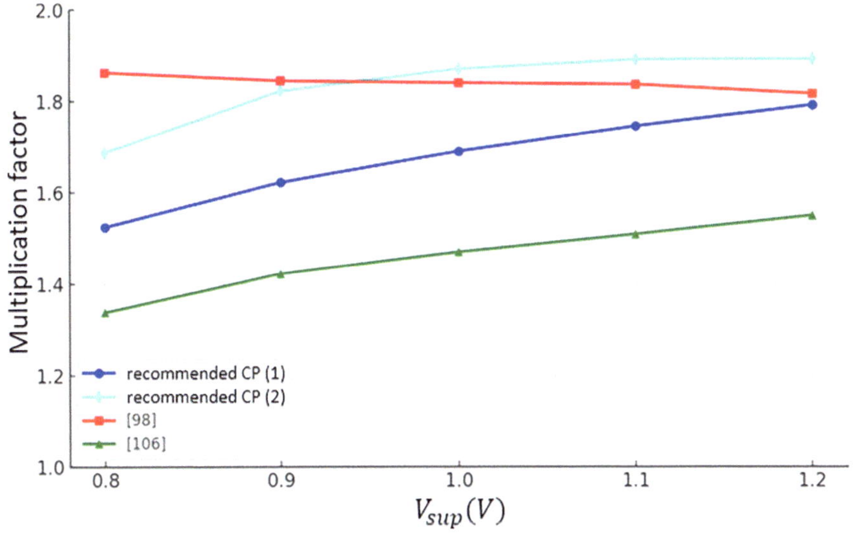

Fig. 3.39 Multiplication factors of the considered CPs

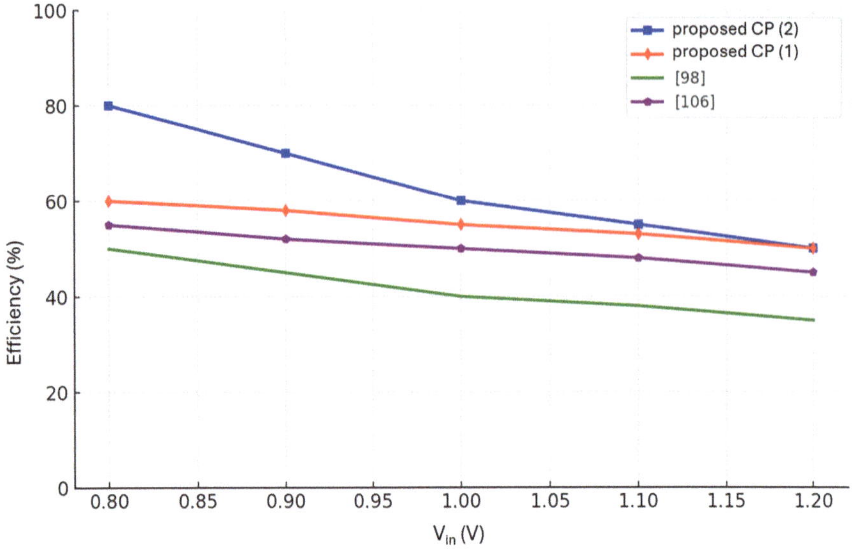

Fig. 3.40 Efficiency of power consumed in the considered CPs

capacitances may arise. The physical design of the proposed second circuit is also more complex, because it is necessary to maintain symmetry and isolation not for one, but for three pairs of clock signals (Fig. 3.41) [70].

To evaluate the protection of circuits from the effects of aging phenomena, simulation has been performed that reflects the state of the circuit both at the beginning of operation and after 1 and 15 years of operation (Fig. 3.42) [70].

From simulation results, it becomes clear that in the absence of NMOS transistors in the CP circuit, the conditions that accelerate aging phenomena are eliminated. The deterioration of the output voltage of the considered circuits with only PMOS transistors over time bears a character close to linear. This indicates that the elements of these circuits do not operate under overvoltage stress conditions, and the small deterioration of the output voltage over time is due only to inevitable aging phenomena, such as the decrease in conductivity of wires and interconnects. In the case of both proposed CPs, this deterioration is smaller than the accepted 10% maximum limit (Table 3.2) [70].

The proposed CP controlled with six-phase clock signals is significantly more efficient in case of low supply voltages (Fig. 3.39) [70] through more complex and precise control of the gates, and at high voltages it is not inferior in terms of both input power and resistance to aging phenomena (Table 3.3) [70].

Thus, in the proposed methods, at the cost of forming more complex clock signals and an increase in area, the reliability problems of the CP are solved without a loss of output power.

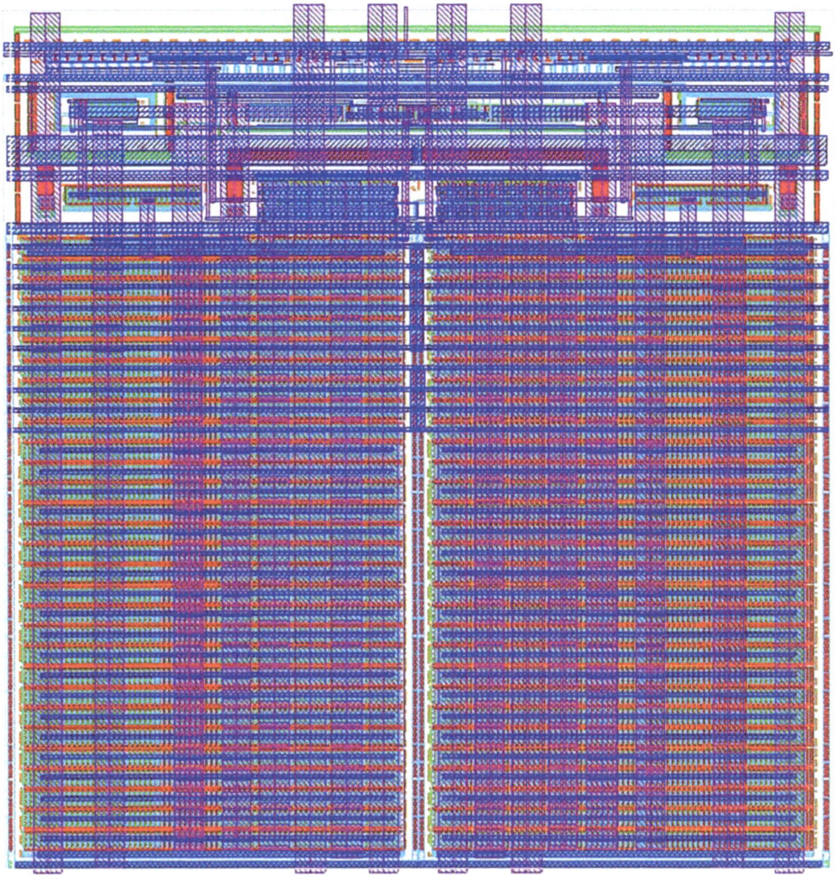

Fig. 3.41 Physical design of the proposed CP

3.2.2 Reduction of Aging Phenomena Through Dynamic Current Limitation of the Output pMOS Transistor in High-Voltage Stabilizers

As noted in Sect. 3.1, the presence of large currents and electric fields in the output stages of VRs with high input voltage levels contributes to an increase in the effects of BTI, HCI, and TDDB aging phenomena. The constant and dynamic current limitations proposed there only partially solve these problems. In the case of constant current limiting, the output PMOS transistor was protected from a large vertical field, and the effect of TDDB phenomena in it was reduced [98]. However, this method does not protect the VR from short circuits and overloads. The dynamic current limiting system proposed in [100] prevented the VR from operating in overload mode, but its volt-ampere characteristic was not entirely within the safe

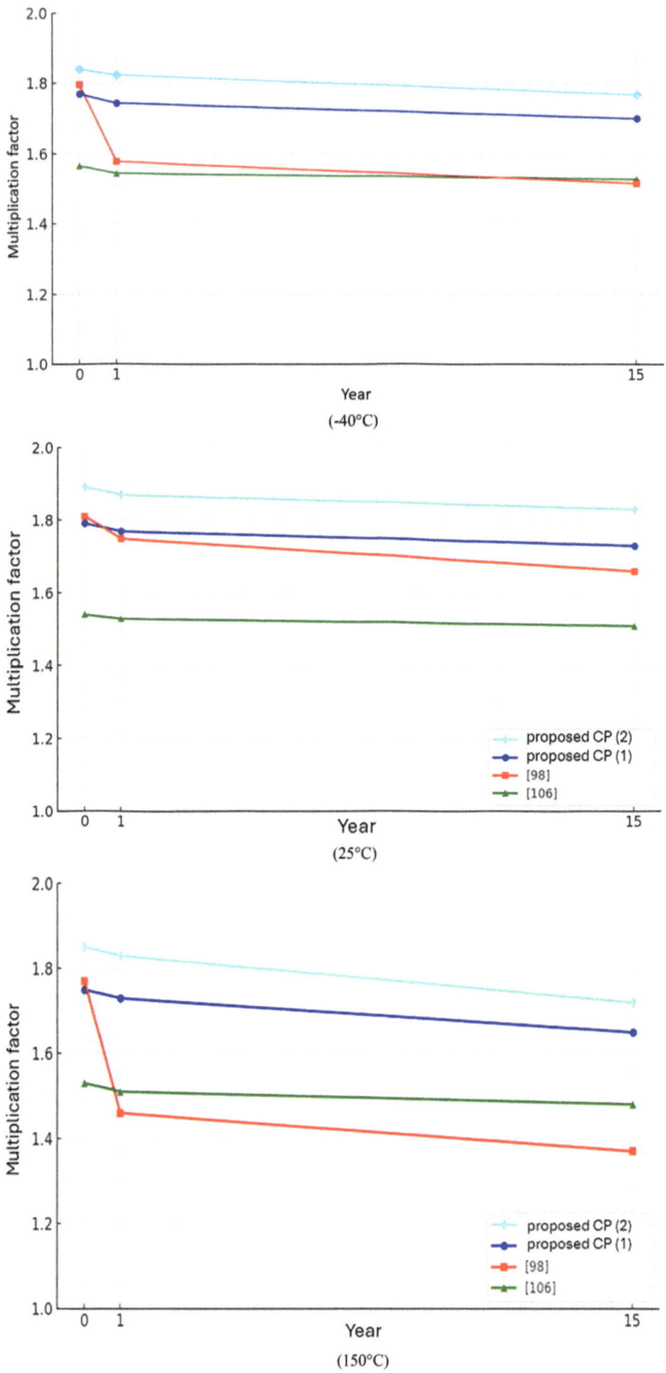

Fig. 3.42 Deterioration of output voltage multiplication factors of CPs because of aging

Table 3.2 Deterioration of the considered CPs

		V_{out}		
T	Architecture	0 years	1 years	15 years
−40 °C	Cross-coupled NMOS [97]	2.16	1.89 (−12,2%)	1.81 (−16,1%)
	PMOS diode [105]	1.86	1.84 (−1%)	1.82 (−2.1%)
	Proposed four-phase CP	2.13	2.1 (−1.4%)	2.04 (−4,3%)
	Proposed six-phase CP	2.19	2.17 (−1%)	2.12 (−3,1%)
25 °C	Cross-coupled NMOS [97]	2.17	2.10 (−3,2%)	1.99 (−8,3%)
	PMOS diode [105]	1.84	1.83 (−0,6%)	1.81 (−1,7%)
	Proposed four-phase CP	2.14	2.12 (−0,8%)	2.08 (−2,8%)
	Proposed six-phase CP	2.26	2.24 (−0,9%)	2.19 (−3,1%)
150 °C	Cross-coupled NMOS [97]	2.12	1.75 (−17,5%)	1.64 (−22,7%)
	PMOS diode [105]	1.83	1.81 (−1,1%)	1.77 (−3,3%)
	Proposed four-phase CP	2.10	2.08 (−1%)	1.98 (−5,7%)
	Proposed six-phase CP	2.22	2.19 (−1,4%)	2.06 (−7,2%)

Table 3.3 Comparison of parameters of the considered CPs

V_{sup} = 1,2 V, Temperature = 150 °C, Aging = 15 years, I_{load} = 100 µA				
Circuit	V_{out} (V)	ΔV_{out} (V)	Area (µm²)	P_{in} (mW)
Cross-coupled NMOS	2,12	0,48	3135	0,58
PMOS diode	1,83	0,06	3229	0,39
Proposed four-phase CP	2,10	0,12	3479	0,42
Proposed six-phase CP	2,22	0,16	3730	0,43

operating area of the output PMOS transistor, because after exceeding the limit, the current decreased linearly (Fig. 3.43) [49].

A new approach to dynamic current limiting is proposed, where the output current, after exceeding the maximum threshold, instead of decreasing in a directly proportional relation to the load, is shut off. To reduce the sensitivity of the system controlling the output current to various PVT deviations and short-term fluctuations, it is proposed to provide hysteresis through dynamic and constant limitations. This will allow protecting the VR from overload and short circuits, while simultaneously using the entire safe operating area of the transistor.

To ensure the described function, it is proposed to set the minimum output value of the OpAmp through the M2 transistor, and to constantly monitor the VR's output voltage through a comparator and after exceeding the set value, shut it off through the M3 transistor (Fig. 3.44) [49].

When the OpAmp's output voltage reaches $V_{bias} - V_{th,M2}$, the M2 transistor begins to charge the gate of M1 and prevent its further drop. In the case of a constant gate–source voltage of the M1 transistor, the output current cannot increase. Therefore, in the case of a decrease in load resistance, the output voltage will also decrease. As a result, the gate-drain voltage will increase, which will lead to an increase in the effect of aging phenomena. Due to the small overlap area of the gate and drain, the

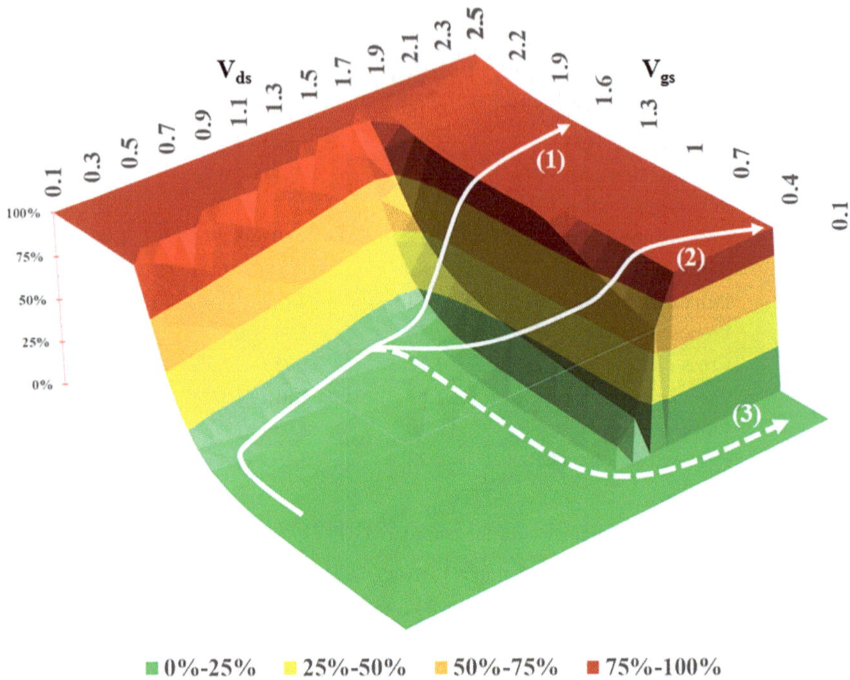

Fig. 3.43 Deterioration of PMOS transistor drain current depending on gate–source and drain–source voltages

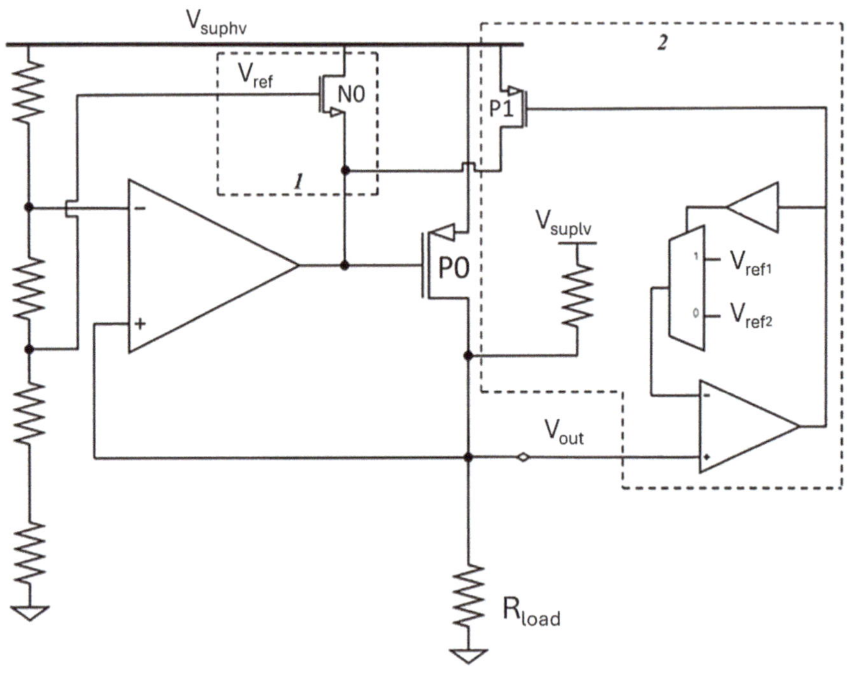

Fig. 3.44 Voltage regulator with proposed dynamic current limitation

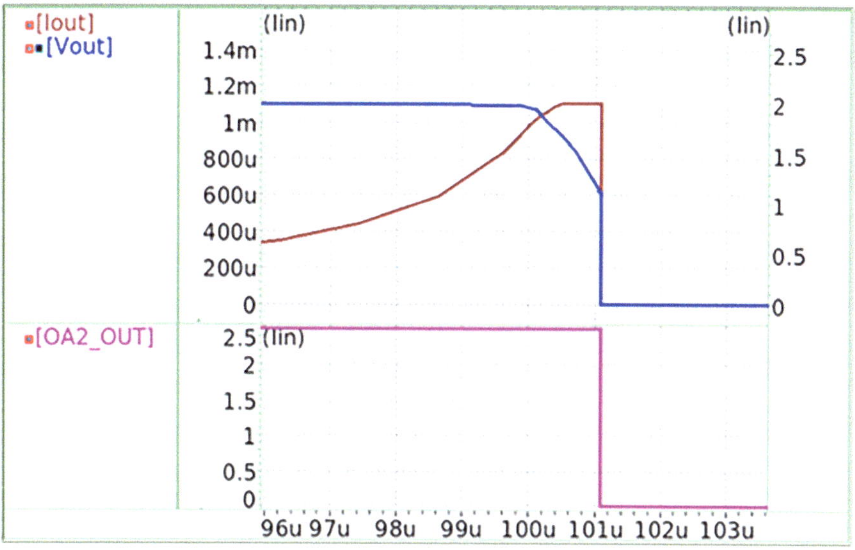

Fig. 3.45 Timing characteristic of VR's operation

effect of the TDDB phenomenon is not noticeable, but in the case of large currents, the vertical and horizontal electric fields create a risk of hot carrier injection.

To avoid HCI phenomena, the VR's output voltage is constantly monitored by a comparator, and in case of exceeding the reference level, the gate of M1 is connected to the supply rail through the M3 transistor. Through this, M1 is closed, and no current flows through its channel under the influence of a large electric field, which protects the gate insulation from "hot" carriers (Fig. 3.45) [49]. When the load resistance returns to the operating range, the control of the M1 transistor's gate is transferred back to the OpAmp through the same comparator, and the VR's negative feedback operation continues.

Simulation was performed both under typical and process and temperature extreme conditions. The output load value of 2 kΩ was chosen as the activation threshold of the first current limiting stage, and 1 kΩ for the second one. For the first simulation, the case is considered where the output load falls below the constant limiting threshold for a short time, but without crossing the second one, it returns to the normal operating range. In the second simulation case, the load returns to the operating range after crossing the minimum threshold (Fig. 3.46) [49].

As a result, the following dependence of the output current on the output load is obtained (Fig. 3.47) [49].

From the results of the performed timing simulation, the relation between the gate–source and drain–source voltages of the proposed VR's output PMOS transistor is derived (Fig. 3.48) [49].

Superimposing the derived curves on the transistor's degradation table, it can be seen that in both typical and extreme cases of VR's operation, under the influence of short circuits of the output load, the PMOS transistor does not leave its safe

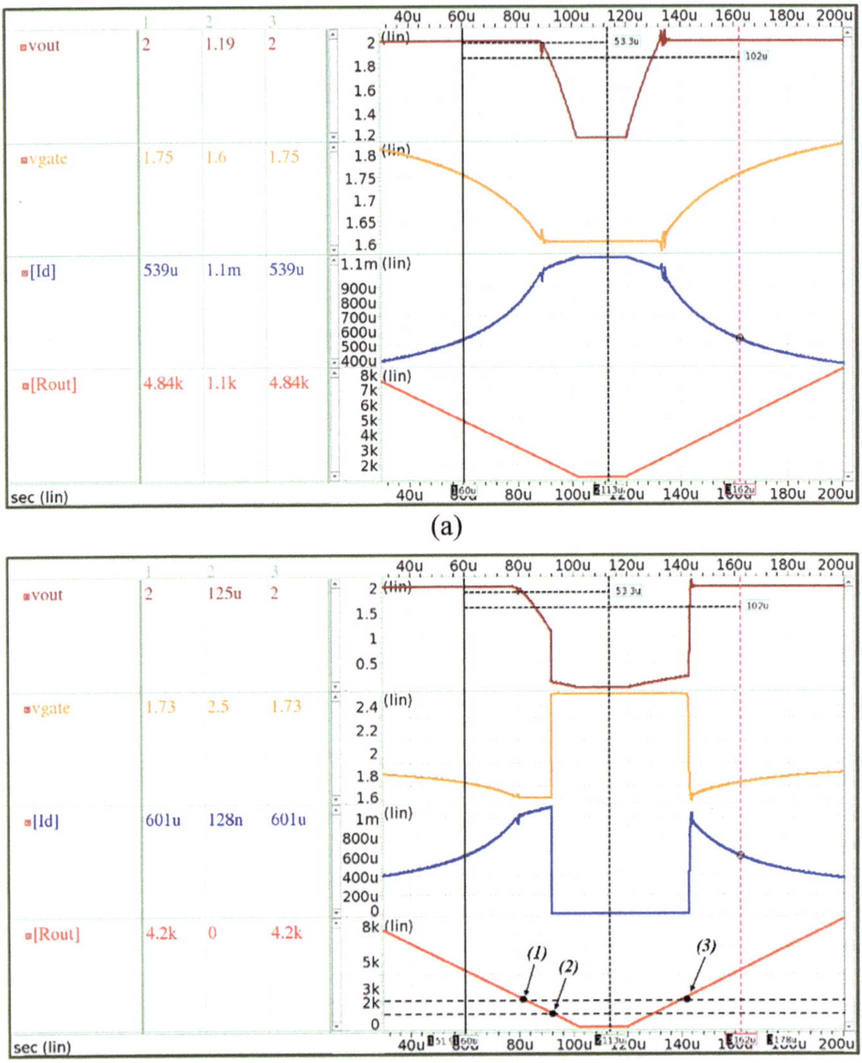

Fig. 3.46 Timing characteristic of VR's operation (**a**—load does not cross lower threshold, **b**—load crosses lower threshold)

operating area (Fig. 3.49) [49]. Thanks to this, it does not lose its most important parameters over time, which are necessary to ensure the accuracy and noise immunity of the output voltage.

The output PMOS transistors constitute a significant part of the VR circuit's area, which must provide a significant small-signal resistance from the supply rail to the

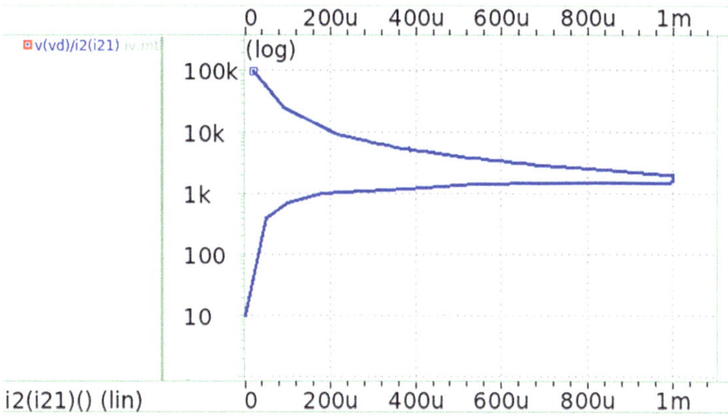

Fig. 3.47 Dependence of VR's output current on load

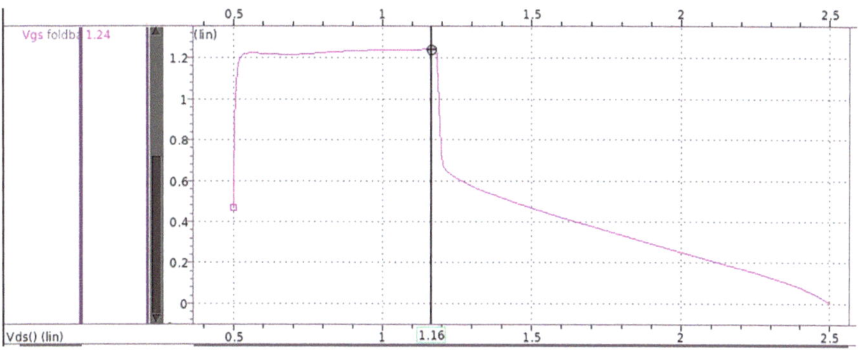

Fig. 3.48 Relation between the gate and drain voltages of a PMOS transistor

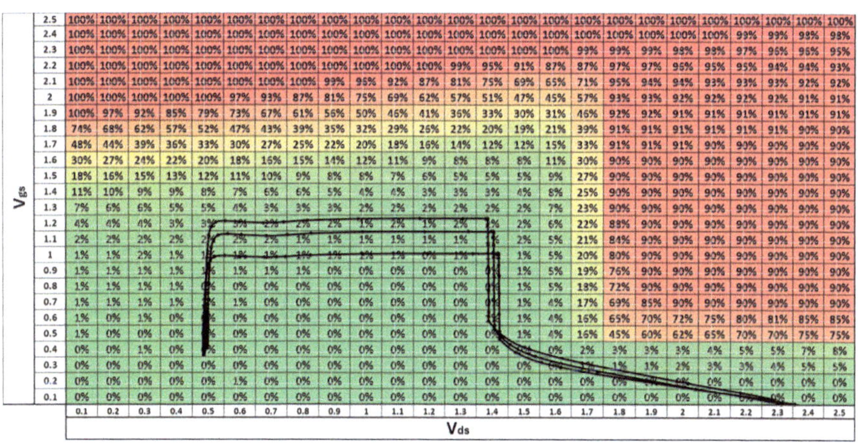

Fig. 3.49 Superimposition of the output PMOS transistor's I–V characteristic and the degradation table

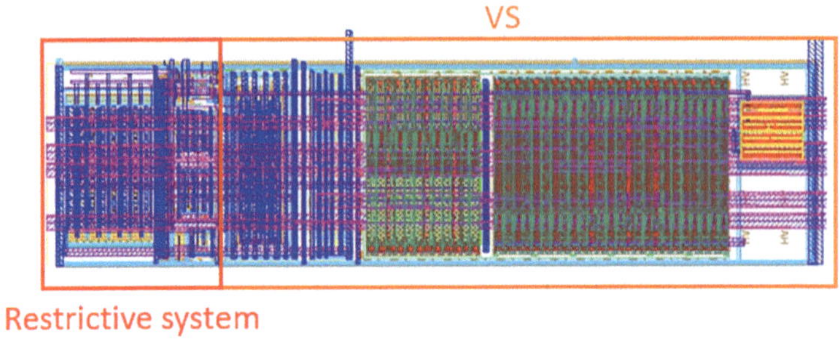

Fig. 3.50 Physical design of the proposed VR

Table 3.4 Comparison of existing and proposed solutions

Circuit	TDDB protection	BTI protection	HCI protection	Area increase
Classical	No	No	No	0%
Constant limit	Yes	No	No	+5%
Dynamic limit	Yes	Partially	Partially	+10%
Proposed	Yes	Yes	Yes	+20%

output for noise immunity, for which a large channel length is necessary. In parallel, it is also necessary to ensure a minimum voltage drop across the channel, for which a large channel width is necessary. As a result, output PMOS transistors occupy 70% of VR's total area, and OpAmp and resistances and capacitances intended for phase-frequency compensation occupy the remaining 30%. After making the changes described in the proposed system, VR's area increases by another 20% (Fig. 3.50) [49].

Thus, thanks to the proposed solutions, the VR is protected from operating in overloaded conditions at the cost of a 20% increase in area, as a result of which it avoids the effects of phenomena that accelerate aging (Table 3.4) [49].

3.2.3 *Elimination of Errors in Level Shifters Through a Supply Voltage Detection Circuit*

As noted above, the operating range of LSs is very sensitive to PVT deviations and has a large distribution. For their reliable operation, control through a supply voltage detection circuit is necessary. The existing supply voltage detection circuits considered above were designed according to the slowest extreme case, and as a result, had a very large margin for faster PVT cases. The proposed detection circuit with a threshold inversely proportional to the temperature partially reduced the over-designed margin, but deviations due to process and high supply voltage continued to remain ignored. A supply voltage detection circuit is proposed that will follow the PVT behavior of the LS (Fig. 3.51) [106].

The area marked in red is the loss of operating range that a system with a constant supply voltage detection threshold incurs. The area marked in purple is the loss

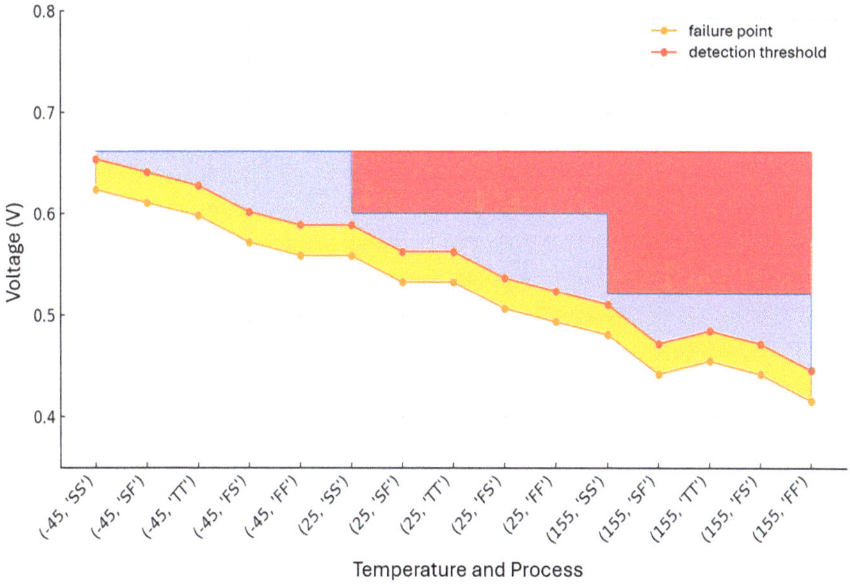

Fig. 3.51 Threshold voltages of existing and proposed supply voltage detection circuits

of operating range in the case of a temperature-sensitive threshold, and the one marked in yellow is for the proposed PVT-sensitive threshold. In the case of a constant threshold, the average margin is 21%, and in the case of a threshold with existing temperature dependence, it is 11%. In obtaining the proposed behavior of the supply voltage detection circuit, a minimum loss of operating range will be obtained. To obtain the proposed behavior, it is necessary to build the supply voltage detection circuit based on the LS circuit used in the IC. The main disadvantage of the LS's operation under low supply conditions is that its output value in the case of failure is unpredictable. It is necessary to ensure a path of low conductivity from the unbuffered output signal to the supply ground. Through this, in the case of failure, the output voltage will always be equal to the supply voltage. Due to asymmetries in the input branches of the proposed circuit, it fails at higher input voltage conditions than in the case of the LS's symmetrical structure. As a result, a circuit performing the LS function is obtained, which fails at a supply voltage several millivolts higher than the classical LS and has a predictable output behavior in the failure state (Fig. 3.52) [106].

The gates of PMOS transistors forming the low conductivity path to the high supply voltage rail are controlled by the low supply voltage. Thanks to this, when the difference between the two supply voltages is large, the conductivity of this branch also increases.

In the performed simulation, two identical LSs were used, the operating state of the first of which is controlled by the proposed supply voltage detection circuit, and the second operates without limitation (Fig. 3.53) [106].

In the obtained results, it can be seen that in the edge range of the operating domain, the uncontrolled LS consumes about 1000 times more current. The

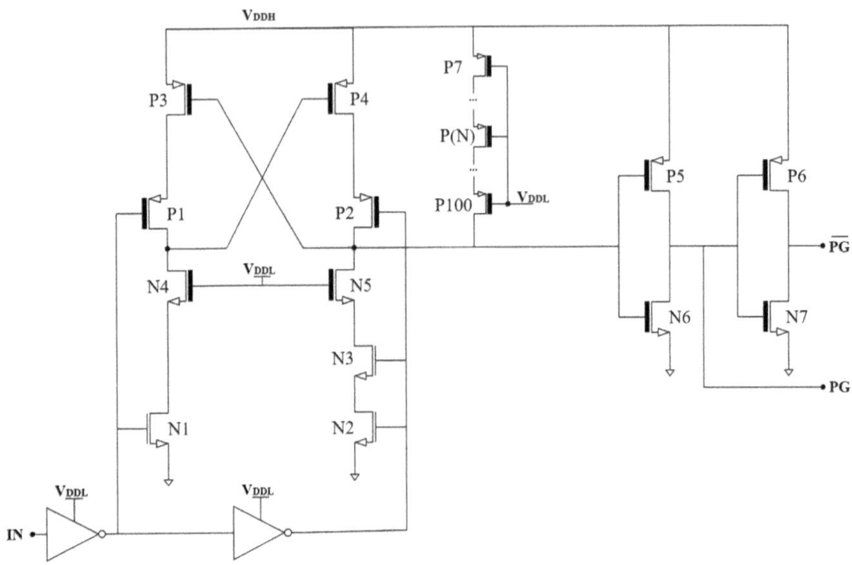

Fig. 3.52 Proposed supply voltage detection circuit

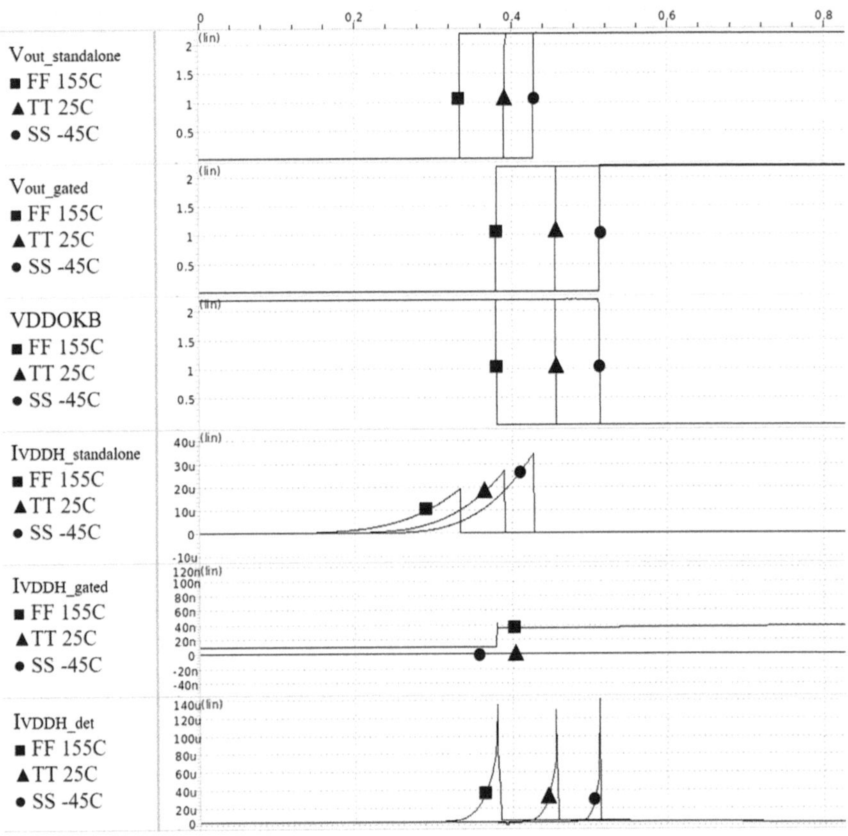

Fig. 3.53 Input–output characteristics of simulated circuits

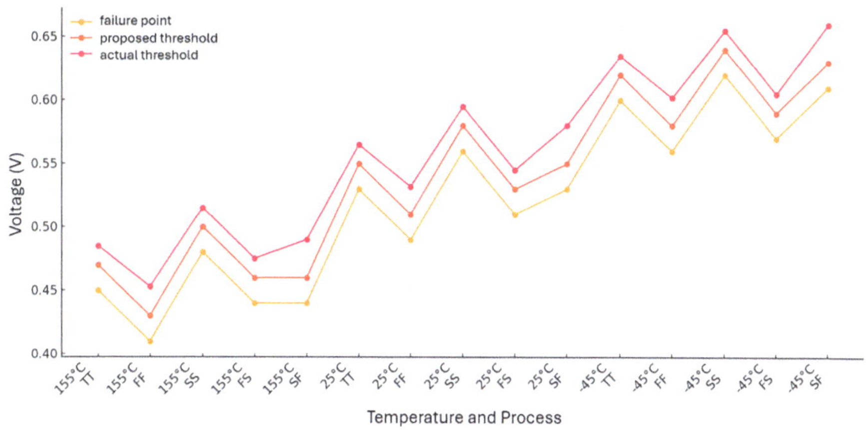

Fig. 3.54 Simulation results of the supply voltage detection threshold circuit

proposed circuit consumes 3–4 times more maximum current in a narrower range. Taking into account the fact that hundreds of LSs are used in the above control systems, the proposed circuit, in addition to ensuring their reliable operation, also ensures a significant reduction in power consumed under extreme conditions.

As a result of the simulation performed for all PVT cases, the following dependence of the supply voltage detection threshold is obtained (Fig. 3.54) [106].

The simulation results satisfy the intended requirements for all PVT cases. It is also necessary to ensure that the proposed circuit maintains the required behavior under manufacturing deviations. The models intended for FF, SS, SF, and FS processes assume that the characteristics of all transistors in the circuit deviate uniformly from the nominal values. However, in a real manufacturing process, these deviations have a Gaussian distribution. Through statistical models, it is possible to obtain, in the result of several hundred simulations, deviations that occur in only one out of billions of IC instances. Only if the circuit operates accurately under these extreme, highly improbable deviation conditions can we be confident that the proposed circuit is suitable for mass production.

As a result of simulation performed with the Monte Carlo method for asymmetries, for a 6σ distribution, an always positive result of the difference between the voltage detection threshold and the LS's failure threshold is obtained (Fig. 3.55) [106].

In case of applying the proposed circuit, it is possible to expand the operating range at the cost of a much smaller area and an increase in power consumption (Table 3.5) [106].

Thus, in the case of using the proposed circuit, the problem of predictability of the failure state of the LSs in the control system embedded in the IMD is solved at the cost of an area of 425 μm^2 and a static power consumption of 4.37 μW.

Fig. 3.55 Difference between voltage detection and LS failure thresholds

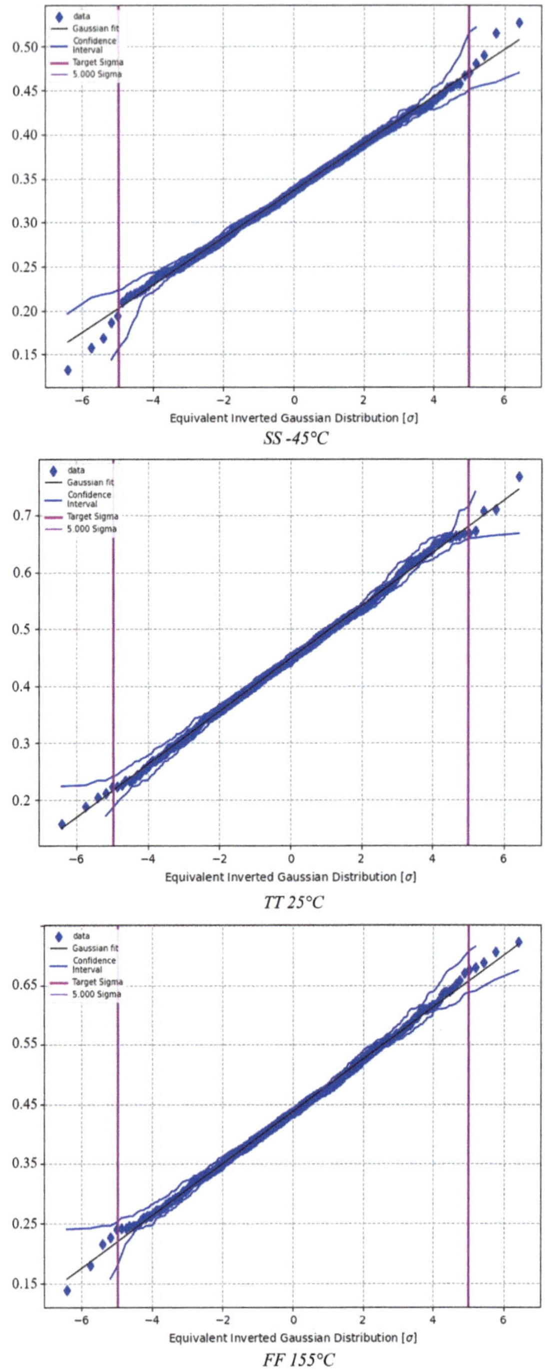

Table 3.5 Comparison of existing and proposed solutions

Detection threshold	Operating range loss	Power	Area
Constant [102]	21%	0,72 μW	110*75
Temperature sensitive [103]	11%	0,814 μW	118*67
Proposed	4%	4,37 μW	25*17

Conclusions

1. Circuits of charge pumps consisting only of PMOS type transistors, controlled by four- and six-phase clock signals, have been developed, in which the effect of aging phenomena has been reduced, ensuring <10% power deviation after 15 years of operation under high temperature conditions, at the cost of a 10% and 15% increase in area, respectively.
2. A method of dynamic current limiting of the voltage regulator has been proposed, whereby the aging-accelerating effects on the PMOS transistor supplying the output stage have been reduced, ensuring <10% power deviation after 15 years of operation under high-temperature conditions, at the cost of a 20% increase in area.
3. A supply voltage detection circuit has been created to control the voltage level shifters in control systems embedded in integrated memory devices, which eliminates their unpredictable failure state under low-voltage conditions, occupying only an area of 425 μm^2.

References

1. G.E. Moore, Cramming more components onto integrated circuits. IEEE Solid-State Circ. Soc. Newsl. **11**(3), 33–35 (2006). https://doi.org/10.1109/N-SSC.2006.4785860
2. R.K. Bassett, *To the Digital Age: Research Labs, Start-up Companies, and the Rise of MOS Technology* (Johns Hopkins University Press, Baltimore, MD, 2003), 440p
3. A. Mocuta, P. Weckx, S. Demuynck, D. Radisic, Y. Oniki, J. Ryckaert, Enabling CMOS scaling towards 3 nm and beyond, in *2018 IEEE Symposium on VLSI Technology. Honolulu, HI, USA*, (IEEE, 2018), pp. 147–148. https://doi.org/10.1109/VLSIT.2018.8510683
4. R. Dulong, *Towards New Memory Paradigms: Integrating Non-Volatile Main Memory and Remote Direct Memory Access in Modern Systems* (Institut Polytechnique de Paris; Université de Neuchâtel, Paris, 2023), 139p
5. B. De Salvo, *Silicon Non-Volatile Memories: Paths of Innovation* (John Wiley & Sons, Hoboken, NJ, 2013), 233p
6. J. Pan, Z. Wang, B. Zhao, J. Yin, P. Guo, Y. Yang, T.-L. Ren, Recent progress of non-volatile memory devices based on two-dimensional materials. Chips **3**, 271–295 (2024)
7. W. Banerjee, Challenges and applications of emerging nonvolatile memory devices. Electronics **9**, 1029 (2020)
8. E. Pérez, Ó. González Ossorio, S. Dueñas, H. Castán, H. García, C. Wenger, Programming pulse width assessment for reliable and low-energy endurance performance in Al:HfO_2-based RRAM arrays. Electronics **9**, 864 (2020)
9. H.H. Radamson, Y. Miao, Z. Zhou, Z. Wu, Z. Kong, J. Gao, et al., CMOS scaling for the 5 nm node and beyond: device, process and technology. Nanomaterials (Basel) **14**(10), 837 (2024). https://doi.org/10.3390/nano14100837

10. H.H. Radamson, X. He, Q. Zhang, J. Liu, H. Cui, et al., Miniaturization of CMOS. Micromachines (Basel) **10**(5), 293 (2019). https://doi.org/10.3390/mi10050293
11. H.H. Radamson, H. Zhu, Z. Wu, X. He, H. Lin, J. Liu, et al., State of the art and future perspectives in advanced CMOS technology. Nanomaterials (Basel) **10**(8), 1555 (2020). https://doi.org/10.3390/nano10081555
12. J. Huang, Research progresses on suppressing the short-channel effects of field-effect transistor. Highlights Sci. Eng. Technol. **27**, 361–367 (2022). https://doi.org/10.54097/hset.v27i.3779
13. S. Radhakrishnan, S.N.S. Vishnu, S.I. Ahmed, R. Thiruvengadathan, Effects of channel length scaling on the electrical characteristics of multilayer MoS_2 field effect transistor. Micromachines (Basel) **14**(2), 275 (2023). https://doi.org/10.3390/mi14020275
14. T. Ishihara, Y. Nakasaki, D. Matsushita, K. Tatsumura, K. Kato, Interface defect engineering for high-performance MOSFETs with novel carrier mobility model: theory and experimental verification. AIP Adv. **10**(5), 055020 (2020). https://doi.org/10.1063/5.0005813
15. K. Singh, J. Pineda de Gyvez, Twenty years of near/sub-threshold design trends and enablement. IEEE Trans Circuits Syst II Express Briefs **68**(1), 5–11 (2021). https://doi.org/10.1109/TCSII.2020.3040970
16. S. Chopra, S. Subramaniam, A review on challenges for MOSFET scaling. Int. J. Innov. Sci. Eng. Technol. **2**(4), 1055–1057 (2015)
17. M. Kumar, Effects of scaling on MOS device performance. IOSR J. VLSI Signal Process. **5**(1 Ver. III), 25–28 (2015)
18. X. Liang, H. Feng, Y. Xiang, J. Sun, Y. Wei, et al., Oxide electric field-induced degradation of SiC MOSFET for heavy-ion irradiation. Electronics **12**, 2886 (2023)
19. H. Kim, J. Park, K. Roy, Y. Kim, Aging compensation with dynamic computation approximation. IEEE Trans. Circ. Syst. I Regul. Pap. **67**(4), 1319–1332 (2020). https://doi.org/10.1109/TCSI.2020.2969462
20. S. Spyridonos, Y. Tsiatouhas, BTI aging influence on charge pump circuits, in *2022 11th International Conference on Modern Circuits and Systems Technologies (MOCAST). Bremen, Germany*, (IEEE, 2022), pp. 1–4. https://doi.org/10.1109/MOCAST54814.2022.9837768
21. M.-D. Ker, S.-L. Chen, C.-S. Tsai, Design of charge pump circuit with consideration of gate-oxide reliability in low-voltage CMOS processes. IEEE J. Solid State Circuits **41**(5), 1100–1107 (2006). https://doi.org/10.1109/JSSC.2006.872704
22. R. Lajmi, S. Ben Dhia, N. Selmi, M. Masmoudi, Characterization of low drop-out during aging and design for yield. Microelectron. Reliab. **76–77**, 92–96 (2017). https://doi.org/10.1016/j.microrel.2017.07.047
23. S. Chowdhury, H. Shen, B. Park, N. Maghari, D. Forte, Aging analysis of low dropout regulator for universal recycled IC detection, in *2019 IEEE Computer Society Annual Symposium on VLSI (ISVLSI). Miami, FL, USA*, (IEEE, 2019), pp. 604–609. https://doi.org/10.1109/ISVLSI.2019.00113
24. J. Cai, B. Halak, D. Rossi, Analysis of BTI aging of level shifters, in *2016 IEEE 22nd International Symposium on On-Line Testing and Robust System Design (IOLTS). Sant Feliu de Guixols, Spain*, (IEEE, 2016), pp. 17–18. https://doi.org/10.1109/IOLTS.2016.7604662
25. B. Halak, V. Tenentes, D. Rossi, The impact of BTI aging on the reliability of level shifters in nano-scale CMOS technology. Microelectron. Reliab. **67**, 74–81 (2016). https://doi.org/10.1016/j.microrel.2016.10.018
26. Y. Cai, S. Ghose, E. Haratsch, Y. Luo, O. Mutlu, Errors in flash-memory-based solid-state drives: analysis, mitigation, and recovery. arXiv (2017). https://doi.org/10.48550/arXiv.1711.11427
27. Raspberry Pi Foundation, *DEF CON 32 Capture the Flag Competition* (DEF CON, Las Vegas, NV, 2024)
28. S. Ghosh, M.N.I. Khan, A. De, J.-W. Jang, Security and privacy threats to on-Chip non-volatile memories and countermeasures, in *2016 IEEE/ACM International Conference on*

Computer-Aided Design (ICCAD). Austin, TX, USA, (IEEE, 2016), pp. 1–6. https://doi.org/10.1145/2966986.2980064
29. N.I. Deligiannis, R. Cantoro, M.S. Reorda, M. Traiola, et al., Towards the integration of reliability and security mechanisms to enhance the fault resilience of neural networks. IEEE Access **9**, 155998–156012 (2021). https://doi.org/10.1109/ACCESS.2021.3129149
30. X. Wang, Y. Song, K. Prakash, et al., Quality-driven design methodology for PUFs in FPGAs for secure IoT, in *2023 24th International Symposium on Quality Electronic Design (ISQED). San Francisco, CA, USA*, (IEEE, 2023), pp. 1–8. https://doi.org/10.1109/ISQED57927.2023.10129405
31. J.-W. Jang, Security of non-volatile memories - attack models, analyses, and countermeasures. PhD dissertation, University of South Florida, Tampa, FL, 2015, 158p. https://digitalcommons.usf.edu/etd/5775
32. M.N.I. Khan, S. Ghosh, Comprehensive study of security and privacy of emerging non-volatile memories. J. Low Power Electron. Appl. **11**(3), 36 (2021). https://doi.org/10.3390/jlpea11030036
33. Tyndall National Institute, *Multi-Function Chip Developed to Enable Next Generation Implantable Device Innovations* (Tyndall National Institute, Cork, 2016) https://www.sciencedaily.com/releases/2016/10/161006084934.htm
34. FPT Semiconductor, *Integrated Circuits: The Backbone of Modern Technology. White Paper* (FPT Corporation, Hanoi, 2024) https://fpt-semiconductor.com/whitepapers
35. Market Research Intellect, *The Brain Behind the Wheel: Automotive Integrated Circuits Fueling the Next Era of Car Technology: Market Research Report MR-AT-1124* (Market Research Intellect, New York, NY, 2024), 45p
36. Automotive Electronics Council, *AEC-Q100-005 Rev-D1: Non-Volatile Memory Program/Erase Endurance, Data Retention, and Operating Life Test. Automotive Electronics Council Standard* (AEC, Detroit, MI, 2012), 28p
37. Automotive Electronics Council, *AEC-Q100 Rev-J: Failure Mechanism Based Stress Test Qualification for Integrated Circuits in Automotive Applications. Automotive Electronics Council Standard* (AEC, Detroit, MI, 2023), 112p
38. A.P. Chandrakasan, N. Verma, D.C. Daly, Ultralow-power electronics for biomedical applications. Annu. Rev. Biomed. Eng. **10**, 247–274 (2008). https://doi.org/10.1146/annurev.bioeng.10.061807.160547
39. R. Oshiro, Fundamentals of AEC-Q100: what 'automotive qualified' really means. Autom. Electron. Mag. **15**(11), 24–28 (2018)
40. T. Derenda, M. Zanne, M. Zöldy, A. Torok, Automatization in road transport: a review. Prod. Eng. Arch. **20**, 3–7 (2018). https://doi.org/10.30657/pea.2018.20.01
41. P.R. Genssler, H.E. Barkam, K. Pandaram, et al., Modeling and predicting transistor aging under workload dependency using machine learning. IEEE Trans. Circ. Syst. I Regul. Pap. **70**(9), 3699–3711 (2023). https://doi.org/10.1109/TCSI.2023.3289325
42. W. Ye, M.B. Alawieh, C.-L. Hsu, et al., Dealing with aging and yield in scaled technologies, in *Embedded Systems Design for High-Speed Data Acquisition*, (Springer, 2020), pp. 409–429. https://doi.org/10.1007/978-3-030-52017-5_17
43. B. Tudor, L. Lin, T. Nirschl, P. Oldiges, MOSRA: an efficient and versatile MOS aging modeling and reliability analysis solution for 45 nm and below, in *2010 IEEE International Conference on Solid-State and Integrated Circuit Technology (ICSICT). Shanghai, China*, (IEEE, 2010), pp. 1645–1647. https://doi.org/10.1109/ICSICT.2010.5667399
44. L.R. Juracy, M.T. Moreira, A.D. Amory, F.G. Moraes, A survey of aging monitors and reconfiguration techniques. arXiv:2007.07829 (2020) https://arxiv.org/abs/2007.07829
45. L. Lanzieri, G. Martino, G. Fey, et al., A review of techniques for aging detection and monitoring on embedded systems. ACM Comput. Surv. **57**(1), 24 (2025). https://doi.org/10.1145/3638054

46. K. Nanbakhsh, A.S. Idil, C. Lamont, et al., On the longevity and inherent hermeticity of silicon-ICs: evaluation of bare-die and PDMS-coated ICs after accelerated aging and implantation studies. Nat. Commun. **16**, 12 (2025). https://doi.org/10.1038/s41467-024-55298-4
47. N. Parihar, N. Goel, S. Mukhopadhyay, S. Mahapatra, BTI analysis tool—modeling of NBTI DC, AC stress and recovery time kinetics, nitrogen impact, and EOL estimation. IEEE Trans. Electron. Devices **65**(2), 392–403 (2018). https://doi.org/10.1109/TED.2017.2780082
48. A. Santana-Andreo, P. Martín-Lloret, E. Roca, et al., Characterization and analysis of BTI and HCI effects in CMOS current mirrors, in *2022 18th International Conference on Synthesis, Modeling, Analysis and Simulation Methods and Applications to Circuit Design (SMACD). Villasimius, Italy*, (IEEE, 2022), pp. 1–4. https://doi.org/10.1109/SMACD55068.2022.9816212
49. H. Sahakyan, Foldback current limiting in low-dropout voltage regulators with aging analysis based operating envelope, in *2023 IEEE East-West Design & Test Symposium (EWDTS). Batumi, Georgia*, (IEEE, 2023), pp. 1–4. https://doi.org/10.1109/EWDTS59469.2023.10297089
50. J. Ji, J.Y. Yang, S. Lee, et al., Heterogeneous integration of high-k complex-oxide gate dielectrics on wide band-gap high-electron-mobility transistors. Commun. Eng. **3**, 15 (2024). https://doi.org/10.1038/s44172-024-00161-z
51. K. Puschkarsky, H. Reisinger, G.A. Rott, C. Schlünder, W. Gustin, T. Grasser, An efficient analog compact NBTI model for stress and recovery based on activation energy maps. IEEE Trans. Electron. Devices **66**(11), 4623–4630 (2019). https://doi.org/10.1109/TED.2019.2941889
52. S. More, Aging degradation and countermeasures in deep-submicrometer analog and mixed-signal integrated circuits. PhD dissertation, Technical University of Munich, Munich, Germany, 2011, 153p
53. Y. Wang, Y. Li, Y. Yang, W. Chen, Hot carrier injection reliability in nanoscale field effect transistors: modeling and simulation methods. Electronics **11**(21), 3601 (2022). https://doi.org/10.3390/electronics11213601
54. S. Tyaginov, E. Bury, A. Grill, et al., Compact physics hot-carrier degradation model valid over a wide bias range. Micromachines (Basel) **14**(11), 2018 (2023). https://doi.org/10.3390/mi14112018
55. A. Gupta, C. Gupta, H. Jatana, A. Dixit, Investigation of hot-carrier degradation in 0.18 μm MOSFETs for the evaluation of device lifetime and digital circuit performance. IEEE Trans. Device Mater. Reliab. **19**(4), 678–685 (2019). https://doi.org/10.1109/TDMR.2019.2938319
56. Y.-H. Huang, C. Lin, W. Chen, Y. Wang, Investigation of hot-carrier-injection assisted TDDB and multi-stage Hot-hole induced leakage current in BCD HV NMOS, in *2017 IEEE International Reliability Physics Symposium (IRPS). Monterey, CA, USA*, (IEEE, 2017), pp. 2D-1.1–2D-1.5. https://doi.org/10.1109/IRPS.2017.7936261
57. J.W. McPherson, Physical model for the frequency dependence of time-dependent dielectric breakdown (TDDB). AIP Adv. **13**(5), 055217 (2023). https://doi.org/10.1063/5.0150268
58. J. Lienig, M. Thiele, *Fundamentals of Electromigration-Aware Integrated Circuit Design* (Springer, Berlin, 2018), 315p. https://doi.org/10.1007/978-3-319-73558-0
59. W.-S. Zhao, R. Zhang, D.-W. Wang, Recent progress in physics-based modeling of electromigration in integrated circuit interconnects. Micromachines **13**(6), 883 (2022). https://doi.org/10.3390/mi13060883
60. Z. Shen, S. Jing, Y., et al., Electromigration in three-dimensional integrated circuits. Appl. Phys. Rev. **10**(2), 021309 (2023). https://doi.org/10.1063/5.0139658
61. F.N. Najm, V. Sukharev, Electromigration simulation and design considerations for integrated circuit power grids. J. Vac. Sci. Technol. B. **38**(6), 063204 (2020). https://doi.org/10.1116/6.0000476
62. O. Prakash, C.K. Dabhi, Y.S. Chauhan, H. Amrouch, Transistor self-heating: the rising challenge for semiconductor testing, in *2021 IEEE 39th VLSI Test Symposium (VTS). San Diego, CA, USA*, (IEEE, 2021), pp. 1–7. https://doi.org/10.1109/VTS50974.2021.9441002

63. S.J. Kang, J.H. Kim, Y.S. Song, et al., Investigation of self-heating effects in vertically stacked GAA MOSFET with wrap-around contact. IEEE Trans. Electron Devices **69**(3), 910–914 (2022). https://doi.org/10.1109/TED.2022.3140283
64. A. Atamuratov, B. Jabbarova, M. Khalilloev, et al., Influence of the gate oxide and back oxide material types on self-heating effect in junctionless FinFET, in *2024 IEEE 25th International Conference of Young Professionals in Electron Devices and Materials (EDM). Altai, Russian Federation*, (IEEE, 2024), pp. 2630–2633. https://doi.org/10.1109/EDM61683.2024.10615156
65. Z.-L. Tang, B.-Y. Cao, Heat generation mechanisms of self-heating effects in SOI-MOS. IEEE J. Electron Devices Soc. **12**, 350–358 (2024). https://doi.org/10.1109/JEDS.2024.3393019
66. P. Shrivastava, Application of FMEA in developing design and reliability verification plan, in *2023 Annual Reliability and Maintainability Symposium (RAMS). Orlando, FL, USA*, (IEEE, 2023), pp. 1–6. https://doi.org/10.1109/RAMS51473.2023.10088221
67. K. Henderson, B. Hulse, M. Styduhar, P. Michelson, Ensuring robust ESD design with comprehensive reliability verification, in *2022 IEEE International Integrated Reliability Workshop (IIRW). South Lake Tahoe, CA, USA*, (IEEE, 2022), pp. 1–4. https://doi.org/10.1109/IIRW56459.2022.10032761
68. C. Li, W. Wang, Research on the integrated design technology of performance verification and qualification test, in *2022 4th International Conference on System Reliability and Safety Engineering (SRSE). Guangzhou, China*, (IEEE, 2022), pp. 226–230. https://doi.org/10.1109/SRSE56746.2022.10067773
69. H. Sahakyan, A. Hayrapetyan, S. Harutyunyan, et al., A low dropout voltage regulator with higher than supply output voltage and load-based frequency control system, in *2020 IEEE 40th International Conference on Electronics and Nanotechnology (ELNANO). Kyiv, Ukraine*, (IEEE, 2020), pp. 366–369. https://doi.org/10.1109/ELNANO50318.2020.9088754
70. H. Gomtsyan, H. Sahakyan, R. Soghomonyan, et al., Reliable all PMOS voltage doublers for low-voltage applications, in *2024 IEEE East-West Design & Test Symposium (EWDTS). Yerevan, Armenia*, (IEEE, 2024), pp. 1–4. https://doi.org/10.1109/EWDTS59469.2023.10297089
71. V. Sahakyan, R. Soghomonyan, S. Harutyunyan, H. Sahakyan, et al., Aging protected precision CMOS current reference with process, supply voltage and temperature compensation, in *2024 IEEE East-West Design & Test Symposium (EWDTS). Yerevan, Armenia*, (IEEE, 2024), pp. 1–4. https://doi.org/10.1109/EWDTS59469.2023.10297089
72. T.-M. Zhou, Y. Zhou, J. Lan, Y. Yao, Performance degradation and reliability evaluation of domestic operational amplifier, in *2023 6th International Conference on Electronics Technology (ICET). Chengdu, China*, (IEEE, 2023), pp. 1002–1006. https://doi.org/10.1109/ICET58434.2023.10212068
73. S. Nitundil, N. Singh, R. Balaji, P. Arora, Design and comparative analysis of a two-stage ultra-low-power subthreshold operational amplifier in 180 nm, 90 nm, and 45 nm technology, in *2021 Devices for Integrated Circuit (DevIC). Kalyani, India*, (IEEE, 2021), pp. 36–40. https://doi.org/10.1109/DevIC50843.2021.9455791
74. R.O. Nunes, R.L. de Orio, Operational amplifier performance degradation and time-to-failure due to electromigration, in *2018 31st Symposium on Integrated Circuits and Systems Design (SBCCI). Bento Gonçalves, Brazil*, (IEEE, 2018), pp. 1–6. https://doi.org/10.1109/SBCCI.2018.8533265
75. D. Maljar, M. Šovčík, D. Arbet, V. Stopjaková, Enhanced reliability of fully differential difference amplifier through on-chip digital calibration, in *2021 24th International Symposium on Design and Diagnostics of Electronic Circuits & Systems (DDECS). Vienna, Austria*, (IEEE, 2021), pp. 119–122. https://doi.org/10.1109/DDECS52668.2021.9417022
76. X. Gui, R. Tang, Y. Zhang, et al., A voltage-controlled ring oscillator with VCO-gain variation compensation. IEEE Microw. Wirel. Compon. Lett. **30**(3), 288–291 (2020). https://doi.org/10.1109/LMWC.2020.2967391

77. O. Petrosyan, R. Soghomonyan, H. Sahakyan, et al., Fully integrated resistance calibration method using reference frequency and MOSFET capacitance, in *2024 IEEE East-West Design & Test Symposium (EWDTS). Yerevan, Armenia*, (IEEE, 2024), pp. 1–4. https://doi.org/10.1109/EWDTS63723.2024.10873752
78. A. Soni, R. Dhiman, R. Chandel, Design and performance analysis of voltage-controlled oscillator, in *2023 3rd International Conference on Intelligent Technologies (CONIT). Hubli, India*, (IEEE, 2023), pp. 1–5. https://doi.org/10.1109/CONIT59222.2023.10205948
79. L. Rodriguez, E. Raygada, C. Silva, J. Saldaña, Design of a CMOS cross-coupled voltage doubler, *2016 IEEE ANDESCON. Arequipa, Peru* (IEEE, 2016), pp. 1–4. https://doi.org/10.1109/ANDESCON.2016.7836258
80. A.R. Guruprasad, A. Akshatha, R. Chetan, Performance analysis of non-overlapping two-phase clock signal generators, in *2023 IEEE 12th International Conference on Communication Systems and Network Technologies (CSNT). Bhopal, India*, (IEEE, 2023), pp. 439–443. https://doi.org/10.1109/CSNT57126.2023.10134682
81. A. Laha, A high voltage boost DC-DC converter using voltage doubler and switched-capacitor technique, in *2020 International Conference on Power Electronics and Renewable Energy Applications (PEREA). Kannur, India*, (IEEE, 2020), pp. 1–5. https://doi.org/10.1109/PEREA51218.2020.9339788
82. M.H.A. Hasan, M.S.R. Roky, M.A.R. Rokon, N.H. Uday, A comparative analysis of voltage doubler and diode rectifier for LCC converter in DC-DC charging systems, in *2024 IEEE Region 10 Symposium (TENSYMP)*, (IEEE, 2024), pp. 1–5. https://doi.org/10.1109/TENSYMP61132.2024.10752320
83. V. Melikyan, M. Martirosyan, A. Melikyan, G. Piliposyan, 14 nm Educational design kit: capabilities, deployment and future, in *Proceedings of the 7th Small Systems Simulation Symposium. Niš, Serbia*, (2018), pp. 37–41
84. J. Tang, Z.R. Shen, B.M. Lei, Research on accelerated test method based on reliability simulation and Arrhenius model, in *2022 4th International Conference on System Reliability and Safety Engineering (SRSE). Guangzhou, China*, (2022), pp. 539–542. https://doi.org/10.1109/SRSE56746.2022.10067691
85. X. Ma, Y. Lu, Q. Li, et al., An NMOS digital LDO With NAND-based analog-assisted loop in 28-nm CMOS. IEEE Trans. Circ. Syst. I Regul. Pap. **67**(11), 4041–4052 (2020). https://doi.org/10.1109/TCSI.2020.3009454
86. M. Máximo, M. Santos, J. Oliveira, P. Toledo, et al., Implementation of NMOS-based LDO using recycling folded-cascode in 16 nm FinFET, in *2024 8th International Young Engineers Forum on Electrical and Computer Engineering (YEF-ECE). Caparica/Lisbon, Portugal*, (IEEE, 2024), pp. 107–110. https://doi.org/10.1109/YEF-ECE62614.2024.10625344
87. B. Lin, P. Qin, X. Yi, et al., A high power supply rejection NMOS LDO with dual-loop and charge pump, in *2024 IEEE 10th International Symposium on Microwave, Antenna, Propagation and EMC Technologies for Wireless Communications (MAPE). Guangzhou, China*, (IEEE, 2024), pp. 1–4. https://doi.org/10.1109/MAPE62875.2024.10813817
88. W.-Y. Hsu, J. Aymerich, X. Yang, et al., A 0-to-35mA NMOS capacitor-less LDO with dual-loop regulation achieving 3 ns response time and 1pF-to-10nF loading range, in *ESSCIRC 2023 - IEEE 49th European Solid State Circuits Conference. Lisbon, Portugal*, (IEEE, 2023), pp. 253–256. https://doi.org/10.1109/ESSCIRC59616.2023.10268699
89. L. Xu, K. Choo, D. Blaauw, D. Sylvester, An analog-assisted digital LDO with single sub-threshold output pMOS achieving 1.44-fs FOM. IEEE Solid-State Circ. Lett. **4**, 154–157 (2021). https://doi.org/10.1109/LSSC.2021.3107870
90. R. Zhang, Z. Liu, X. Wang, A capacitor-less LDO with nested miller compensation and bulk-driven techniques in 90 nm CMOS, in *2021 4th International Conference on Circuits, Systems and Simulation (ICCSS). Kuala Lumpur, Malaysia*, (IEEE, 2021), pp. 51–55. https://doi.org/10.1109/ICCSS51193.2021.9464211
91. X.-C. Gong, J.-J. Kuang, X. Ming, et al., A fast transient PMOS LDO with AP3 buffer and shaped-hybrid-bias EA techniques achieving 8.15ps FOM, in *2024 IEEE International*

Symposium on Circuits and Systems (ISCAS). Singapore, (IEEE, 2024), pp. 1–5. https://doi.org/10.1109/ISCAS58744.2024.10558628

92. B. Saurab, A.P. Chavan, Ultra-low power, area efficient and high-speed voltage level shifter based on wilson current mirror, in *2021 IEEE Mysore Sub Section International Conference (MysuruCon). Hassan, India*, (IEEE, 2021), pp. 108–113. https://doi.org/10.1109/MysuruCon52639.2021.9641657
93. R.N. Ray, M.M. Tripathi, C.I. Kumar, High performance energy efficient CMOS voltage level shifter design, in *2022 2nd International Conference on Intelligent Technologies (CONIT). Hubli, India*, (IEEE, 2022), pp. 1–5. https://doi.org/10.1109/CONIT55038.2022.9847677
94. A. Kapoor, A. Thapar, C.S. Jha, C.I. Kumar, A high performance and low power subthreshold voltage level shifter design, in *2024 37th International Conference on VLSI Design and 2024 23rd International Conference on Embedded Systems (VLSID). Kolkata, India*, (IEEE, 2024), pp. 623–627. https://doi.org/10.1109/VLSID60093.2024.00110
95. Z. Zhou, M. Tan, A 20-V pulse driver based on all-nMOS charge pump without reversion loss and overstress in 65-nm standard CMOS technology. IEEE Trans. VLSI Syst. **32**(10), 1812–1821 (2024). https://doi.org/10.1109/TVLSI.2024.3435974
96. N. Lilic, R. Kappel, G. Roehrer, H. Zimmermann, Methods for the charge pump power efficiency improvement in triple well technologies, in *2020 27th IEEE International Conference on Electronics, Circuits and Systems (ICECS). Glasgow, UK*, (IEEE, 2020), pp. 1–4. https://doi.org/10.1109/ICECS49266.2020.9294893
97. J. Zhang, Y. Zhao, S. Peng, et al., A high-efficiency and low-ripple boost charge pump, in *2023 5th International Conference on Circuits and Systems (ICCS). Huzhou, China*, (IEEE, 2023), pp. 263–267. https://doi.org/10.1109/ICCS59502.2023.10367225
98. N. Wang, Y. Zhang, E. Qi, Y. Cheng, An LDO current limit protection circuit with hiccup mode, in *2023 8th International Conference on Integrated Circuits and Microsystems (ICICM). Nanjing, China*, (IEEE, 2023), pp. 209–213. https://doi.org/10.1109/ICICM59499.2023.10365917
99. E.-A. Marian, M.-D. Farcaş, R.-C. Oneţ, Comparison of three overcurrent protection circuits for LDOs, in *2023 International Semiconductor Conference (CAS). Sinaia, Romania*, (IEEE, 2023), pp. 171–174. https://doi.org/10.1109/CAS59036.2023.10303716
100. Y. Wang, X. Wang, H. Yao, Design of a foldback current limit circuit in LDO, in *2022 14th International Conference on Measuring Technology and Mechatronics Automation (ICMTMA). Changsha, China*, (IEEE, 2022), pp. 62–65. https://doi.org/10.1109/ICMTMA54903.2022.00019
101. W. Zhang, Foldback current limiting design and optimization for hot-swap and E-fuse ICs, in *PCIM Asia 2020: International Exhibition and Conference for Power Electronics, Intelligent Motion, Renewable Energy and Energy Management. Shanghai, China*, (2020), pp. 1–5
102. Y. Lin, K. Xu, A temperature-compensated power-on-reset circuit in 40 nm CMOS, in *2019 IEEE International Conference on Electron Devices and Solid-State Circuits (EDSSC). Xi'an, China*, (IEEE, 2019), pp. 1–2. https://doi.org/10.1109/EDSSC.2019.8754457
103. S. Delshadpour, M. Steger, S. Singh, A 0.7 μW explicit bandgap less POR circuit with brownout detection, in *2020 IEEE 11th Latin American Symposium on Circuits & Systems (LASCAS). San Jose, Costa Rica*, (IEEE, 2020), pp. 1–4. https://doi.org/10.1109/LASCAS45839.2020.9069023
104. A.A. Antonov, M.S. Karpovich, V.Y. Vasilyev, Power-on reset IC module with brownout detection and fast transients immunity, in *2021 XV International Scientific-Technical Conference on Actual Problems of Electronic Instrument Engineering (APEIE). Novosibirsk, Russian Federation*, (IEEE, 2021), pp. 273–277. https://doi.org/10.1109/APEIE52976.2021.9647518
105. S.S. Harutyunyan, H.T. Kostanyan, M.T. Grigoryan, et al., A reliable PMOS-based charge pump architecture. Proc. RA NAS NPUA Ser. Tech. Sci. **LXXIII**(2), 181–187 (2020)
106. H. Sahakyan, A novel PVT-tracking power supply detection circuit for reliable level shifter operation in multi-voltage ICs. Proc. RA NAS NPUA Ser. Tech. Sci. **LXXVII**(N3), 349–356 (2024)

Chapter 4
Development of Means for Reliability Improvement in High-Speed Serial Data Receivers

4.1 General Issues of Developing Means for Reliability Improvement in High-Speed Serial Data Receivers

4.1.1 Need to Develop Means for Reliability Improvement in High-Speed Serial DataReceivers

Integrated circuits (ICs) are widely used in modern electronic systems. Their fabrication has historically utilized two primary transistor technologies. In the initial period, until the 1970s, ICs used bipolar transistors with channel lengths of up to 20 μm [1, 2]. Subsequently, a transition was made to metal-oxide-semiconductor (MOS) transistors. This change was made to meet the requirements of low power consumption, small area, and manufacturing simplicity. Over time, the channel length of MOS transistors has been scaled down to 3 nm [3]. The reduction in channel length has made it possible to design more functional ICs. Parallel to the increase in the functionality of modern ICs, the requirements for speed have also increased [4]. The main factor limiting speed is the input-output (I/O) nodes, due to their large physical size and external load, which is tens of times greater than the load inside an IC [5]. These nodes also perform signal parameter conversion to make them more readable for receivers.

I/O nodes perform both data reception and transmission functions [6]. In ICs with built-in I/O nodes, data transmission can be implemented parallelly and serially [7–9]. For parallel data transmission, the transmitter and receivernodes are connected by multiple transmission lines [7]. In this case, each bit of data is transmitted over a separate transmission line, which further increases the speed of the I/O system. However, along with the increase in speed, power consumption and heat dissipation increase, and synchronization problems arise, which complicate the design process [8]. In contrast to parallel transmission, one of the main disadvantages of serial transmission was the low speed [9]. To avoid the difficulties arising during the

V. Melikyan, K. Li, *Parametric Optimization Techniques of Integrated Circuits*,
https://doi.org/10.1007/978-3-032-20400-4_4

design of I/O nodes with parallel data transmission, ICs with serial data transmission I/O nodes are now widely used (Fig. 4.1) [10].

In such I/O node structures, parallelly transmitted data is converted to serial and fed as an input signal to the transmitter node [10].

Data passes from the transmitter to the receiver node via the transmission line. During transmission, signal attenuation occurs [11, 12]. This is due to the distributed parameter structure of the transmission line's equivalent circuit. The transmission line is represented by numerous cascaded connections of a circuit consisting of elements R_0, L_0, G_0, and C_0 (Fig. 4.2) [13–16], where R_0 is the resistance of each cascade, L_0 is the inductance, G_0 is the conductance, and C_0 is the capacitance.

The considered transmission line is called homogeneous because all similar elements of cascaded circuits are equal [15]. To obtain the equation for the impedance Z_0 of such a transmission line, it is necessary to solve the following differential equations derived from Ohm's and Kirchhoff's laws [13]:

$$\frac{\partial V(x,t)}{\partial x} = -R_0 I(x,t) - L\frac{\partial I(x,t)}{\partial t}, \tag{4.1}$$

$$\frac{\partial I(x,t)}{\partial x} = -G_0 I(x,t) - C\frac{\partial I(x,t)}{\partial t}: \tag{4.2}$$

As a result, the impedance of a homogeneous transmission line is calculated as follows [13]:

$$Z_0 = \frac{V(x)}{I(x)} = \sqrt{\frac{R_0 + j\omega L}{G_0 + j\omega C}}. \tag{4.3}$$

From expressions (4.1) and (4.2), it becomes clear that the transmitted signal has a wave nature, since the voltage and current functions are characterized by two variables [16]. This means that signal reflection can be observed when the transmission line length is greater than the wavelength of the data [16, 17]. Due to this phenomenon, the useful component of the signal is distorted during transmission over the line. Distortion, in turn, causes unreliable data reception [18]. The reflection

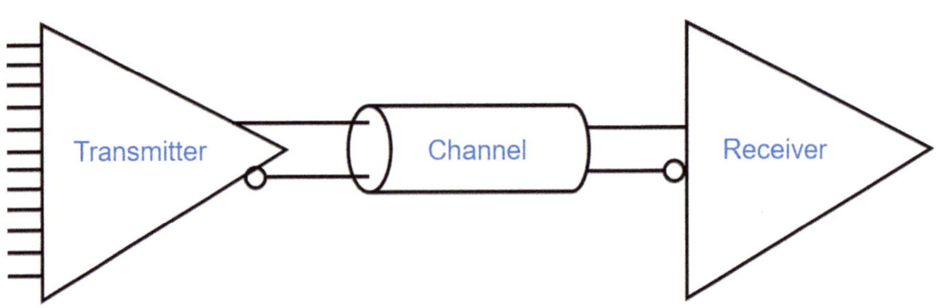

Fig. 4.1 Structure of an I/O node with serial data transmission

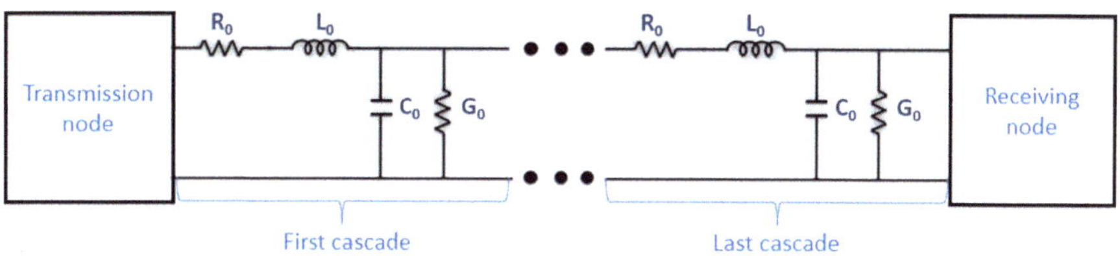

Fig. 4.2 Transmission line structure

Fig. 4.3. Amplitude–frequency response of a transmission line

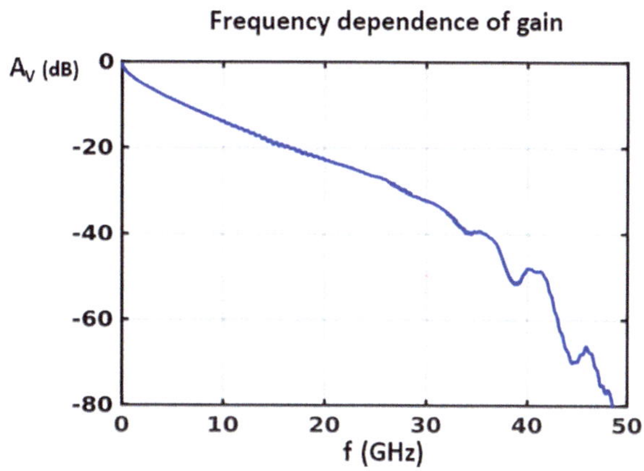

coefficient for the transmitter (K_{tr}) and receiver (K_r) nodes is determined by the following formulas [13]:

$$K_{tr} = \frac{R_{out} - Z_0}{R_{out} + Z_0}, \tag{4.4}$$

$$K_r = \frac{R_{in} - Z_0}{R_{in} + Z_0}. \tag{4.5}$$

They become most vivid at high frequencies (Fig. 4.3) [19].

From formulas (4.4) and (4.5), it follows that to avoid signal reflection and distortion, a Resistance Coordination System (RCS) is necessary [20, 21].

In I/O nodes, RCS is used to match the output impedance of the transmitter and the input impedance of a receiver according to the characteristic impedance. Its calculation is performed empirically [20]. As a result of experiments, those resistance values are recorded for which minimum power consumption and minimum voltage drop per unit length are observed as a result of data propagation through the transmission line [20]. As a result, the characteristic impedance calculation is performed by selecting the best compromise value from the algebraic and geometric means of these two recorded values. For example, according to the Universal Serial Bus (USB) standard, the minimum voltage drop per unit length of the line was observed at 77 Ω, and the minimum power consumption at 30 Ω resistances (Fig. 4.4) [21].

As a result, calculations of algebraic and geometric means yielded 53.5 Ω and 48 Ω, respectively. The best compromise value chosen from these two was 50 Ω [21].

As a result of matching the input impedance of a receiver node, the transmitted data, without distortion of the useful component but attenuated, is passed to the Equalizer (E) [22, 23]. The function of E is to obtain the same amplitude value for all frequency components of the data at the output. Implementing such a function means providing amplification corresponding to the attenuation factor for the entire high-frequency component (Fig. 4.5) [24]. Since the data transmission speed of I/O nodes has increased over time, particularly reaching up to 20 GHz in USB systems [25], 16 GHz in PCIe [26], and up to 56 GHz in XSR [27], it follows that for the same transmission line, the attenuation of data at the maximum allowable frequency components will be greater.

As a result of high attenuation, the amplitude value of the data in those components will tend to 0. This means that to preserve signal integrity (SI), the designed equalizer circuit (EC) must perform equalization for components with amplitude values reaching down to tens of millivolts [28].

To ensure uninterrupted system operation, it is necessary to perform SI checks at the input and output of every node. For SI checking, Eye diagrams (ED) of the signals obtained at the input of the receiver node and the output of the equalizer circuit are derived [29–31]. EDs are obtained by isolating and overlaying signals over a unit period. The following are measured from the ED:

- Horizontal and vertical eye openings.
- Signal jitter.

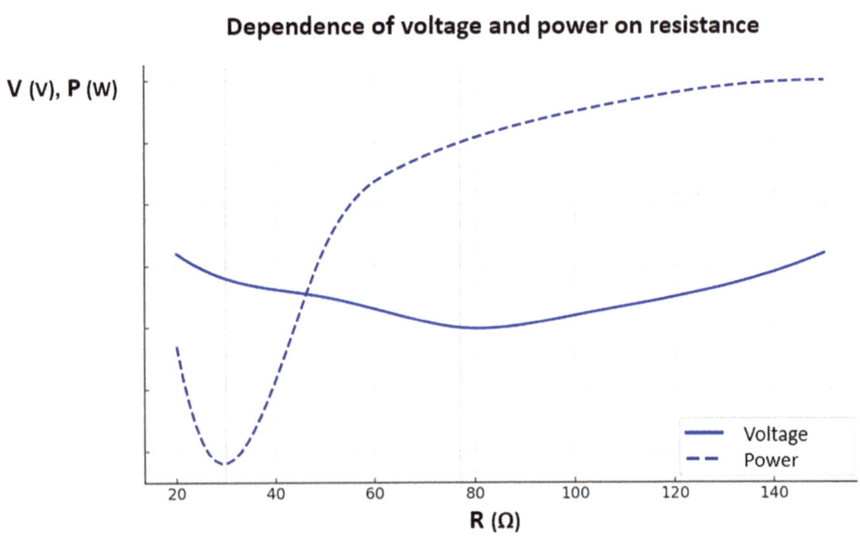

Fig. 4.4 Dependence of transmission line resistance, power, and voltage drops

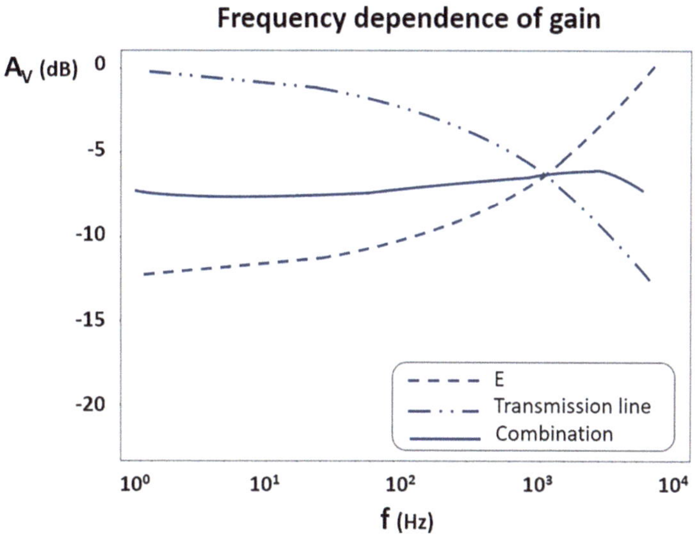

Fig. 4.5 Amplitude–frequency response of the transmission line, equalizer circuit, and the overall system

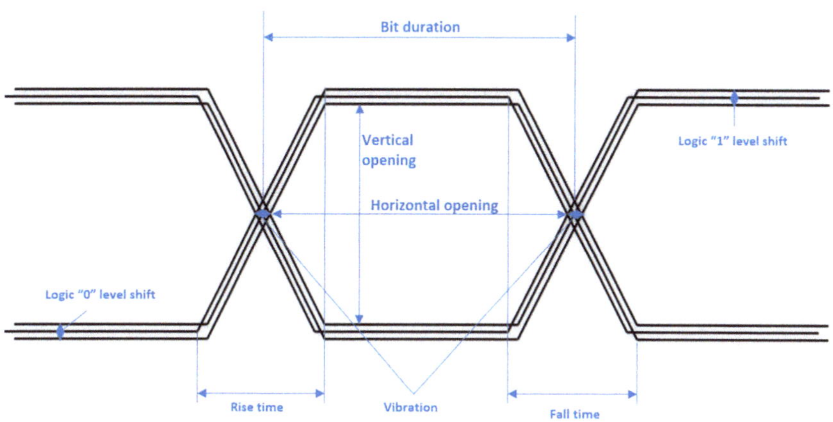

Fig. 4.6 Parameters measured via eye diagram

- Deviations of logical 1 and 0 levels.
- Rise and fall times.
- Bit duration (Fig. 4.6) [31].

EDs are open or closed [29]. EDs are considered open if their vertical opening constitutes 50–70% of the data voltage amplitude, and the horizontal opening constitutes 80–90% of the bit duration (Fig. 4.7) [30].

EDs are considered closed if their vertical and horizontal openings do not meet the above requirements (Fig. 4.8) [30].

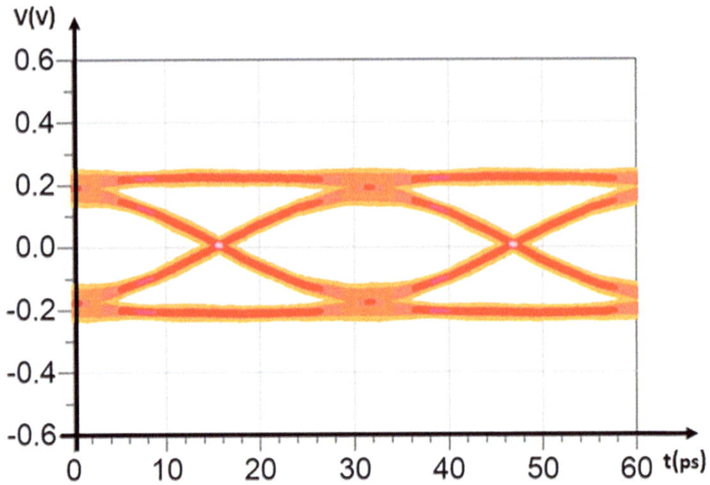

Fig. 4.7 Open eye diagram

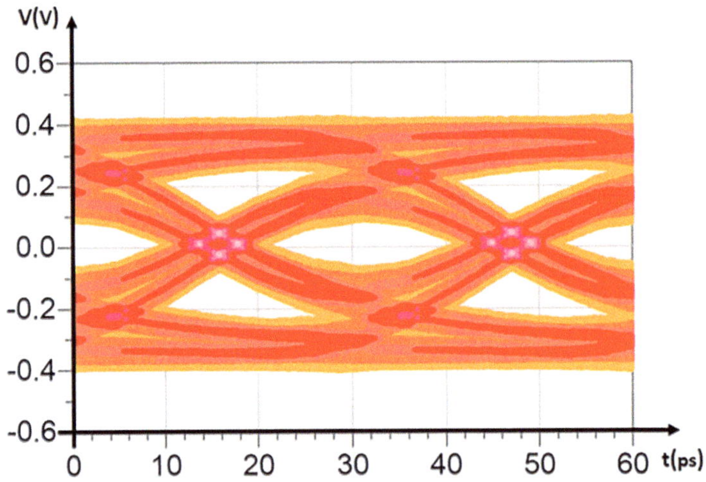

Fig. 4.8. Closed eye diagram

EDs become closed when rise and fall times increase, deviations of logical 0 and 1 levels increase, and signal jitter increases [31]. The large values of these parameters are a consequence of a number of phenomena that contribute to reliability degradation. Due to these phenomena, degradation of key parameters is observed in the semiconductor elements used in circuits designed in various nodes of ICs [32]. Specifically, degradation of interconnects and the transistor gate oxide layer [33], reduction in conductivity, which can result in changes in the threshold voltages of semiconductor elements [34], leakage currents exceeding the maximum allowable value, voltage drops in power rails, high power consumption, and thermal shutdowns [35]. As a result of the long-term effects of these phenomena, breaks in interconnects and punctures in semiconductor elements can occur [36]. As a result of the listed undesirable phenomena, SI is violated, and data read failure occurs [37].

Several reputable companies, such as Broadcom and Qualcomm, have faced such problems. In 2015, in Qualcomm's Snapdragon 810 processors used in mobile phones, a series of functional failures occurred due to phenomena affecting IC reliability. As a result, there were customer complaints, and the company implemented special measures aimed at reducing the impact of this phenomenon [38]. In 2017, Ethernet Switches designed and manufactured by Broadcom, which are widely used in various network equipment, had reliability issues in the receiver node, resulting in significant failures in a number of data centers and enterprise networks [39].

To avoid the above cases, IC design companies perform special tests where a number of operating conditions are established. These include supply voltage values with ±10% deviation from the nominal value and an operating temperature range from −55 °C to +150 °C [40]. To perform simulations including such phenomena, several software tools exist. These include PrimeSim and HSPICE developed by Synopsys [41], Spectre [42], and PSpice [43] from Cadence, and ELDO [44] from Siemens. As a result of such tests, it is possible to calculate the effects of phenomena disrupting the long-term uninterrupted operation of ICs during the design phase.

Thus, the involvement of modern electronic communication systems in important fields, such as medicine and mechanical engineering, implies a sharp increase in reliability criteria to ensure long-term uninterrupted operation. Therefore, there is a need to develop means for improving the reliability of high-speed serial data reception systems.

4.1.2 *Key Factors Determining Deterioration of Reliability of Integrated Circuits*

The main phenomena contributing to the deterioration of IC reliability are crosstalk [45–48], IR drop [49–52], electromigration [53, 54], self-heating [55–58], and aging [59–66].

The crosstalk phenomenon occurs when interconnects of two or more different frequency signals are routed in close proximity [45] during the physical design of ICs. The effect of this phenomenon is maximized when they are located quite close to each other. In this case, a parasitic capacitance arises between two metals. The reactive resistance of parasitic capacitance for high-frequency signals decreases, resulting in the distortion of signals on those interconnects under the influence of alternating currents (Fig. 4.9) [46].

The interconnect whose signal switching distorts the others is called the aggressor. The interconnect whose signal is distorted due to the crosstalk phenomenon is

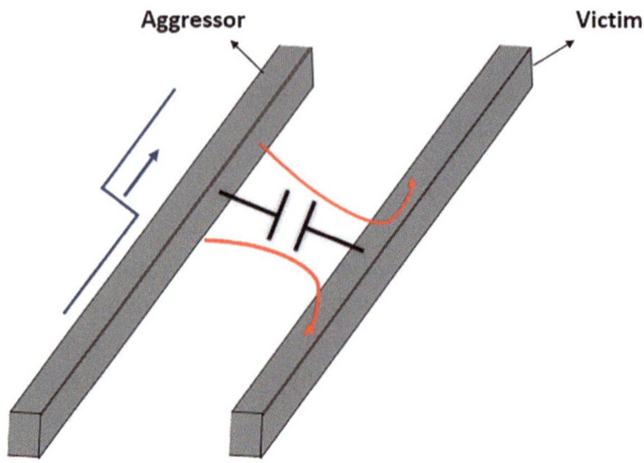

Fig. 4.9 Effect of crosstalk phenomenon

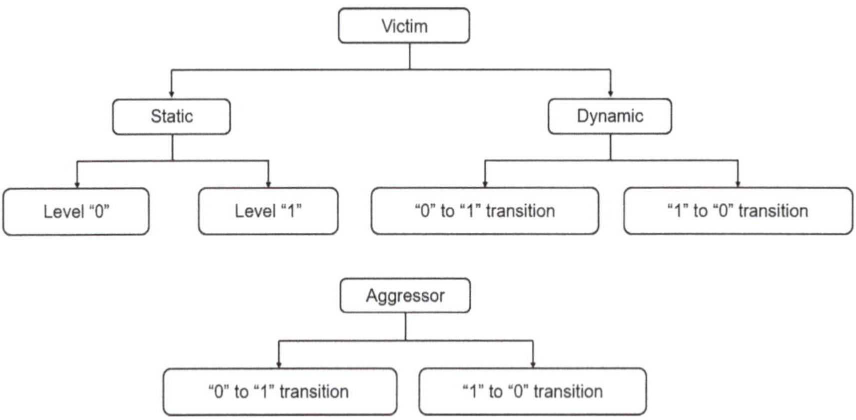

Fig. 4.10 Possible states of victim and aggressor

called the victim. The victim's signals can be in four, and the aggressor's signals can be in two possible states (Fig. 4.10) [47].

In this case, eight different crosstalk scenarios are possible (Fig. 4.11) [47].

One of the known methods developed to reduce the crosstalk phenomenon is to add an interconnect between the victim and the aggressor during physical design, which is grounded or connected to the supply voltage. In this case, the parasitic capacitance between the aggressor and the victim decreases [48].

IR drop is a phenomenon resulting from the physical design of ICs, where voltage drops occur on the power rails due to currents flowing through the circuits (Fig. 4.12) [49].

The IR drop phenomenon has a significant impact on the voltage distribution from the power rails to individual sections of ICs (Fig. 4.13) [50].

During IC operation, currents flow through the power rails, which cause voltage drops in the interconnects, resulting in a decrease in the supply voltage values reaching individual sections of the ICs according to Ohm's law. In the literature, two

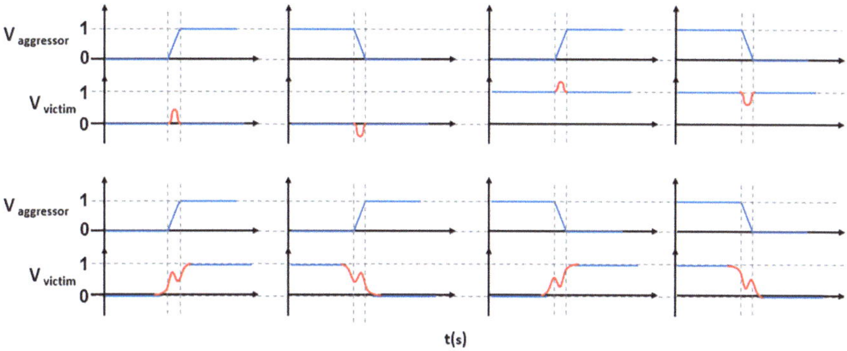

Fig. 4.11 Possible scenarios of crosstalk phenomenon

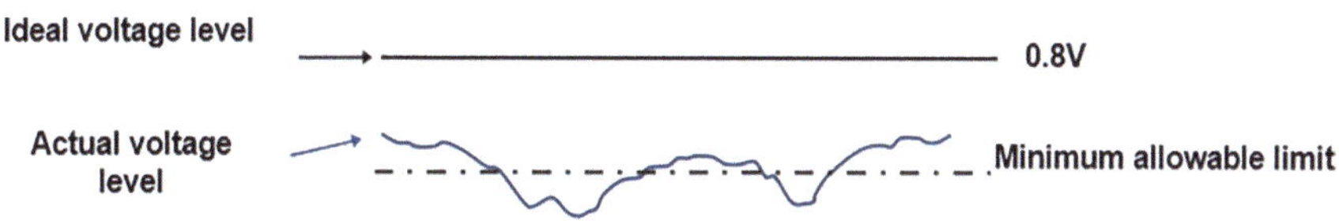

Fig. 4.12 IR drop

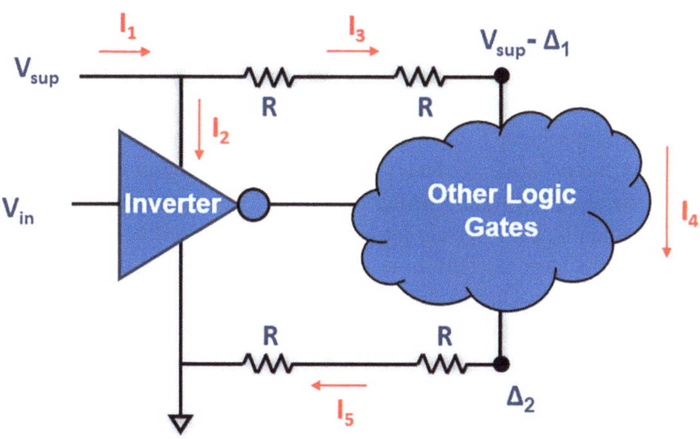

Fig. 4.13 Change in supply voltage in individual sections of ICs

types of IR drop are distinguished: static and dynamic [50, 51]. Static IR drop is observed with constant current flowing through the power rails. Dynamic IR drop is observed as a result of alternating currents caused by numerous transistor switching. Using the example of a Complementary-MOS (CMOS) inverter—during the output signal change, the resistances of the transistors change, and during this time, the current flowing through them changes (Fig. 4.14) [52, 67].

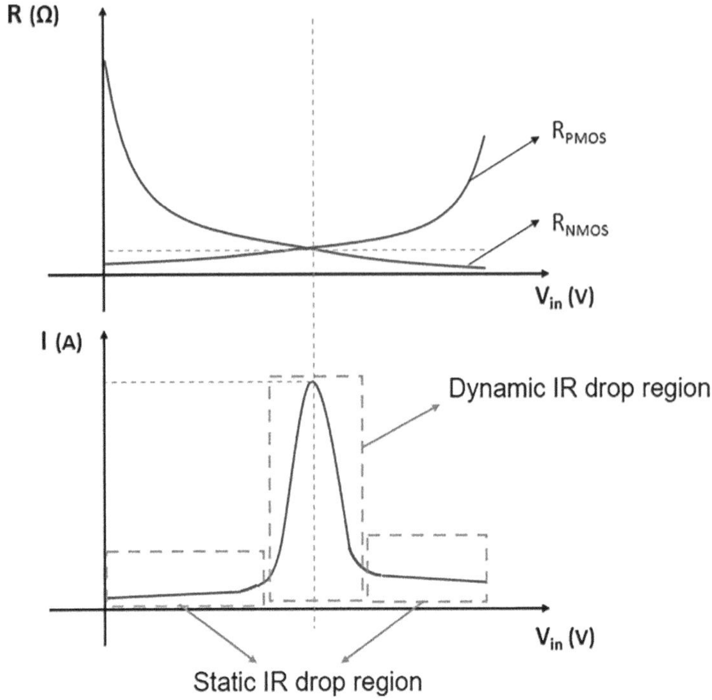

Fig. 4.14 Resistance of CMOS inverter elements and the current flowing through them

When a constant logical 1 or 0 signal is formed at the inverter output, one of the transistors is in the off state, so leakage current flows through it, causing static IR drop in the power rails. Since during switching the transistor resistances do not change proportionally, for a certain input voltage, they have the smallest total resistance. Due to this, a large current flows through the transistors, causing dynamic IR drop in the power rails. Due to this phenomenon, as a result of a decrease in the supply voltage level observed in individual sections of the IC, the parameters of the circuits located in that section change. To reduce the effect of the IR drop phenomenon, optimizations of the arrangement of power rails and interconnects are performed [49, 50].

Another phenomenon that promotes the reduction of IC reliability is electromigration. In this phenomenon, electrons collide with metal ions during their movement, causing displacements that lead to changes in the structure of IC interconnects [53, 54]. Displacements of metal ions in interconnects cause breaks, voids, and hillocks, which can cause interruptions of current flowing through that section or short circuits, leading to functional failures of circuits (Fig. 4.15) [53].

For the mean time to failure (MTTF) of interconnects under the influence of this phenomenon, certain calculations are performed using Black's equation [53]:

$$\mathrm{MTTF} = \frac{A}{J^n} \cdot e^{\frac{E_a}{kT}}, \tag{4.6}$$

Fig. 4.15. Effect of electromigration phenomenon in metal interconnects

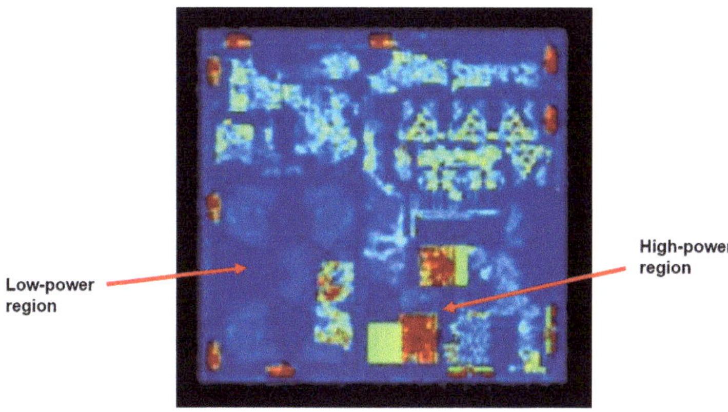

Fig. 4.16 Thermal map of ICs

where A is a constant for the material, J is the current density, n is the current density coefficient, E_a is the activation energy for electromigration, k is Boltzmann's constant, and T is the absolute temperature in Kelvin.

The self-heating phenomenon occurs in ICs during operation as a result of power consumption [56]. During operation, currents caused by transistor switching occur, and power consumption takes place, resulting in heat dissipation in various sections of ICs (Fig. 4.16) [55].

As a result of circuit heating, the leakage currents flowing in them increase, which in turn causes additional heat dissipation. High temperature accelerates system aging by changing charge carrier mobility and degrades material properties. This phenomenon is especially seen in ICs designed with small technology processes [57].

To avoid this phenomenon, it is proposed to manufacture ICs with elements that have higher thermal conductivity to reduce thermal resistance. Design optimization is also proposed [58].

Another means of increasing reliability is the reduction of aging phenomena. Aging phenomena occur during the continuous operation of ICs and cause the degradation of key parameters of semiconductor elements used in them and circuits composed of them [59]. During the aging of these elements, some physical properties of semiconductor material change, resulting in parametric degradation. Three mechanisms contributing to IC aging are known in the literature: bias temperature instability (BTI) [59, 61, 65, 68, 69], hot carrier injection (HCI) [65, 66, 68, 69], and time-dependent dielectric breakdown (TDDB) [68–71].

BTI is a phenomenon of reliability degradation in MOS transistors that affects both NMOS and PMOS transistors. This phenomenon manifests itself as a gradual change in the threshold voltage of transistors. These changes occur as a result of prolonged voltage application to the transistor gate (Fig. 4.17) [68].

There are negative and positive BTIs. Negative BTI (NBTI) occurs in PMOS transistors when a negative voltage, or, which is the same, a voltage corresponding to a low logical level in the context of ICs, is applied to the gate-substrate at high temperatures (100–150 °C) [59]. This "stress" causes hole accumulation in the gate oxide and silicon-oxide interface, leading to the formation of interface states and fixed charges. Positive BTI (PBTI) occurs in NMOS transistors when a positive voltage, or, which is the same, a voltage corresponding to a high logical level in ICs, is applied to the gate-substrate under the aforementioned temperature conditions [61]. Electrons accumulate in the gate oxide, causing the aforementioned degradation effects. As a result of the effects of these phenomena, threshold voltage change occurs over time, resulting in a decrease in drain–source current, an increase in leakage currents, and a decrease in transistor switching speed [61, 65].

As a result of these phenomena, the change in the transistor's threshold voltage is given by the following expression [69]:

$$\Delta V_{\text{th}} = A \cdot t^n \cdot e^{\frac{-E_a}{kT}}, \tag{4.7}$$

where A is the pre-exponential empirical coefficient, t is the voltage application time, n is the time exponent (usually from 0.1 to 0.3), E_a is the activation energy, k is Boltzmann's constant, and T is the absolute temperature.

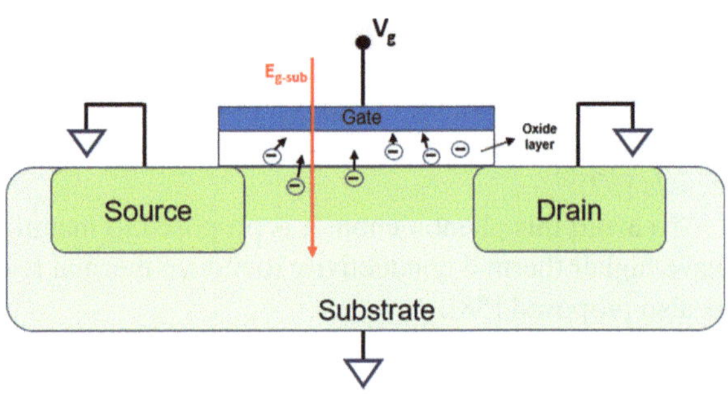

Fig. 4.17 Reflection of BTI phenomenon

To avoid this phenomenon, it is proposed to use high-k insulators and metal gates, or to reduce the voltage applied to the gate [59, 61].

The other aging phenomenon, HCI, occurs when the main charge carriers in transistors, under the influence of high electric fields between gate–source and drain–source, acquire high kinetic energy and become injected into the gate oxide layer (Fig. 4.18) [68].

As a result of this phenomenon, damage to the oxide layer occurs, which increases the transistor's threshold voltage. Consequently, the drain current and transistor conductivity decrease [65]. The threshold voltage change due to the effect of this phenomenon is calculated by the following expression [69]:

$$\Delta V_{\mathrm{th}} = B \cdot I_d^{\,m} \cdot e^{\frac{-E_a}{kT}}, \tag{4.8}$$

where B and m are empirical coefficients, I_{d} is the drain current, E_{a} is the activation energy, k is Boltzmann's constant, and T is the absolute temperature.

To avoid this phenomenon, it is proposed to lightly dope the drain region or reduce the electric fields [65, 66].

The effect of the TDDB phenomenon is due to the long-term application of a high electric field between gate and drain. As a result of the application of such fields, the main charge carriers of transistors accumulate in the gate oxide and silicon-oxide interface, which causes their gradual degradation over time (Fig. 4.19) [68].

Due to the effect of this phenomenon, a large number of traps increase in the oxide layer, resulting in current flow through them due to tunneling [70]. Over time, the increase in the number of traps causes large currents, resulting in the final breakdown of the oxide layer. The breakdown time of the oxide layer as a result of this phenomenon's effect is calculated by the following expression [69]:

$$t_{\mathrm{breakdown}} = A \cdot e^{\frac{E_a}{kT}} \cdot \left(\frac{1}{E}\right)^{n}, \tag{4.9}$$

where A and n are the empirical coefficients, E_a is the activation energy, E is the electric field, k is Boltzmann's constant, and T is the absolute temperature.

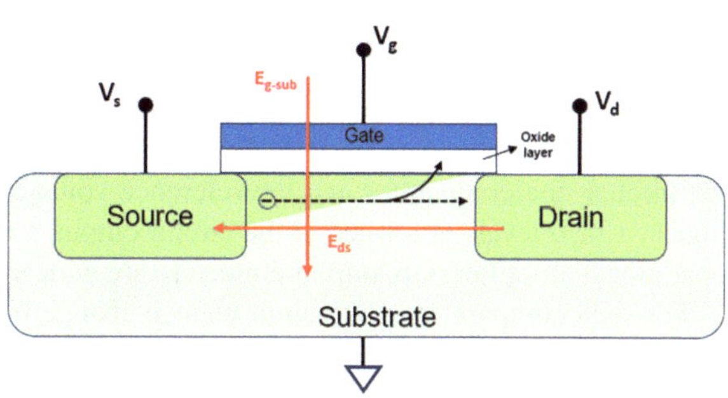

Fig. 4.18 Reflection of HCI phenomenon

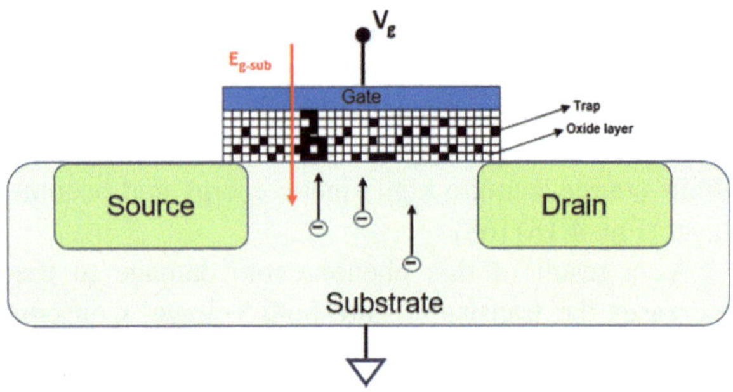

Fig. 4.19 Reflection of TDDB phenomenon

To avoid these phenomena, it is proposed to use high-k insulators, perform process optimization to reduce trap formation in the gate oxide, or develop special design techniques [70, 71].

Thus, to avoid the main factors causing the aforementioned reliability degradation, it is necessary to apply methods that will reduce their effect on circuit parameters. These methods, along with the reduction of effects, may lead to degradation of area, power, and other parameters.

4.1.3 Existing Solutions and Key Issues to Increase Reliability of High-Speed Serial Data Receivers

High-speed receiver systems consist of a large number of analog and mixed-signal circuits, which are considered the most sensitive circuits in ICs. Therefore, these circuits are the main victims of reliability phenomena, causing SI failures in systems.

The functionality of high-speed receiver nodes is implemented through equalizer circuits, voltage regulators (VR), charge pump (CP) circuits, and comparators. All the listed circuits are highly sensitive circuits. The development of methods for improving the reliability of such circuits is based on compromise solutions, given the strict limitations on area, functionality, power consumption, and speed of modern ICs. The development of these methods has become a serious challenge, considering the already high frequencies and reliability problems.

Existing Methods for Increasing the Reliability of Comparators

Comparators are widely used in high-speed receiver nodes. They can equalize high and low levels of EC output signal to voltage levels corresponding to logical 0 and 1. One of the necessary inputs for comparator's operation is the reference voltage, the other is the output data from the EC. As a result of comparing the data transmitted through the comparator and the reference voltage, a voltage corresponding to logical 1 or 0 levels is formed at the circuit output. Comparators based on CMOS operational amplifier (OpAmp) architecture are widely used in ICs (Fig. 4.20) [72].

For such comparators, the output voltage change from input signals is given by the following formula [73]:

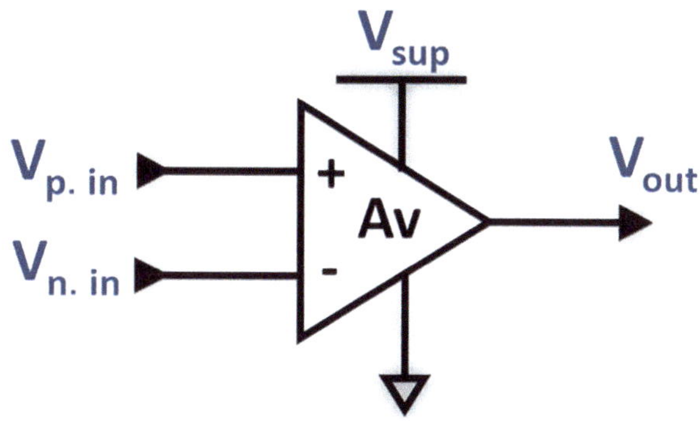

Fig. 4.20 Comparator based on CMOS operational amplifier

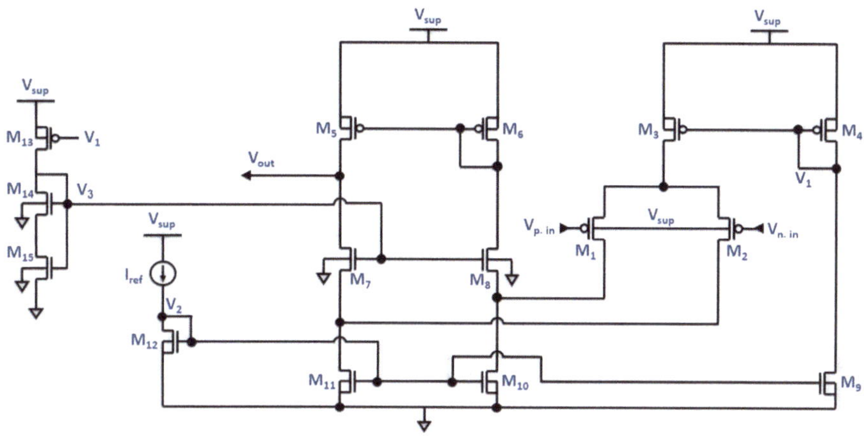

Fig. 4.21 Comparator based on folded cascode OpAmp

$$V_{\text{out}} = \left(V_{p.\text{in}} - V_{n.\text{in}}\right) A_V, \tag{4.10}$$

where $V_{p.\text{in}}$ and $V_{n.\text{in}}$ are the positive and negative inputs, respectively, and A_V is the circuit's gain.

Important parameters of such comparators are gain, sensitivity, speed, and noise immunity. The degradation of the above parameters may be due to the effect of aging phenomena on the circuit. Parametric degradation occurs both as a result of long-term continuous operation and when switching from long-term off mode to operating mode. Comparators based on folded cascode OpAmps are widely used in ICs (Fig. 4.21) [74].

In basic folded cascode OpAmp-based comparators, T_{A1}, T_{A2}, T_{A3}, T_{A4}, T_{A5}, and T_{A6} transistors are also added to ensure off mode (Fig. 4.22) [75].

Experiments were conducted using simulation programs to calculate aging phenomena over 10 years for the two aforementioned operating modes (Table 4.1) [75].

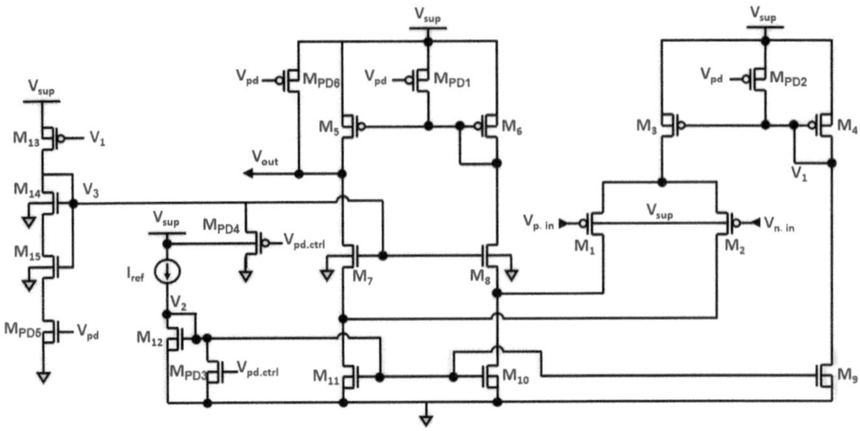

Fig. 4.22. Comparator with transistors providing off mode

Table 4.1 Maximum deviations of I_d and V_{th} of transistors as a result of 10-year aging of folded cascode OpAmp-based comparator

	Operating mode		Off mode	
Transistor name	ΔV_{th} (mV)	ΔI_d (%)	ΔV_{th} (mV)	ΔI_d (%)
T_7, T_8	24	3,1	192	19,6
T_9	20	2,8	198	20,1
T_{A1}, T_{A2}, T_{A6}	0,1	0,1	252	24,2
T_{A3}, T_{A4}	0,1	0,1	254	24,3
T_{A5}	148	14,2	0,1	0,1

As seen from the table, the parametric deviations obtained as a result of simulation for the effect of HCI aging in long-term operating mode and TDDB aging in off mode exceed the allowable 10% value.

In one of the known approaches in the literature, T_{A7} and T_{A8} transistors have been added to the circuit to reduce the effects of aging phenomena (Fig. 4.23) [76].

Some authors also noted parametric degradation of input transistors due to the effect of aging phenomena when there are deviations in the voltages applied to their gates. Therefore, CMOS switches B1 and B2 and transistors T_{A4}, T_{A5}, T_{A6} were added to the input transistor gates to apply the same voltage to the input transistor gates in off mode. Experiments were conducted using simulation programs to calculate 10-year aging phenomena (Table 4.2) [76].

Existing Methods for Increasing the Reliability of Charge Pump Circuits

For a number of analog and mixed-signal nodes in high-speed receiver systems, such as equalizer and comparator circuits, it is necessary to provide large gains for processing the signal transmitted over the transmission line. To provide such gain, voltage regulators (VR) exceeding the supply level are designed [77]. The CP circuit is an integral part of such VRs. With the help of the CP circuit, it is possible to

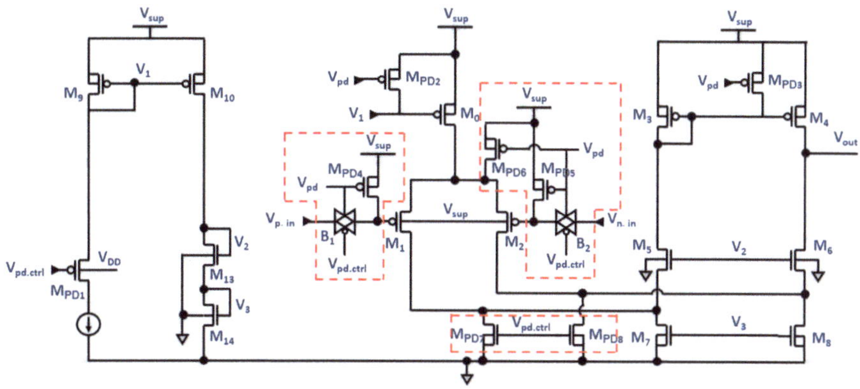

Fig. 4.23 Proposed comparator to reduce aging phenomena

Table 4.2 Folded cascode OpAmp-based comparator before and after aging protection

Parameter	Before aging protection	After aging protection
Constant gain (dB)	66,2	80,01
Largest ΔV_{th} (mV)	254	0,01
Largest ΔI_d (%)	24,3	0,1
Largest voltage deviation (mV)	35	3

obtain a voltage exceeding the supply level by 1.5–2 times. This voltage, in turn, serves as the supply for the VR. As a result, it is possible to obtain a stable voltage exceeding the supply level. Cross-coupled CP circuits are widely used in the literature (Fig. 4.24a–c) [78, 79].

The circuits consist of C_1 and C_2, capacitances connected to T_1, T_2 input transistors and T_3, T_4 switch transistors, respectively. All three circuits have two inputs to which anti-phase Clocks (clk) are connected. The clks switch from logical 0 to 1 levels. To obtain a value at the CP circuit output that is twice the supply voltage, the logical 1 level must correspond to the V_{sup} voltage. As a result of clk switching and the cross-coupling of T_1, T_2 transistor gates, $2V_{sup}$ voltages are formed at nodes A and B every half-cycle and transferred to the output through T3 and T4 [78–81].

The functions of these circuits are the same, but the effects hindering reliable operation are different. In all three circuits, there are leakage currents from output to input (V_{sup}) due to simultaneous clk switching. For these currents, the authors propose adding delays in the clks to prevent simultaneous switching (Fig. 4.25) [78, 79].

In the first circuit, there are leakage currents flowing from nodes A and B to the input, caused by the leakage currents between the drain and bulk of transistors T_1 and T_2. Since the bulks of these transistors are connected to the source, as a result, no potential difference exceeding the maximum allowable value arises between any two of their pins [79]. Therefore, there cannot be an aging effect in those transistors. However, this does not apply to transistors T_3 and T_4, because when switching from

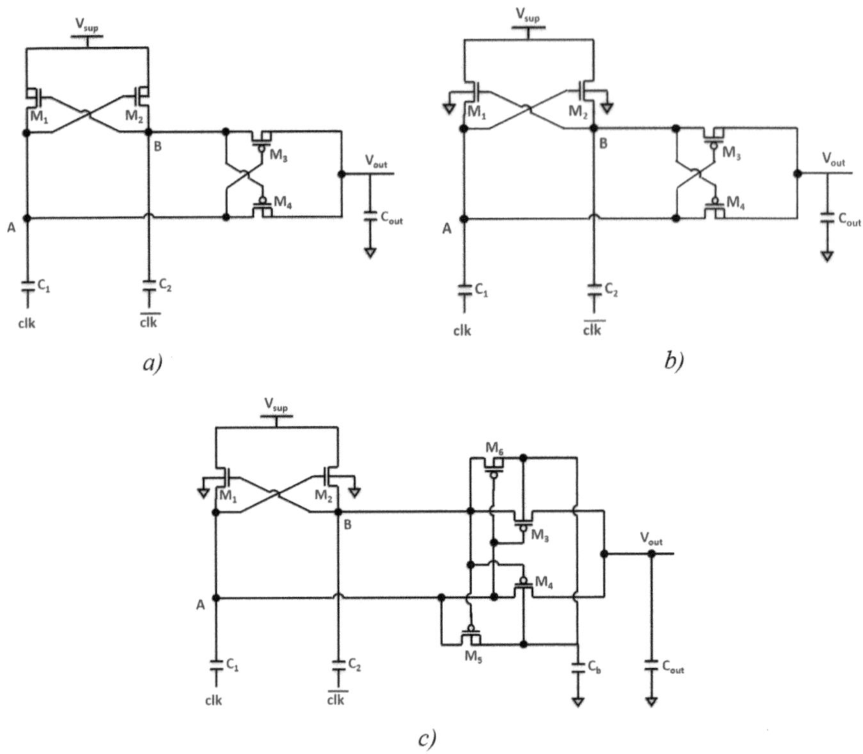

Fig. 4.24 (**a–c**) Cross-coupled CP circuits

Fig. 4.25 Anti-phase clks with added delays

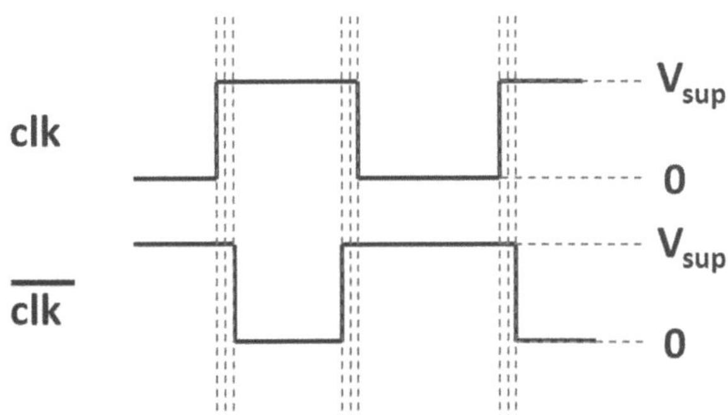

circuit's off mode to operating mode, the gate-bulk and gate–drain voltages exceed the maximum allowable value. Therefore, HCI aging occurs. Experiments were conducted using simulation programs to evaluate the effect of HCI (Table 4.3).

In the second circuit, unlike the first one, it is proposed to ground the bulks of input transistors T_1 and T_2. With this connection, the leakage currents flowing from nodes A and B to the input are eliminated. However, reverse-biased P-N junction currents from drain and source to bulk arise, and a potential difference exceeding

Table 4.3 Maximum deviations of I_d and V_{th} of transistors as a result of 10-year aging of CP circuit shown in Fig. 4.20a

Transistor name	Operating mode	
	ΔV_{th} (mV)	ΔI_d (%)
T_1 and T_2	10	0,1
T_3 and T_4	120	19,3

Table 4.4 Maximum deviations of I_d and V_{th} of transistors as a result of 10-year aging of CP circuit shown in Fig. 4.20b

Transistor name	Operating mode	
	ΔV_{th} (mV)	ΔI_d (%)
T_1 and T_2	136	20,3
T_3 and T_4	119	19,2

Table 4.5 Maximum deviations of I_d and V_{th} of transistors as a result of 10-year aging of CP circuit shown in Fig. 4.20c

Transistor name	Operating mode	
	ΔV_{th} (mV)	ΔI_d (%)
T_1 and T_2	139	20,9
T_3 and T_4	18	0,3

the maximum allowable value between gate and bulk, resulting in HCI aging in those transistors [80]. Experiments were conducted using simulation programs to evaluate the HCI effect in the second circuit (Table 4.4).

In the third circuit, capacitance C_{bulk} and transistors T_5 and T_6 have been added to reduce the aging phenomena in transistors T_3 and T_4. The added transistors charge the C_{bulk} capacitance up to $2V_{sup}$ voltage during circuit operation, which does not discharge in both off and operating modes of the circuit [78, 81]. Since the bulks of transistors T_3 and T_4 are connected to the anode of C_{bulk}, no voltage exceeding the maximum allowable value arises between any two of their pins. Experiments were conducted using simulation programs to evaluate the effect of aging phenomena in the third circuit (Table 4.5).

Existing Methods for Increasing the Reliability of Equalizers

ECs play a crucial role in high-speed receiver systems. ECs compensate for the losses caused by the transmission line to preserve the SI of the transmitted data. In high-speed receiver systems, the voltage amplitude value of the data passed through the transmission line can reach up to several tens of mV. For ECs to be able to preserve data SI, it is necessary to reduce the effects of phenomena that promote reliability degradation. Passive and active element ECs are found in the literature. Passive element ECs are composed of a combination of high- and low-frequency filters (Fig. 4.26) [82, 83].

The transfer function of the passive element EC is given as follows [83].

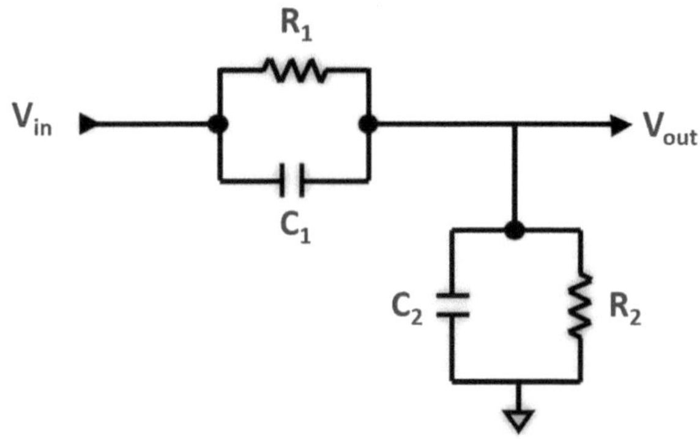

Fig. 4.26 Passive element EC

$$H(s) = \frac{R_2}{R_2 + R_1} * \frac{1 + C_1 R_1 s}{1 + \frac{R_1 R_2}{R_1 + R_2}(C_1 + C_2)s}, \tag{4.11}$$

where R_1 and R_2 are the resistances of the EC, and C_1 and C_2 are the capacitances. The zero (ω_{zero}) and pole (ω_{pole}) frequencies from the transfer function are calculated by the following expressions [83]:

$$\omega_{zero} = \frac{1}{R_1 C_1}, \omega_{\mathbf{pole}} = \frac{1}{\frac{R_1 R_2}{R_1 + R_2}(C_1 + C_2)}: \tag{4.12}$$

The gains of the passive EC, before the zero frequency ($A_{toz.}$) and after the pole frequency ($A_{p.f.}$), are given as follows [83]:

$$A_{\mathbf{toz.}} = \frac{R_2}{R_1 + R_2}, A_{\mathbf{p.f.}} = \frac{C_1}{C_1 + C_2} \tag{4.13}$$

Active element ECs are composed of two differential branches connected to each other by variable resistances and capacitances (Fig. 4.27) [83, 84].

The transfer function of the active element EC is given as follows [84]:

$$H(s) = \frac{-g_m R_1 (1 + sR_2 C_2)}{1 + \frac{g_m R}{2} + sR_2 C_2}, \tag{4.14}$$

where R_1 is the amplifier load, g_m is the transconductance of the input transistors, and C_2 and R_2 are the feedback capacitance and resistance.

From the transfer function of the active element EC, two poles and one zero arise, which are calculated by the following expressions [85]:

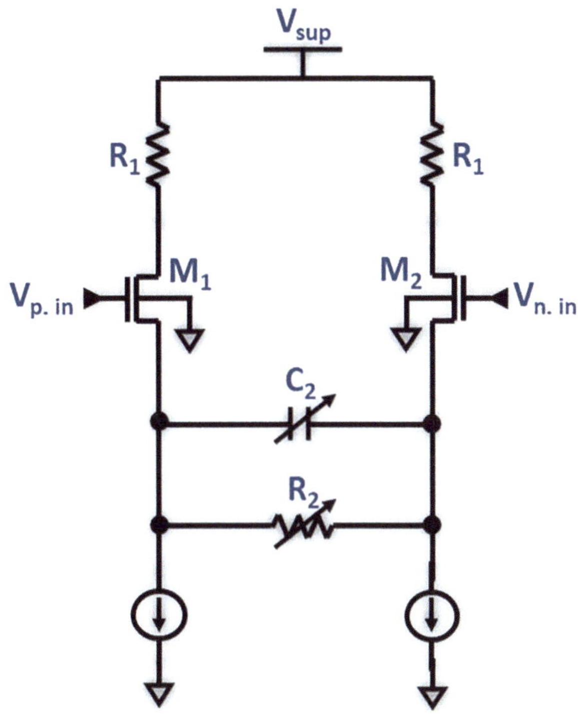

Fig. 4.27 Active element EC

$$\omega_{\text{pole}} = \frac{1 + g_m R_2 / 2}{R_2 C_2}, \omega_{\text{pole}} = \frac{1}{R_1 C_2}, \omega_{\text{zero}} = \frac{1}{R_2 C_2}. \tag{4.15}$$

The EC's gain coefficient before the zero frequency and the maximum gain coefficient are given as follows [85]:

$$A_{\text{toz.}} = \frac{g_m R_1}{1 + g_m R_2 / 2}, A_{\text{peak}} = g_m R_1. \tag{4.16}$$

Active element ICs are widely used in high-speed receiver systems.

In one known approach, it is proposed to add second-order low-frequency filters at the inputs and a high-frequency filter at the output for the active element EC (Fig. 4.28) [86].

The basis of the proposal is reliability improvement through distortion filtering, as a result of which the vertical and horizontal eye openings in the data ED at the system output increase (Fig. 4.29) [86].

As a result of adding filters, it was possible to reduce the effect of distortions at frequencies providing the maximum gain coefficient by increasing the quality factor. As a result of distortions, the quality factor decreases, which changes the amplitude–frequency response of the ECs (Fig. 4.30) [86].

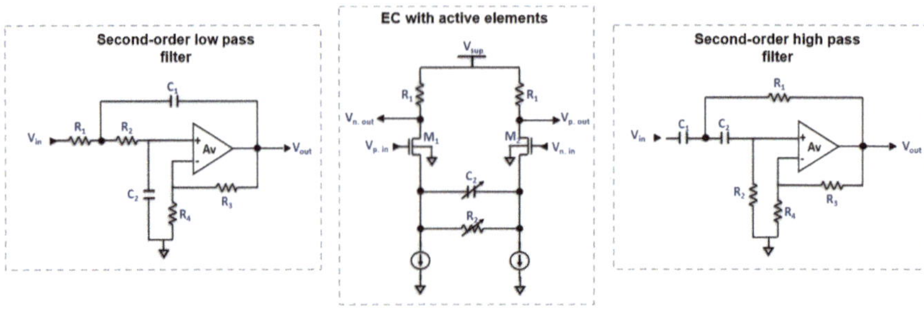

Fig. 4.28 Representation of EC using filters

Fig. 4.29 EDs obtained before and after filter application

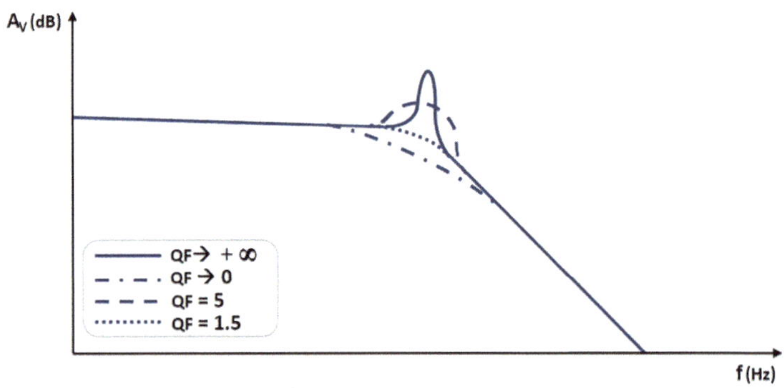

Fig. 4.30 Amplitude–frequency response of the EC for different quality factor values

Considering the impossibility of designing ideal elements in ICs, transistors replace ideal current sources and load resistances in active element ECs. Therefore, besides protection from distortions, it becomes necessary to apply methods for reducing aging phenomena. Experiments were conducted using simulation programs to evaluate the effect of 10-year aging phenomena for active element ECs with current sources and resistances replaced by transistors (Table 4.6) [87].

Table 4.6 Changes in parameters of the equalizer under different Process-voltage-temperature (PVT) conditions before and after 10-year aging

	Before 10-year aging		After 10-year aging	
PVT	Gain (dB)	Corresponding Freq (Hz)	Gain (dB)	Corresponding Freq (Hz)
SS	11,7	4,5	6,29	2,86
TT	12,5	5,1	6,5	3,16
FF	12,9	5,4	7,1	3,68

Thus, by applying known methods from the literature, it is possible to reduce the effects of phenomena that promote reliability degradation, which cause signal distortions, loss currents, and in some cases, even signal loss. However, as the results of 10-year aging of circuits playing an important role in ICs shown in the above tables indicate, it is necessary to develop methods to ensure long-term uninterrupted operation.

4.1.4 *Proposed Approaches to Increase Reliability of High-Speed Serial Data Receivers*

As mentioned in Sect. 4.1, the existing methods for improving the reliability of high-speed serial data reception systems do not sufficiently reduce the parametric degradation due to the effects of aging phenomena. Therefore, there is a need to develop new methods for improving reliability. Based on this, the following approaches have been proposed:

1. Designation of high-resistance points (HRP) in off mode in comparator circuits based on folded cascode OpAmps (Fig. 4.31).

As a result of HRP designation, the pins of transistors are connected to a constant potential during long-term off mode. Therefore, the potential difference between any two pins of transistors does not exceed the maximum allowable aging threshold and remains unchanged. As a result of applying the proposed approach, the parametric degradation due to potential differences between transistor pins is reduced, making it possible to obtain a larger gain and smaller offset voltage compared to the results obtained by applying the method mentioned above, at the cost of increased area. The area increase is due to the addition of transistors necessary for applying the proposed method.

2. Changing the load in circuits with different load structures (Fig. 4.32).

To address the issues described above, it is proposed to replace the cross-coupled NMOS transistors with diode-connected PMOS transistors. As a result of this change, the potential difference between any two pins of the transistors is reduced from $2*V_{sup}$ to V_{sup}. As a result, it becomes possible to reduce the parametric degradation of transistors in CP circuits due to the effects of aging phenomena. As a result

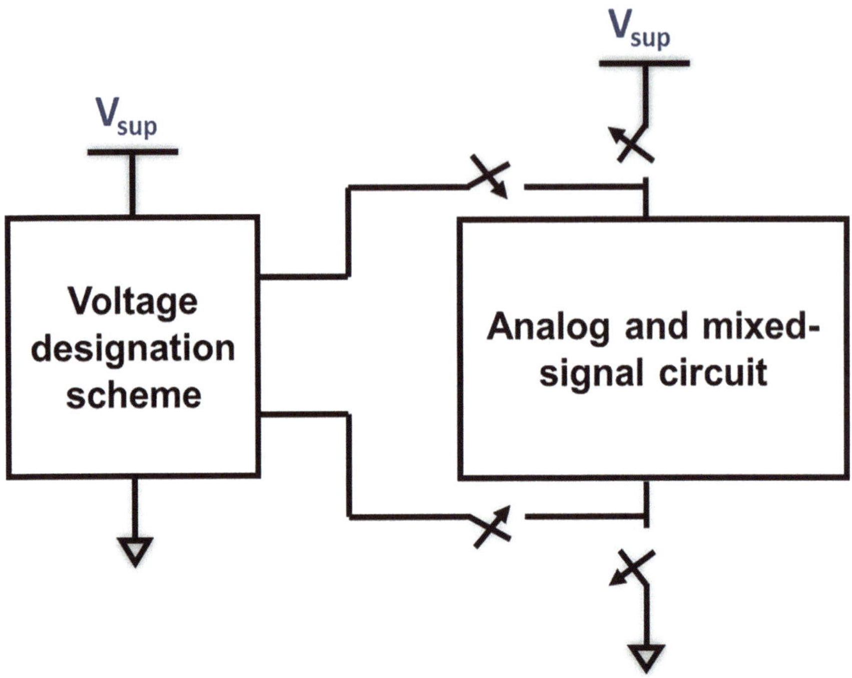

Fig. 4.31 Control of HRPs via voltage designation scheme

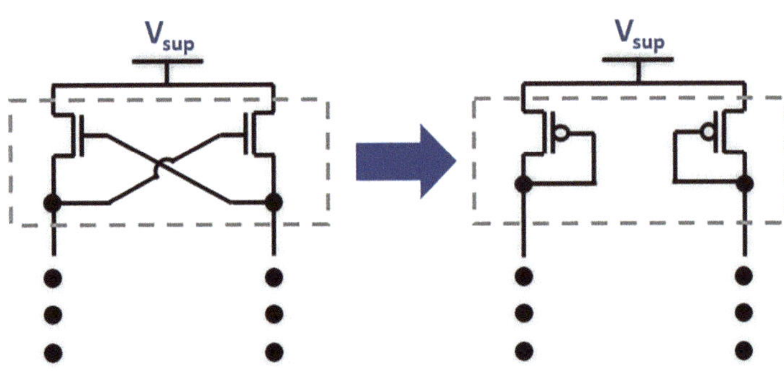

Fig. 4.32 Load structure with NMOS and PMOS transistors

of the proposed approach, it is possible to ensure 10 years of uninterrupted operation of the CP circuit in a smaller occupied area.

3. Adding negative capacitance (Neg-C) in the high-speed equalizer node and applying methods to reduce parametric degradation of transistors due to aging phenomena (Fig. 4.33).

The reliability improvement of the EC, based on distortion filtering, does not reduce the parametric degradation of transistors due to aging phenomena. Therefore, it is

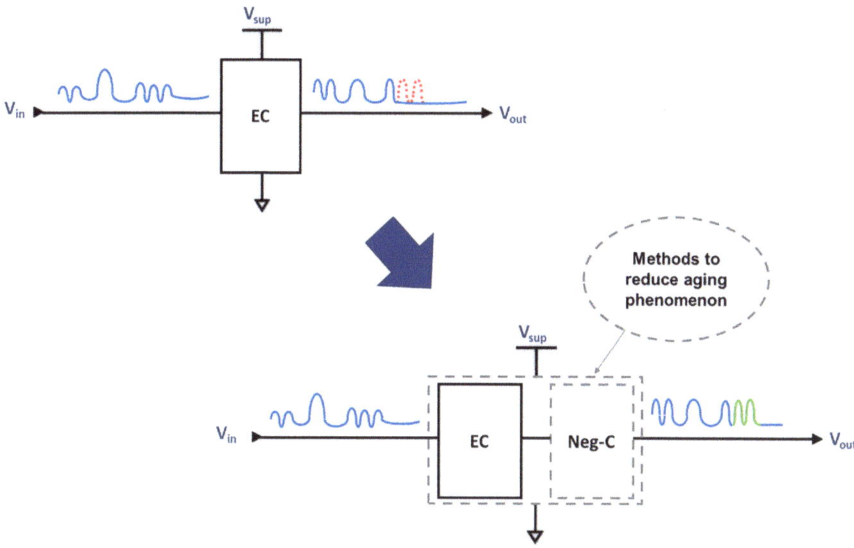

Fig. 4.33 Structure of combined use of EC and Neg-C

proposed to add Neg-C, with which it is possible to perform current division between the EC and Neg-C, preserving the maximum gain and reducing HCI aging. It is also proposed to apply three more methods for reducing the effects of aging phenomena in both the EC and Neg-C to reduce the parametric degradation of transistors when switching from long-term off mode to operating mode, to avoid SI failures.

Conclusions

1. Parallel to scaling of modern technological processes, reliability degradation due to the effects of aging phenomena has become the most important fundamental issue in the design of high-speed serial data reception systems. As a result of reliability degradation in such systems, signal distortions can appear, causing failures up to the point of stopping their operation. Therefore, the development of methods to reduce the effects of aging phenomena has become an extremely important task.
2. The existing methods to improve reliability of high-speed serial data reception systems that account for aging phenomena do not satisfy the solution of current problems, resulting in the need to develop new methods.
3. Approaches to reduce aging phenomena in circuits that are part of receiver nodes of integrated circuits have been proposed, which satisfy current requirements and, at the cost of increased occupied area and simulation time duration, provide necessary operating conditions for them.

4.2 Proposed Solutions to Increase Reliability of High-Speed Serial Data Receivers

The problems described in Sect. 4.1 highlight the lack or inadequacy of existing methods for mitigating aging effects of aging phenomena in highly sensitive circuits that are part of high-speed serial data receiver nodes. The requirements set for ensuring the long-term uninterrupted operation of modern ICs have become stricter, due to sharp changes in their usage environment—voltage and temperature—in various fields. Therefore, there is a need to develop new methods or solve the problems of existing ones, which will take into account the problems promoting reliability degradation caused by the effects of IC aging phenomena. Below are the proposed methods with consideration of aging phenomena in ICs for different PVTs, which solve the shortcomings present in existing methods, aligning them with current requirements.

4.2.1 Method to Reduce the Aging of Key Transistors When Switching to Operating State by Adjusting the Initial State of High-Resistance Points in Off Mode of Integrated Circuits

As mentioned in Sect. 4.1, the degradation of the main parameters of comparators occurs when switching from long-term off mode to operating mode. Therefore, the method of reducing aging of key transistors when switching to the operating state by adjusting the initial state of high-resistance points in off mode was developed and applied to prevent parametric degradation (Fig. 4.34) [75].

Special circuits were designed to condition HRPs. The first circuit consists of series-connected resistances and capacitances connected in parallel to them (Fig. 4.35) [88].

The potential required to condition the HRPs is obtained through the above circuit. However, directly connecting V_1 to the HRPs can lead to a change in that

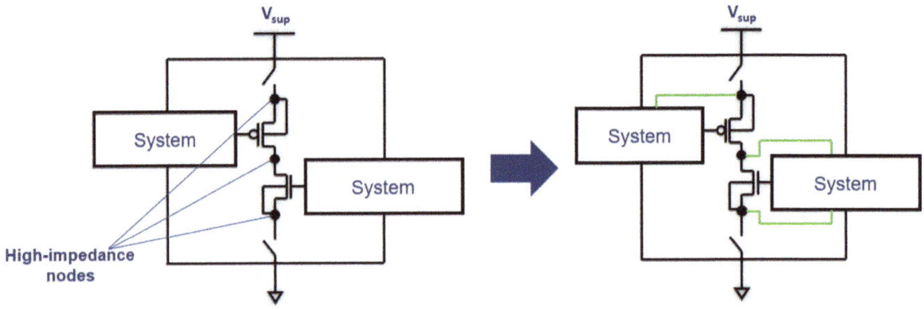

Fig. 4.34 Application of the proposed method

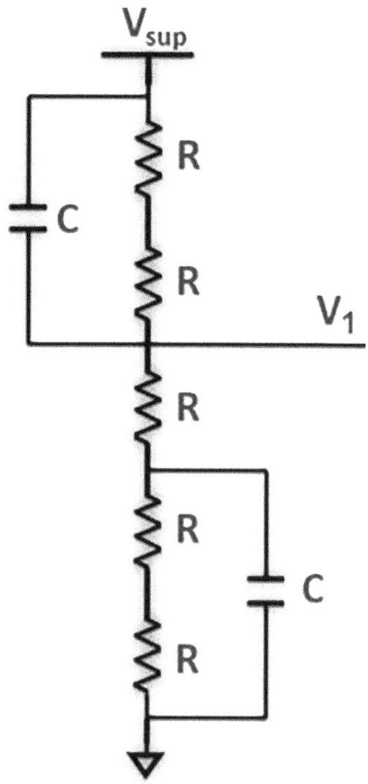

Fig. 4.35 Circuit for obtaining V_1 with resistances and capacitances

voltage, so a constant current generation circuit [89] and a circuit for obtaining V_2 from V_1 are designed (Fig. 4.36) [90].

Through the constant current generating circuit, transistor T_2 mirrors the current flowing through T1. Through a current mirror circuit, this current is transferred via T_3 to transistors T_4, T_8, T_5, and T_6. Since the V_1 potential is connected to the source of the diode-connected transistor T_7, the V_2 voltage will be one threshold voltage higher. To reduce this one threshold voltage, a source follower circuit is connected to the output of this circuit (Fig. 4.37) [91].

As a result, the obtained V_3 voltage value is one threshold voltage less than V_2 and equal to the V_1 voltage. The HRPs present in the circuits are conditioned by this voltage, which varies only with the supply voltage.

As a result of applying the above method in comparators based on folded cascode operational amplifiers, transistors D_1, D_2, D_3, and D_4 have been added to the base circuit, through which nodes A, B, and C are conditioned with the V_3 potential (Fig. 4.38) [75].

The high-resistance points present in the circuit are nodes A, B, C, because in off mode they are disconnected from all points of the circuit. Transistors T_{A1}, T_{A2}, T_{A3}, T_{A4}, and T_{A5} are added to the base circuit to ensure off mode. As a result of their addition, in off mode, transistors T_{12}, T_{11}, T_{10}, T_9, T_5, T_6, T_3, and T_4 are turned off, and T_7, T_8, T_1, and T_2 are connected to nodes A, B, and C [75].

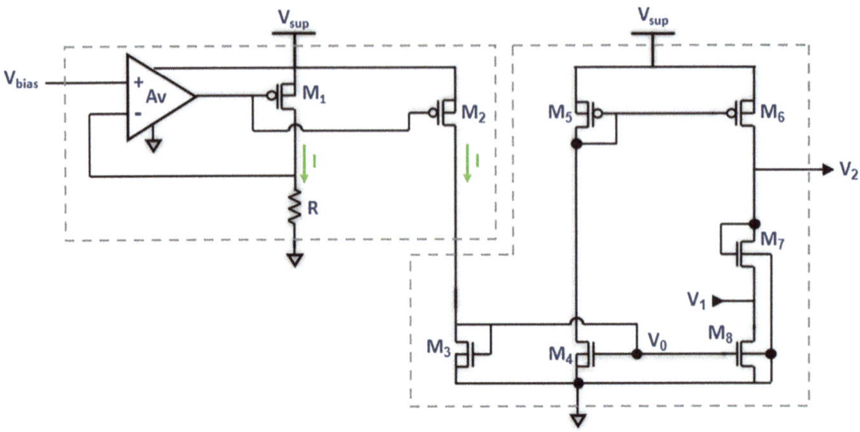

Fig. 4.36 Constant current generation and V_1 to V_2 obtaining circuit

Fig. 4.37 Source follower circuit

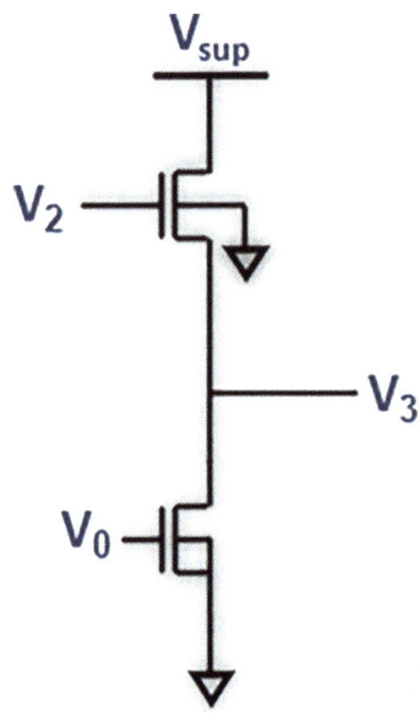

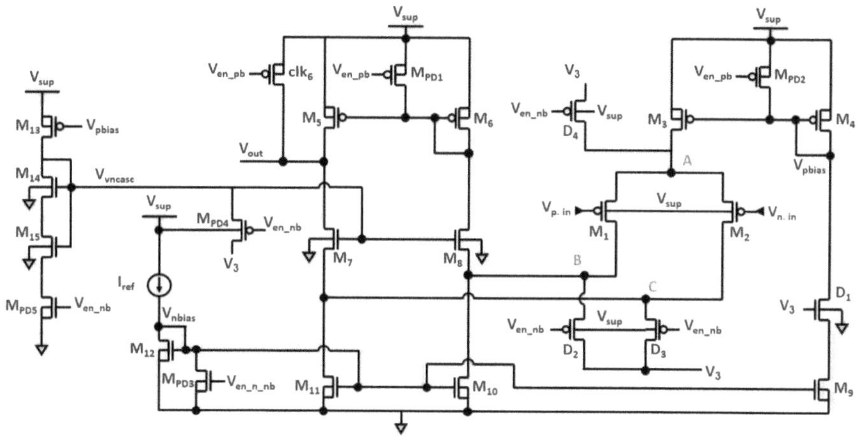

Fig. 4.38 Comparator circuit with application of the proposed method

Table 4.7 Maximum deviations of I_d and V_{th} of transistors in the comparator circuit under different operating modes as a result of 10-year aging

	Operating mode		Off mode	
Transistor name	ΔV_{th} (mV)	ΔI_d (%)	ΔV_{th} (mV)	ΔI_d (%)
T_7, T_8	22	2,9	0,1	0,1
T_9	18	2,6	0,1	0,1
T_1, T_2	14	1,3	0,1	0,1

Without this method, the potentials at these nodes are undefined, which can cause aging effects in the key transistors.

Using simulation tools, the changes in the threshold voltages and drain–source current of the key transistors due to the effects of aging phenomena were measured (Table 4.7) [75].

By mitigating the aging effects on these transistors, the changes in the comparator's key parameters gain, unit gain bandwidth (UGB), and offset voltage can be reduced. To evaluate the changes in the above parameters due to the effects of aging phenomena, the transfer function ($H(s)$) of this circuit was derived [91]:

$$H(s) = \frac{g_{m1} R_{out}}{1 + s R_{out} C_{out}}, \tag{4.17}$$

where g_{m1} is the transconductance of the input transistor, R_{out} and C_{out} are the output resistance and capacitance of the circuit, respectively, and s is the frequency component.

The R_{out} output resistance used in the above formula is calculated for a section of the circuit, since the circuit consists of two similar single-stage amplifiers [91]:

$$R_{\text{out}} = \left[(1 + g_{m8} \left(r_{o10} \| r_{o1} \right) r_{o8} + r_{o10}) \| r_{o6} \right], \tag{4.18}$$

where: g_{m8} is the transconductance of transistor T_8, r_{o10}, r_{o1}, r_{o8}, and r_{o6} are the channel resistances of transistors T_{10}, T_1, T_8, and T_6, respectively.

From (4.17), it can be seen that at low frequencies, the operational amplifier's gain is given by the following expression [91]:

$$A_V = g_{m1} \left[(1 + g_{m8} \left(r_{o10} \| r_{o1} \right) r_{o8} + r_{o10}) \| r_{o6} \right]. \tag{4.19}$$

As a result of aging, the g_m of transistors T_1 and T_2 decreases due to the change in threshold voltages. For the same reason, the transconductances and channel resistances of T_8 and T_7 also decrease. As a result of the change in resistances and transconductances, the output resistance also decreases, which causes a decrease in UGB. UGB is determined by the following expression [91]:

$$\text{UGB} = g_{m1} / C_{\text{out}}. \tag{4.20}$$

The output capacitance contains the drain-bulk and gate-drain capacitances of transistors T_5, T_7, T_{A6}. As a result of the effects of aging phenomena, these capacitances decrease. They decrease due to the increase in the depletion region in the drain area under the influence of the HCI phenomenon. Amplitude–frequency simulations were performed to evaluate the changes in the above parameters at the beginning of operation and after 10 years of aging (Fig. 4.39) [75].

With the applied method, after 10 years of aging, the gain decreased by 2.7% and the UGB by 18.6%. Another important parameter of comparators is the input offset voltage. With the application of the proposed method, the source and drain pins of T_1 and T_2 were connected to the same potential, as a result of which the gate–drain, gate–source, and gate-bulk voltages did not change during long-term off mode. Therefore, in case where the input voltages are connected to the same potential in off mode, the offset voltage between them is equal to 1 mV when switching to operating mode.

And in case of maximum difference in input voltages, when the supply voltage is applied to the input of one transistor and the other's gate is grounded, the offset voltage becomes equal to 2 mV. The input offset voltage values were obtained as a result of Monte-Carlo [92] simulation (Figs. 4.40 and 4.41).

With the application of the proposed method, the physical design of the circuit increased by 5.1% compared to the base (Figs. 4.42 and 4.43; Table 4.8) [75].

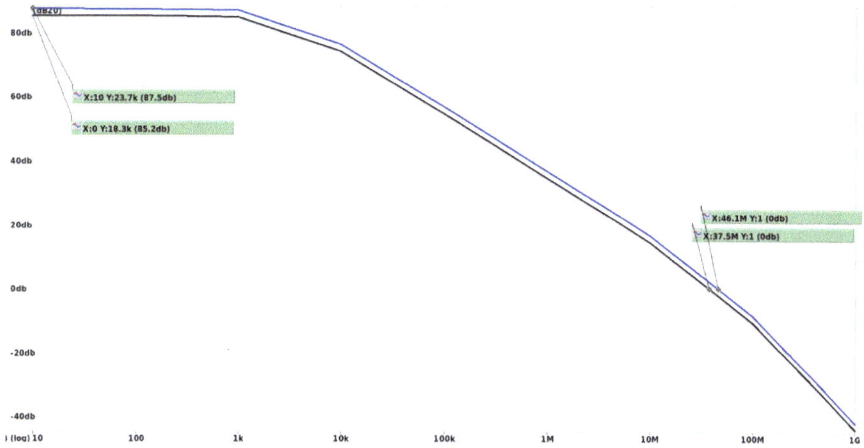

Fig. 4.39 Up to 10 years of aging and the resulting amplitude–frequency response of the comparator

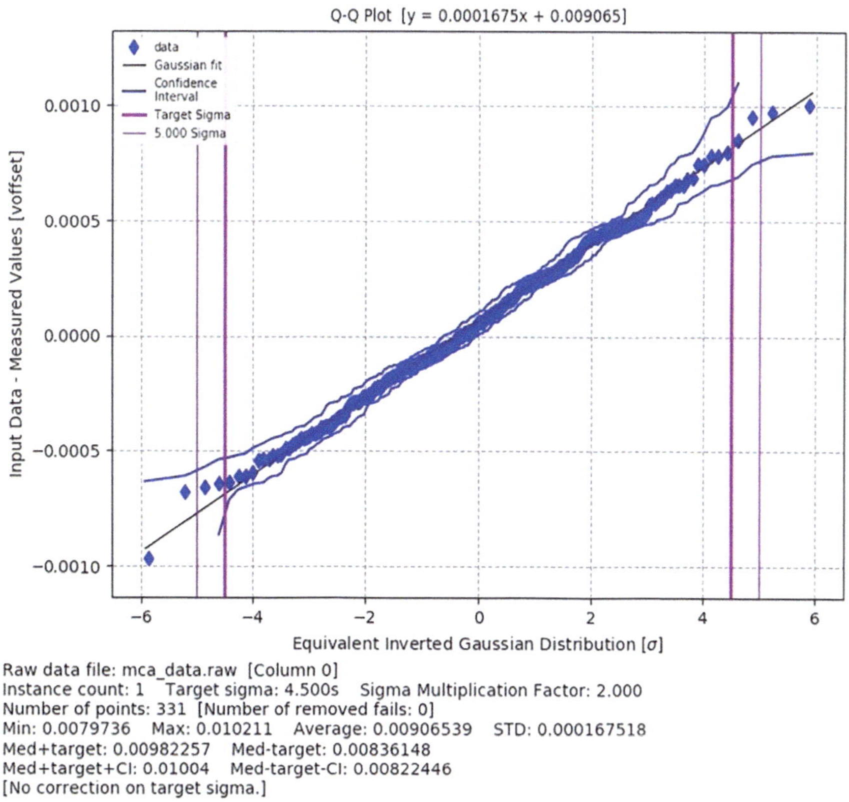

Fig. 4.40 Distribution of offset voltage for the comparator with the same input voltages

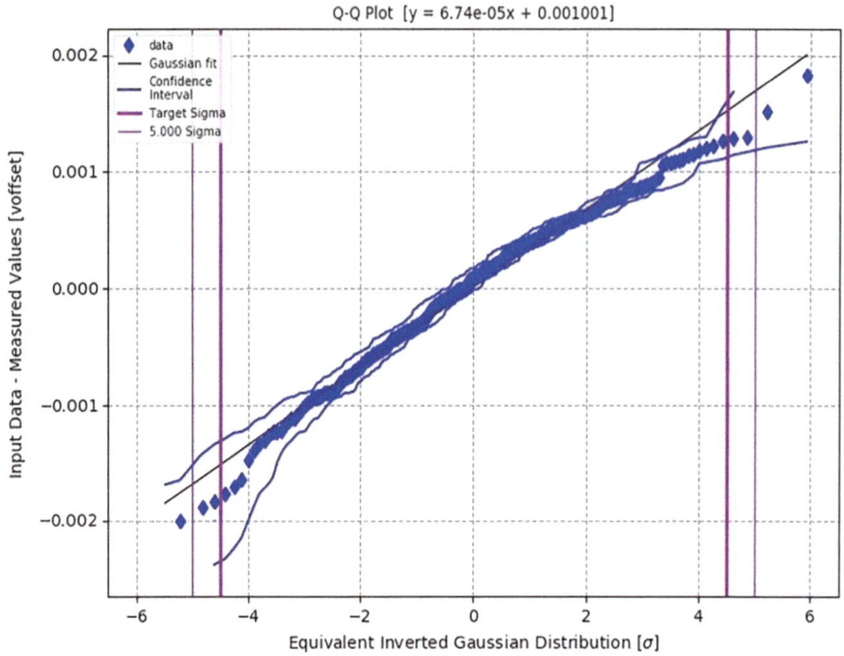

Fig. 4.41 Distribution of offset voltage for the comparator with the largest difference in input voltages

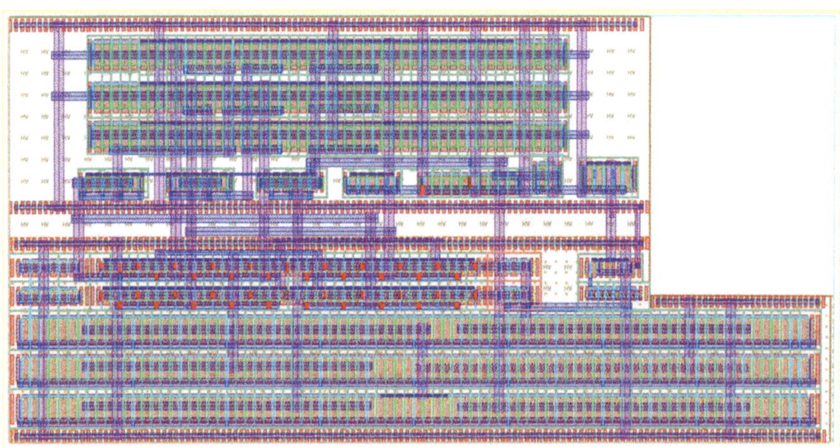

Fig. 4.42 Physical design of the base comparator

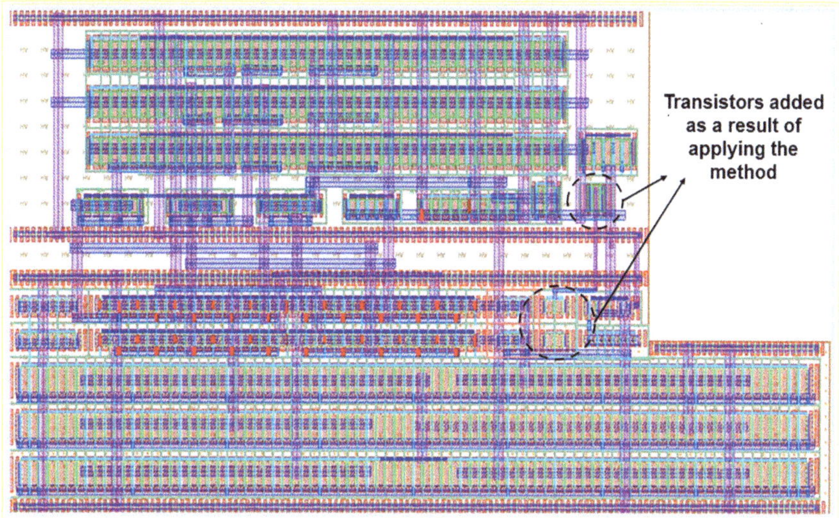

Fig. 4.43 Physical design of the comparator with application of the proposed method

Table 4.8 Changes in parameters of the comparator under different PVT conditions before and after 10-year aging

Parameter	Proposed circuit	Existing circuit	Base circuit
Maximum threshold voltage deviation (mV)	22	0,01	273
Offset voltage (mV)	2	3	35
Gain (dB)	85,2	80,01	76,3

In summary, the proposed method achieved a gain of 85.2 dB after 10-year aging simulation, representing an increase of 6.4% and 11.6% compared to the existing and base circuits, respectively. The physical design area increased by 5.1% compared to the base, which is 0.28% larger than the area of the existing circuit. With the application of the method, the offset voltage decreased by 1.5 times compared to the existing circuit, and by 17.5 times compared to the base.

4.2.2 Method to Increase System Reliability by Changing the Load in Circuits with Different Load Structures

As mentioned above, in CP circuits, deteriorations of transistor parameters occur due to the effects of aging phenomena. Due to them, for years, as a result of simulating an output load requiring up to 500 μA current, functional failures occur in circuits, which manifest as incomplete formation of output signals under different PVT conditions (Fig. 4.44).

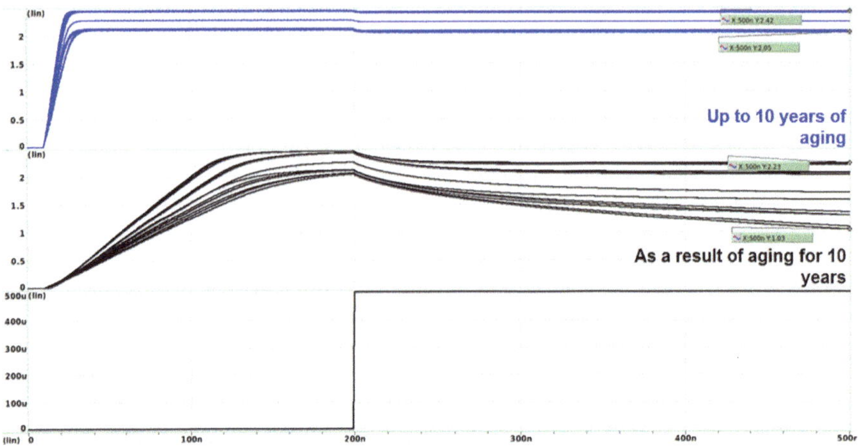

Fig. 4.44. Results of timing simulation of the base CP circuit

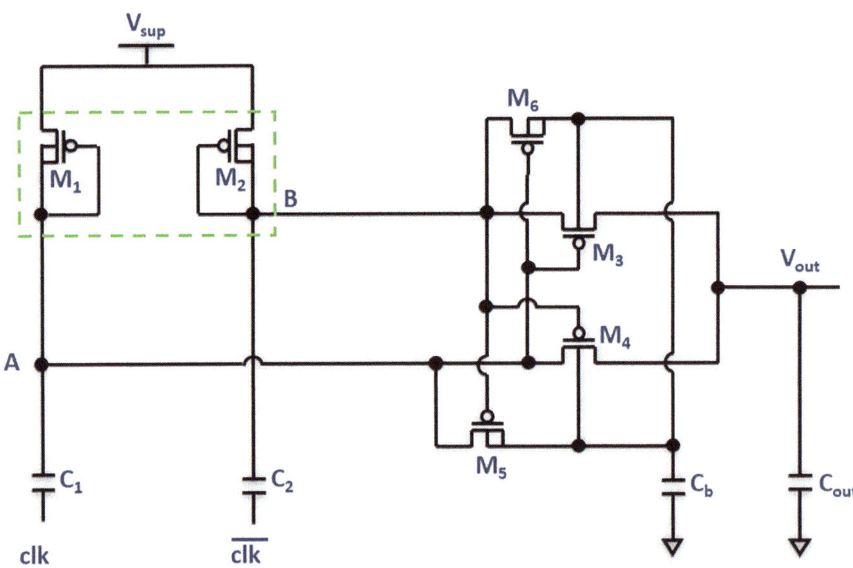

Fig. 4.45 CP circuit with application of the proposed method

To avoid such failures, it is proposed to change the circuit architecture according to the load structure. In the proposed circuit, the cross-coupled NMOS transistors are replaced with diode-connected PMOS type of transistors (Fig. 4.45) [93].

This change ensures that the potential difference between any two terminals of the PMOS transistors does not exceed $2*V_{sup}$, thereby staying within the maximum allowable limit. As a result of 10-year aging, the change in the threshold voltages and drain currents of the transistors does not exceed the allowable 10% (Fig. 4.46).

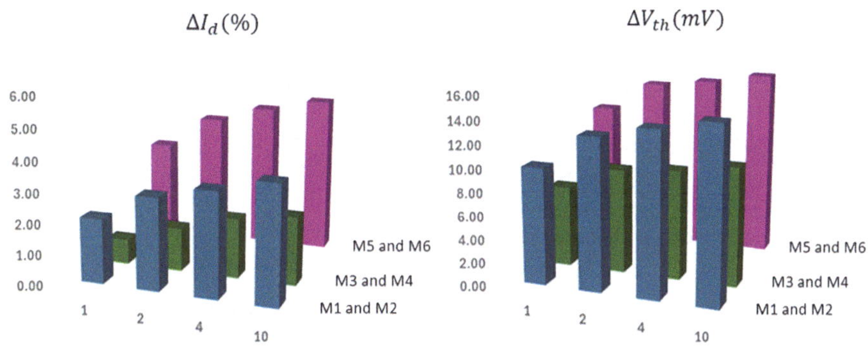

Fig. 4.46 Maximum deviations of I_d and V_{th} of transistors as a result of 10-year aging of the CP circuit with the proposed method applied

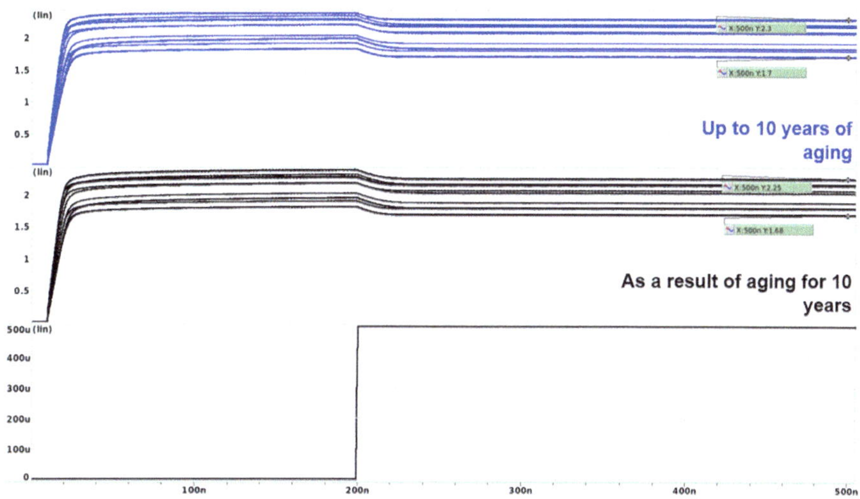

Fig. 4.47 Results of timing simulation of the CP circuit with application of the proposed method

From the results of 10-year aging simulation, it can be seen that parametric degradation is mainly observed in T_5 and T_6 transistors. They perform a switch role, which open every half-cycle and charge the C_{bulk} capacitance up to $V_{sup} - V_{th}$. These transistors are intended to keep the bulk potentials of T_3 and T_4 transistors constant and to avoid additional drain–source voltage drop [93].

With the application of the proposed method, with simulation of an output load consuming up to 500 μA current, no CP circuit operation failures occurred under different PVT conditions (Fig. 4.47).

A parameter comparison was performed for CP circuits with different load architectures at the beginning of operation and as a result of 10-year aging (Table 4.9).

From the results it follows that both before load connection and after load connection, the output voltage of the PMOS architecture CP circuit is significantly

Table 4.9 Changes in parameters of CP circuits with different load architectures at the beginning of operation and as a result of 10-year aging

Parameters under different PVT conditions	Before 10-year aging		After 10-year aging	
	NMOS Arch	PMOS Arch	NMOS Arch	PMOS Arch
Maximum output voltage value, before load connection (V)	2,57	2,38	2,36	2,17
Minimum output voltage value, before load connection (V)	2,09	1,83	2,02	1,79
Maximum output voltage value, after load connection (V)	2,42	2,3	2,23	2,25
Minimum output voltage value, after load connection (V)	2,05	1,7	1,03	1,68
Difference between max and min output voltages, before load connection (V)	0,34	0,38	0,36	0,38
Difference between max and min output voltages, after load connection (V)	0,37	0,6	1,2	0,57
Maximum output setup time (ns)	16,7	13,1	149,2	14,1

lower than that of NMOS, in case before 10-year aging. This is due to the differences in the operating principles of these circuits.

In the proposed circuit, the supply voltage is transferred to nodes A and B through two PMOS diode-connected transistors. Therefore, a potential one threshold voltage ($V_{\text{p.th}}$) less than the supply voltage is formed at those nodes.

Then, through two anti-phase clks, an additional potential equal to the supply voltage is added to nodes A and B every half-cycle. In parallel, the circuit output is connected to nodes A and B through transistors T_3 and T_4. However, transistors T_3 and T_4 also form a voltage drop (V_{ds}) between their source–drain pins due to channel resistance [93].

Thus, the output voltage formation of CP circuit with PMOS architecture can be defined by the following expression:

$$V_{\text{out}} = 2V_{\text{sup}} - V_{\text{p.th}} - V_{\text{sd}}. \tag{4.21}$$

In CP circuits with NMOS architecture, the supply voltage is transferred to nodes A and B through two cross-coupled transistors, as a result of which, due to clk switching, $2{*}V_{\text{sup}}$ voltage is formed at nodes A and B. Therefore, the output voltage formation of the CP circuit with NMOS architecture will be defined by the following expression:

$$V_{\text{out}} = 2V_{\text{sup}} - V_{\text{sd}}. \tag{4.22}$$

From the obtained results, it becomes clear that the setup time of the output of the CP circuit with PMOS architecture is significantly smaller compared to NMOS. Therefore, the rise rates also differ (Figs. 4.48 and 4.49).

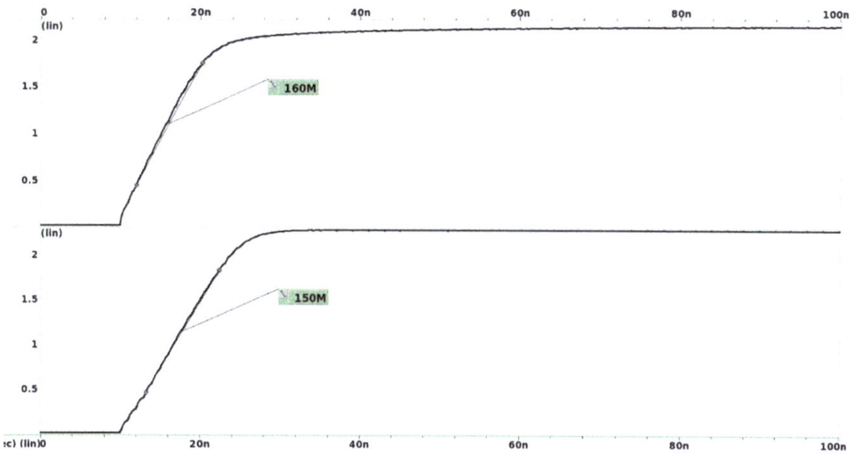

Fig. 4.48 Output rise rates of circuits with base and proposed method application, before 10-year aging

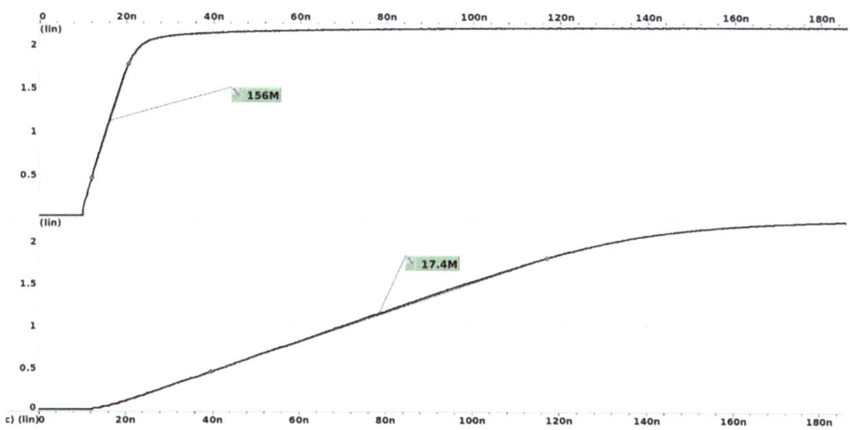

Fig. 4.49 Output rise rates of circuits as a result of 10-year aging with base and proposed method application

This is due to the connection of PMOS transistor bulks. In the proposed circuit, the bulks of T_1 and T_2 transistors are connected to nodes A and B. Therefore, during charging half-cycle, a forward-biased diode is formed between the source and bulk in T_1 and T_2 transistors, which contributes to the fast charging of C_1 and C_2 capacitances. And in CP circuits with NMOS architecture, the source-bulk and drain-bulk diode connections of cross-coupled transistors are always reverse-biased. With their grounded bulks, during charging half-cycle, leakage currents flow, which prolong the setup time of the CP circuit output (Fig. 4.50).

As a result of applying the proposed method, to reduce the effective value of the threshold voltages of PMOS transistors, the sizes of T_1 and T_2 transistors were

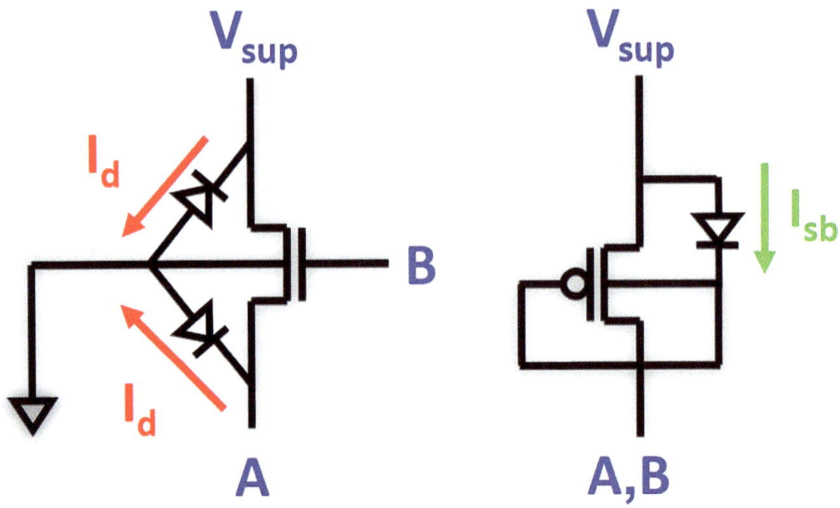

Fig. 4.50 Leakage currents of transistors due to load change in the CP circuit

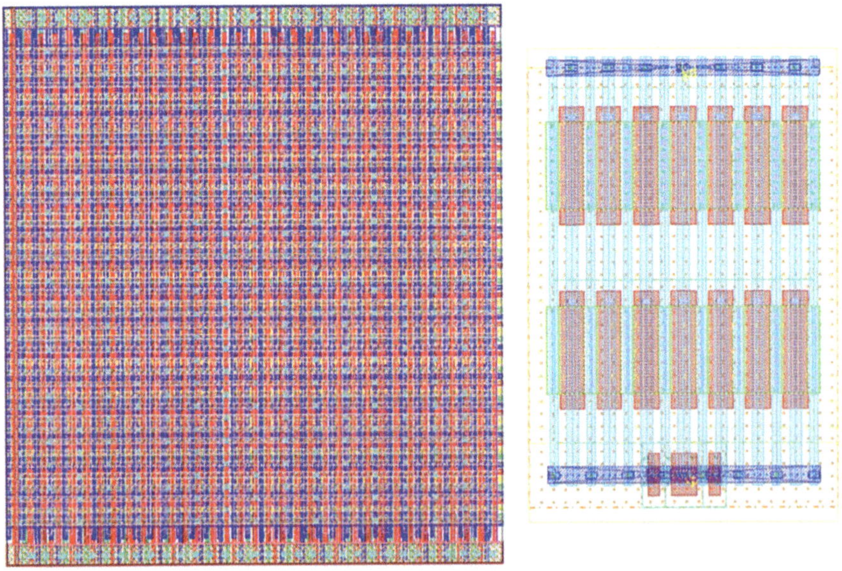

Fig. 4.51 Physical designs of MOM and MOS capacitances

increased by three times compared to NMOS architecture CP circuits. As a result, physical dimensions of CP circuit with PMOS architecture increased by 5.2% compared to NMOS. Since there are strict area limitations in modern ICs, it is proposed to replace the metal-oxide-metal (MOM) [94] capacitances present in the circuit with MOS capacitances (Fig. 4.51).

Unlike MOM, MOS capacitances are nonlinear and are voltage-dependent (Fig. 4.52) [95].

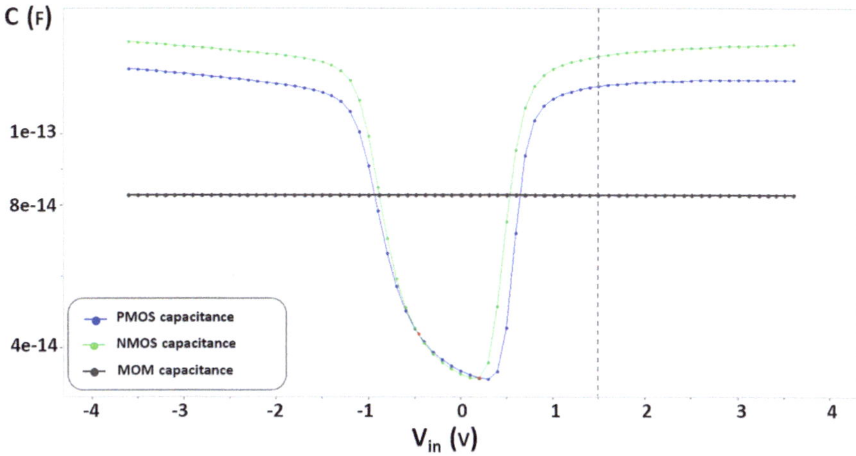

Fig. 4.52 Volt-farad characteristics of PMOS, NMOS, and MOM capacitances

In the stable operating state of the CP circuit, the voltage drops across the capacitances are close to V_{sup}. From the volt-farad characteristic, it can be seen that in the required range, the MOS capacitance value is saturated and exceeds MOM by 1.5 times. As a result, after replacing the capacitances, the area of the proposed circuit will decrease by 26.9% (Fig. 4.53).

Thus, as a result of applying the proposed method, thanks to the designed architecture, it became possible to neutralize the output voltage drop of more than 1 V and the nine-fold increase in output setup time due to aging, at the cost of obtaining a value one threshold voltage less at the circuit output in the main operating state. The area of the CP circuit also decreased by 26.9%, due to the replacement of metal-oxide-metal capacitances with the metal-oxide-semiconductor type.

4.2.3 *Method to Reduce Aging and Increase Reliability in High-Speed Equalizers*

As previously stated, improving the reliability of active equalizer circuits requires not only minimizing distortion but also developing methods to mitigate agingAs a first step, it is proposed to design a Neg-C for the active element EC (Fig. 4.54) [87, 96].

The simplest Neg-C circuit consists of cross-coupled transistors T_1 and T_2, a capacitance connected between the two branches, and current sources. Through the two cross-coupled transistors of thc Ncg-C circuit, it is possible to obtain a negative complex impedance value. The expression for the negative complex impedance (Z) is given [87, 96]:

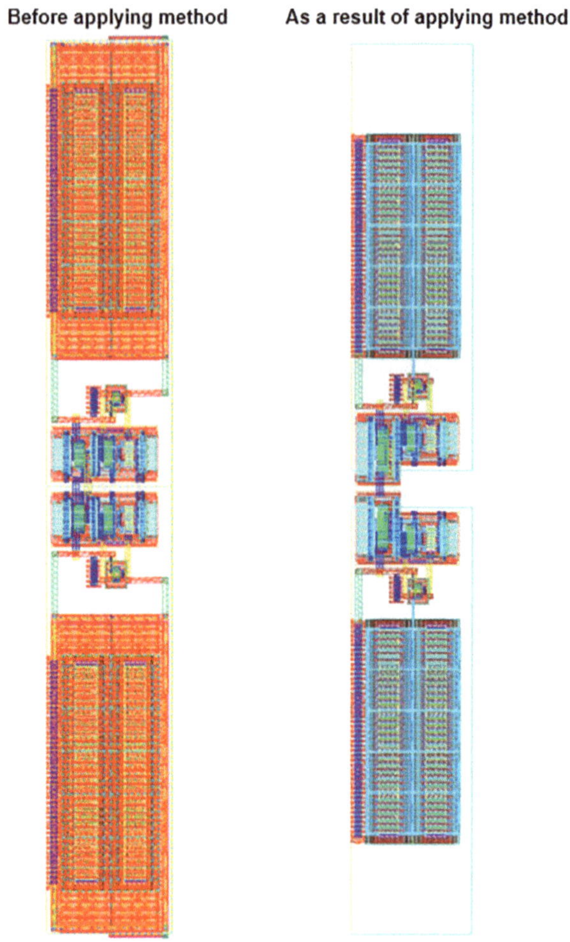

Fig. 4.53 Physical designs of the CP circuit with base and proposed method application

$$Z = -\frac{1}{sC} * \frac{g_m + s\left(C_{sd} + 2C\right)}{g_m - sC_{sd}}, \tag{4.23}$$

where g_m and C_{sd} are the transconductance and source–drain capacitance of transistors, and C is the capacitance connected between two branches.

By changing the value of the current sources in the Neg-C, it is possible to control g_m, as a result of which the negative complex impedance value will change. In the overall system, increasing the negative complex impedance value of the Neg-C allows increasing the bandwidth, the constant region, and the maximum gains of that system (Fig. 4.55) [87, 96].

The basis of the proposal is the reduction of the HCI aging phenomenon effect. As a result of designing the Neg-C, to obtain the same maximum gain, the currents flowing through the EC will be divided between the Neg-C and the EC. Therefore,

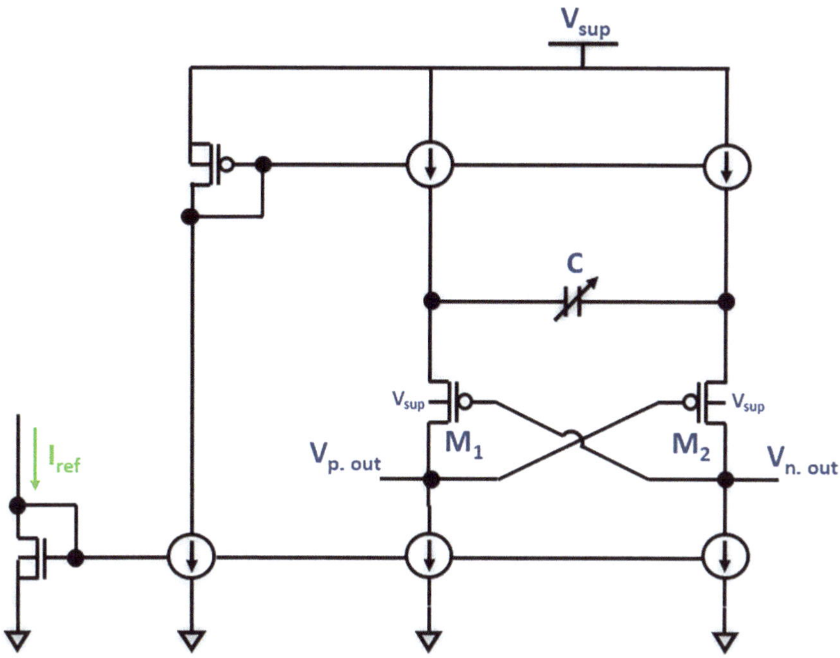

Fig. 4.54 Simplest negative capacitance circuit

Fig. 4.55 Amplitude–frequency response of the overall system

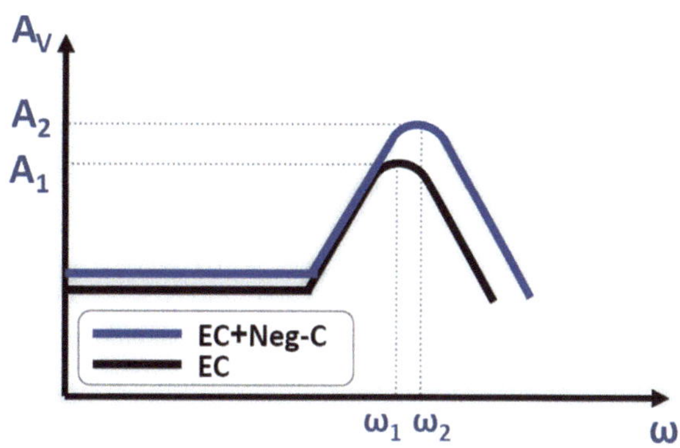

before the Neg-C design, the current flowing through the EC and the currents flowing through the overall system will be equal. With this approach, the degradation of transistor parameters during long-term operation is reduced. To solve the problems of parametric degradation of transistors when switching from long-term off mode to operating mode, it is proposed to apply several methods for reducing aging phenomena for both the EC and the Neg-C [87].

The first method involves conditioning all high-resistance points [75]. According to the second method, the input transistors are isolated [87]. According to the third

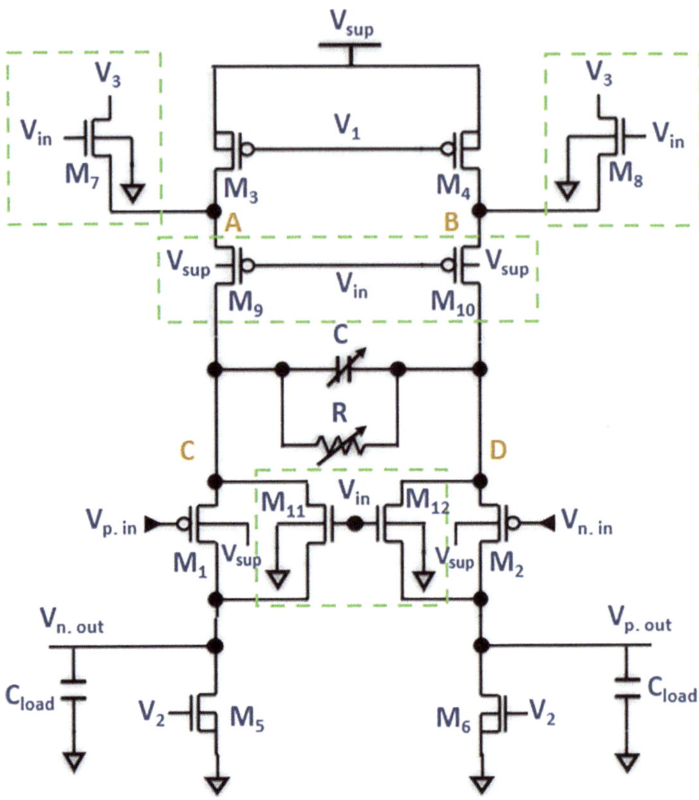

Fig. 4.56 EC with application of the proposed method

method, transistors are added to the circuits that serve as switches in off mode [97, 98]. By applying the above methods, it is proposed to add transistors T_7, T_8, T_9, T_{10}, T_{11}, and T_{12} to the EC circuit (Fig. 4.56) [87].

According to the first method, T_7 and T_8 have been added to condition the A and B high-resistance points. They protect transistors T_3 and T_4, which act as current sources. According to the second method, T_{11} and T_{12} have been added to isolate the T_1 and T_2 input transistors. According to the third method, T_8 and T_9 have been added to disconnect node A from C and B from D, to avoid loss currents. Following the same logic, the application of aging reduction methods is proposed to be performed in the Neg-C (Fig. 4.57) [87].

As a result of applying the proposed methods, for the obtained circuits, using modeling tools, the changes in the threshold voltages and drain–source current of the key transistors due to the effects of aging phenomena were measured (Tables 4.10 and 4.11).

According to (4.14) and (4.23), as a result of protecting the EC and Neg-C input transistors from parametric degradation, it was possible to reduce the change in both the constant region and the maximum gain coefficients of the overall system. In the

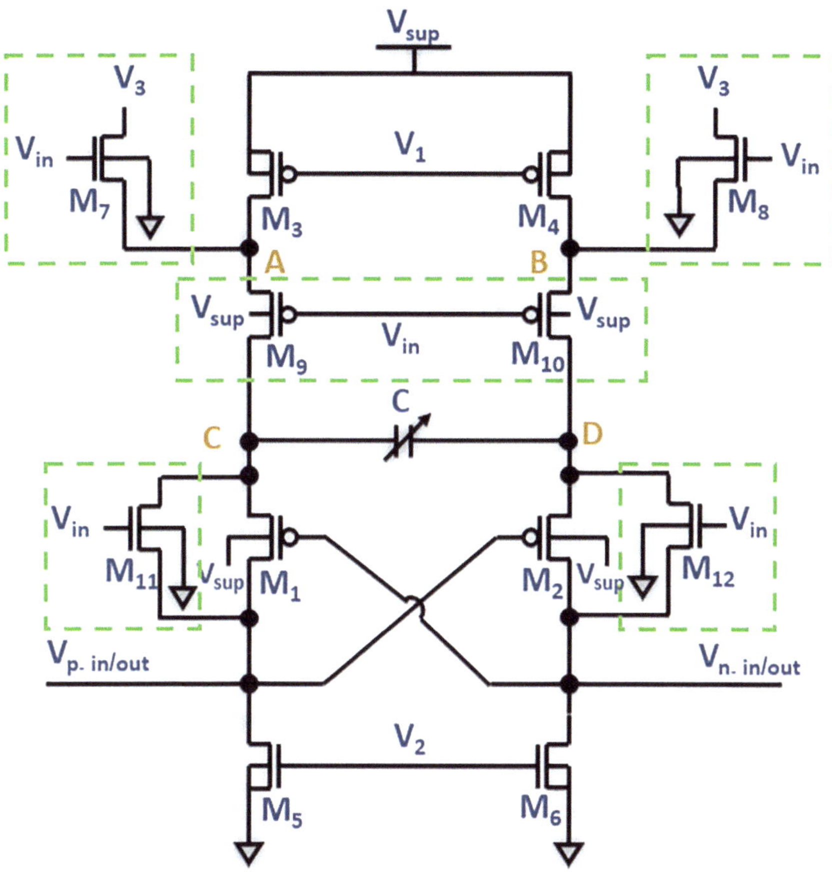

Fig. 4.57 Neg-C with application of the proposed method

Table 4.10 Maximum deviations of I_d and V_{th} of transistors as a result of 10-year aging of the EC

	Before applying proposed methods		As a result of method application	
Transistor name	ΔV_{th} (mV)	ΔI_d (%)	ΔV_{th} (mV)	ΔI_d (%)
T1 and T2	231	22,1	16	0,84
T3 and T4	103	13,3	9	0,53
T5 and T6	207	18,5	11	0,61

EC, as a result of protecting transistors T_5 and T_6 from aging, the changes in the zero and cutoff frequencies have also decreased, according to (4.15) (Fig. 4.58) [87].

Since the EC and Neg-C are composed of differential pairs, it is necessary to follow a special procedure during their physical design. The importance of following the procedure is that the external effects on the semiconductor elements of two branches of circuits are symmetrical. This allows keeping the circuit free from offset voltages and distortions. To maintain symmetry of external effects, matching

Table 4.11 Maximum deviations of I_d and V_{th} of transistors as a result of 10-year aging of the Neg-C

	Before applying proposed methods		As a result of method application	
Transistor name	ΔV_{th} (mV)	ΔI_d (%)	ΔV_{th} (mV)	ΔI_d (%)
T1 and T2	243	23,6	18	0,89
T3 and T4	104	13,3	10	0,57
T5 and T6	210	18,6	13	0,72

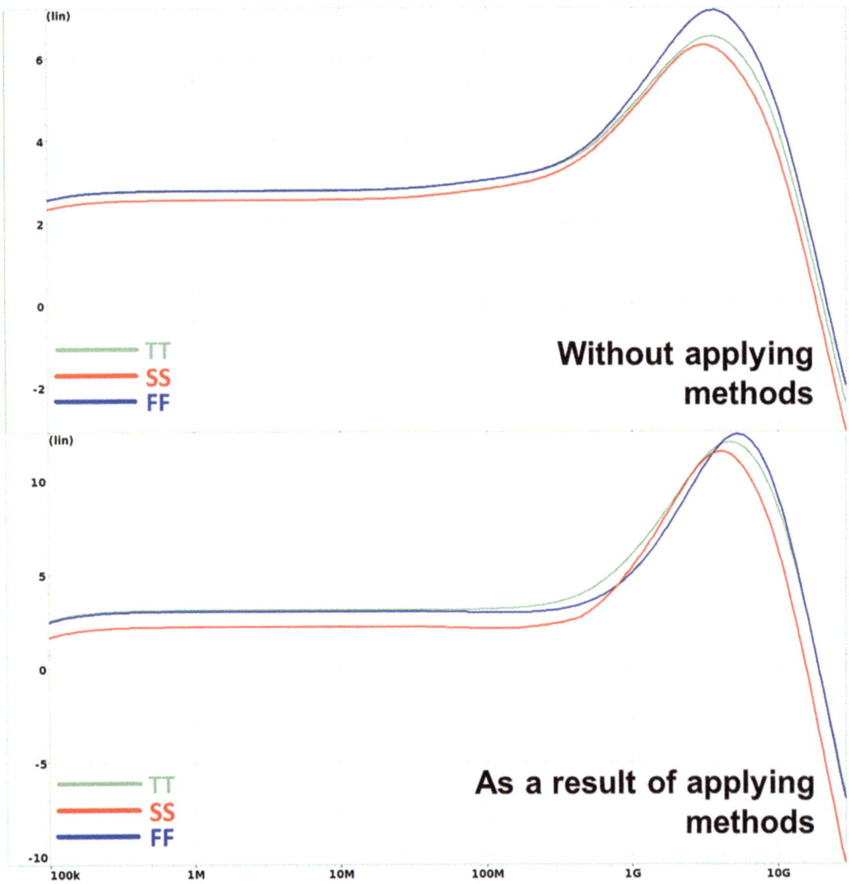

Fig. 4.58 Amplitude–frequency response of the overall system without applying the proposed methods and as a result of their application

method is used during the physical design of ICs [99]. As a result of applying this method, the process variation size and the changes caused during lithography, as well as the effect of temperature distribution, are reduced for two symmetrical branches. The lengths and number of interconnects also change.

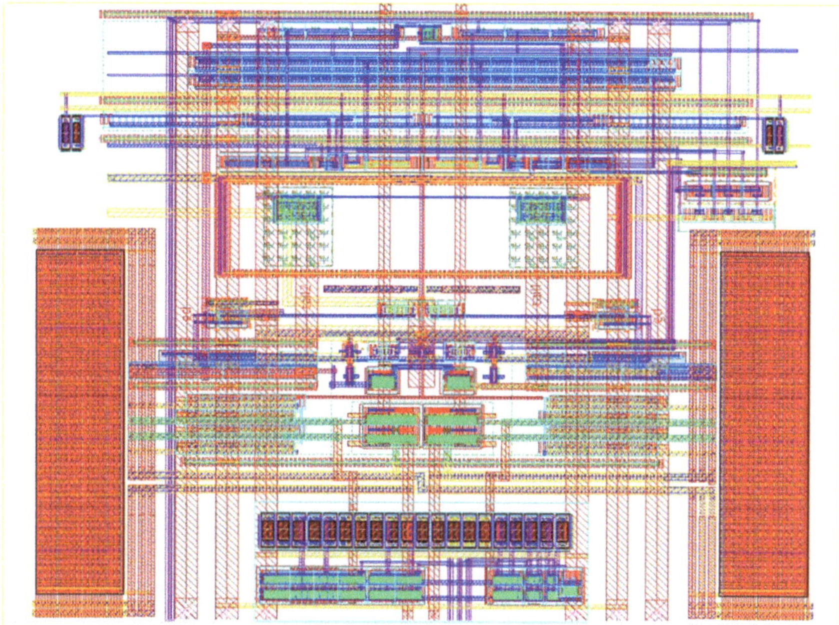

Fig. 4.59 Physical design of the overall system before applying the proposed methods

Therefore, as a result of the physical design, the matching method was applied for both transistors and resistances and capacitances, as a result of which the area of the overall system, due to the application of aging reduction methods, increased by 13% before method application (Figs. 4.59 and 4.60).

With the application of the proposed methods, as a result of 10-year aging, it was possible to obtain a minimum value of the maximum gain of 11.5 dB for different PVTs, instead of the previous 6.29 dB (Table 4.12) [87].

Thus, as a result of 10-year aging of the proposed overall system, it was possible to reduce the change in the maximum gain and frequency from 48% to 3.8% and from 36.4% to 4.2%, respectively, at the cost of a 13% area increase.

Conclusions

1. A method to reduce the effect of aging phenomena in comparators based on folded cascode operational amplifiers has been proposed, which, through the addition of voltage-conditioning transistors, increases the gain compared to the existing and base circuits by 6.4% and 11.6%, respectively, and a decrease in offset voltage by 1.5 and 17.5 times, under the condition of an increase in occupied area by 0.28% and 5.1%.
2. A method to reduce the effect of aging phenomena in the charge pump circuit has been created, which, due to a change in the circuit's load structure, neutralized the output voltage drop of more than 1 V and the nine-fold increase in output settling time, at the cost of a three-fold increase in input transistor size. As a

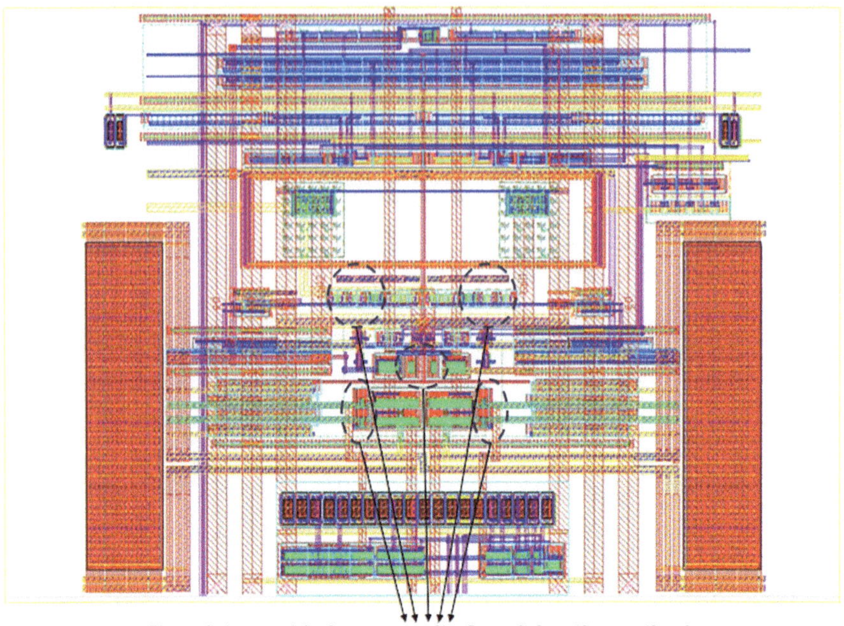

Fig. 4.60 Physical design of the overall system with application of the proposed methods

Table 4.12 Changes in parameters of the overall system under different PVT conditions before and after 10-year aging

Process	Before 10-year aging		After 10-year aging	
	Max gain (dB)	Frequency (GHz)	Max gain (dB)	Frequency (GHz)
SS	11,7	4,5	11,5	4,36
TT	12,5	5,1	12,1	4,99
FF	12,9	5,4	12,4	5,17

result of the method's application, the occupied area decreased by 26.9%, due to the replacement of metal-oxide-metal capacitances with the metal-oxide-semiconductor type.

3. A method to reduce the effect of aging phenomena in the equalizer circuit has been developed, which, due to the addition of a negative capacitance circuit, reduced the effect of the HCI aging phenomenon, ensuring a reduction in the changes of the overall system's maximum gain coefficient and frequency to 3.8% and 4.2%, respectively, at the cost of a 13% area increase.

References

1. H. Jin, The history, current applications and future of integrated circuit. Highlights Sci. Eng. Technol. **31**, 232–238 (2023). https://doi.org/10.54097/hset.v31i.5146

2. Z. Li, Analysis of the development history, key technologies, and future development of integrated circuits. Highlights Sci. Eng. Technol. **81**, 313–317 (2024). https://doi.org/10.54097/bbqvc047
3. S. Kim, H. Park, E. Choi, Y.H. Kim, et al., Reliability assessment of 3 nm GAA logic technology featuring multi-bridge-channel FETs, in *2023 IEEE International Reliability Physics Symposium (IRPS). Monterey, CA, USA*, (IEEE, 2023), pp. 1–8. https://doi.org/10.1109/IRPS48203.2023.10117953
4. S. Sutardja, 1.2 The future of IC design innovation, in *IEEE International Solid-State Circuits Conference (ISSCC) Digest of Technical Papers. San Francisco, CA, USA*, (IEEE, 2015), pp. 1–6. https://doi.org/10.1109/ISSCC.2015.7062847
5. S. Abdennadher, I.O. High-Speed, IP process technology & design interactions: problems & solutions, in *2023 IEEE International Conference on Design, Test and Technology of Integrated Systems (DTTIS). Gammarth, Tunisia*, (IEEE, 2023), pp. 1–6. https://doi.org/10.1109/DTTIS59576.2023.10348358
6. H.X. Jiang, X.H. Zhang, W. Zhao, J.Y. Lu, Study on signal integrity simulation for high-speed serial rapid IO. Appl. Mech. Mater. **340**, 881–886 (2013). https://doi.org/10.4028/www.scientific.net/AMM.340.881
7. M. Ehlert, Different approaches of high-speed data transmission standards. Adv. Radio Sci. **2**, 187–200 (2004). https://doi.org/10.5194/ars-2-187-2004
8. N. Wirth, Serial data transmission, in *Digital Circuit Design for Computer Science Students*, (Springer, Berlin, Heidelberg, 1995), pp. 147–148. https://doi.org/10.1007/978-3-642-57780-2_12
9. R. Dobkin, A. Morgenshtein, A. Kolodny, R. Ginosar, Parallel vs. serial on-chip communication, in *International workshop on system level interconnect prediction (SLIP)*, (ACM, 2008), pp. 43–50. https://doi.org/10.1145/1353610.1353620
10. G. Katz, S. Zlotsky, I. Vinitzky, A. Babecovand, et al., Inclusive performance analysis of 100 Gbps PAM-4 at SerDes using digital equalizers, in *2024 Asian Conference on Communication and Networks (ASIANComNet). Bangkok, Thailand*, (IEEE, 2024), pp. 1–4. https://doi.org/10.1109/ASIANComNet63184.2024.10811094
11. P. Aziz, A. Healey, C. Liu, et al., SerDes design and modeling over 25+ Gb/s serial link. Int. Symp. Microelectron. **2011**, 000044–000060 (2011). https://doi.org/10.4071/isom-2011-TA2-Paper1
12. M.A. Dolatsara, Equalization optimization for SerDes channels with constrained bayesian optimization, in *2022 IEEE International Symposium on Electromagnetic Compatibility & Signal/Power Integrity (EMCSI). Spokane, WA, USA*, (IEEE, 2022), pp. 293–293. https://doi.org/10.1109/EMCSI39492.2022.10050240
13. V.G. Oklobdzija, R.K. Krishnamurthy, *High-Performance Energy-Efficient Microprocessor Design* (Springer Science & Business Media, 2006), 338p
14. Y. Wang, H. Yan, C. Pan, S. Liu, Measurement-based analysis of characteristics of fast moving underwater acoustic communication channel, in *2022 International Conference on Frontiers of Information Technology (FIT). Islamabad, Pakistan*, (IEEE, 2022), pp. 47–52. https://doi.org/10.1109/FIT57066.2022.00019
15. A. Maffucci, A. Tamburrino, S. Ventre, Optimal homogeneous transmission line model for graphene interconnects, in *2019 IEEE 5th International Forum on Research and Technology for Society and Industry (RTSI). Florence, Italy*, (IEEE, 2019), pp. 172–175. https://doi.org/10.1109/RTSI.2019.8895600
16. J. Fadey, E.D. Markus, K. Kanzumba, A Review of the impact of reflection of transient signals on TEM transmission lines, in *2018 Open Innovations Conference (OI). Johannesburg, South Africa*, (IEEE, 2018), pp. 26–33. https://doi.org/10.1109/OI.2018.8535769
17. B. Wu, H. Zheng, X. Zhang, Effect of signal reflection on the performance of high-density ceramic package transmission lines. J. Phys. Conf. Ser. **1325**, 012170 (2019). https://doi.org/10.1088/1742-6596/1325/1/012170

18. L. Tang, O.P. Mahela, B. Khan, Y. Miro, Current- and voltage-actuated transmission line protection scheme using a hybrid combination of signal processing techniques. Sustainability **15**(7), 5715 (2023). https://doi.org/10.3390/su15075715
19. https://www.ieee802.org/3/ck/public/19_09/li_3ck_02_0919.pdf
20. *Inter-Chip USB Supplement to the USB 3.1 Specification*. July 2013
21. *Inter-Chip Supplement to the USB Revision 3.0 Specification*. February 2013
22. S. Ibrahim, B. Razavi, Low-power CMOS equalizer design for 20-Gb/s systems. IEEE J. Solid State Circuits **46**, 1321–1336 (2011). https://doi.org/10.1109/JSSC.2011.2134450
23. M. Maeng, F. Bien, Y. Hur, H. Kim, et al., 0.18-μm CMOS equalization techniques for 10-Gb/s fiber optical communication links. IEEE Trans. Microw. Theory Tech. **53**, 3509–3519 (2005). https://doi.org/10.1109/TMTT.2005.857108
24. P. Naidu, C.-W. Lu, Receiver analog front-end cascading transimpedance amplifier and continuous-time linear equalizer for signals of 5 to 30 Gb/s. Electronics **11**, 1546 (2022). https://doi.org/10.3390/electronics11101546
25. https://usb.org/sites/default/files/usb_data_performance_language_usage_guidelines_jan_2024.pdf
26. http://sabrepc.com/blog/computer-hardware/does-pcie-5-0-matter
27. https://www.synopsys.com/dw/ipdir.php?ds=dwc_usr-xsr_phy
28. C.H. Lee, M.T. Mustaffa, K.H. Chan, Comparison of receiver equalization using first-order and second-order Continuous-Time Linear Equalizer in 45 nm process technology, in *2012 IEEE International Conference on Industrial Applications (ICIAS)*, (IEEE, 2012), pp. 795–800. https://doi.org/10.1109/ICIAS.2012.6306122
29. S.N. Ahmadyan, C. Gu, S. Natarajan, E. Chiprout, et al., Fast eye diagram analysis for high-speed CMOS circuits, in *Design, Automation & Test in Europe Conference & Exhibition*, (EDA Consortium, 2015), pp. 1377–1382
30. https://docs.keysight.com/eesofapps/post-6-eye-opening-experience-with-ctle-678068358.html
31. V.K. Sharma, J.N. Tripathi, R.K. Nagpal, et al., Analysis and estimation of jitter sub-components. Int. J. VLSI Des. Commun. Syst. **5**(2), 15–24 (2014)
32. B. Malozyomov, N. Martyushev, N.N. Bryukhanova, et al., Reliability study of metal-oxide semiconductors in integrated circuits. Micromachines **15**, 561 (2024). https://doi.org/10.3390/mi15050561
33. A. Rogozhin, O. Glaz, Interconnects materials for integrated circuit technology below 5 nm node. Microelectronics **53**, 102–116 (2024). https://doi.org/10.31857/S0544126924010111
34. T. Rudenko, V. Kilchytska, M.K. Md Arshad, et al., On the MOSFET threshold voltage extraction by transconductance and transconductance-to-current ratio change methods: part II - effect of drain voltage. IEEE Trans. Electron Devices **58**, 4180–4188 (2012). https://doi.org/10.1109/TED.2011.2168227
35. A. Olanite, M.S. Musa, Power dissipation of adiabatic logic circuit for FinFET and MOSFET transistors, in *2021 6th IEEE International Conference on Recent Advances and Innovations in Engineering (ICRAIE). Kedah, Malaysia*, (IEEE, 2021), pp. 1–6. https://doi.org/10.1109/ICRAIE52900.2021.9704038
36. Y. Chen, Y. Song, B. Wu, et al., Optimization of impact ionization in metal-oxide-semiconductor field-effect transistors for improvement of breakdown voltage and specific on-resistance. Electronics **13**(20), 4101 (2024). https://doi.org/10.3390/electronics13204101
37. C. Liu, S.K. Lim, A study of signal integrity issues in through-silicon-via-based 3D ICs, in *2010 IEEE International Interconnect Technology Conference. Burlingame, CA, USA*, (IEEE, 2010), pp. 1–3. https://doi.org/10.1109/IITC.2010.5510738
38. https://arstechnica.com/gadgets/2015/04/in-depth-with-the-snapdragon-810s-heat-problems/
39. https://docs.broadcom.com/doc/Upgrade-to-G610
40. *The Gold Standard for Accurate Circuit Simulation* (Synopsys, 2019), 4p
41. *PrimeSim™ HSPICE Reference Manual: Commands and Control Options* (Synopsys Inc., 2020), 823p
42. https://picture.iczhiku.com/resource/eetop/WHIroZOqiUaWEmMX.pdf

43. https://www.seas.upenn.edu/~jan/spice/PSpice_UserguideOrCAD.pdf
44. https://www.eet.bme.hu/~szalai/Download/eldo_ur.pdf
45. R.C. Williamson, P.W. Juodawlkis, J.L. Wasserman, G.E. Betts, et al., Effects of crosstalk in demultiplexers for photonic analog-to-digital converters. J. Lightwave Technol. **19**(2), 230–236 (2001). https://doi.org/10.1109/50.917890
46. https://pdfreshers.blogspot.com/2018/09/cts.html
47. https://teamvlsi.com/2020/06/crosstalk-noise-and-crosstalk-delay-effects-of-crosstalk.html
48. G. Trucco, G. Boselli, V. Liberali, A study of crosstalk through bonding and package parasitics in CMOS mixed analog-digital circuits, in *Integrated Circuit and System Design. Power and Timing Modeling, Optimization and Simulation (PATMOS 2004)*, (Springer, Berlin, Heidelberg, 2004), pp. 138–147. https://doi.org/10.1007/978-3-540-30205-6_16
49. R. Chaudhry, D. Blaauw, R. Panda, T. Edwards, Current signature compression for IR-drop analysis, in *Design Automation Conference*, (Computer Society, Los Angeles, CA, 2000), pp. 162–167. https://doi.org/10.1109/DAC.2000.855296
50. https://hardwarebee.com/WikiBee/ir-drop/
51. M. Abestano, J. Hora, O.J. Gerasta, M.J. de Guzman, Optimizing design power integrity using IR-aware placement, in *2022 37th International Technical Conference on Circuits/Systems, Computers and Communications (ITC-CSCC). Phuket, Thailand*, (2022), pp. 1–3. https://doi.org/10.1109/ITC-CSCC55581.2022.9894858
52. H. Azatyan, H. Sahakyan, A. Galstyan, S. Harutyunyan, IO glitch filter, in *2020 IEEE 40th International Conference on Electronics and Nanotechnology (ELNANO). Kyiv, Ukraine*, (IEEE, 2020), pp. 259–262. https://doi.org/10.1109/ELNANO50318.2020.9088824
53. https://www.synopsys.com/glossary/what-is-electromigration.html
54. R. Martins, N. Lourenço, A. Canelas, N. Horta, Electromigration-aware and IR-drop avoidance routing in analog multiport pin structures, in *2014 Design, Automation & Test in Europe Conference & Exhibition (DATE). Dresden, Germany*, (IEEE, 2014), pp. 1–6. https://doi.org/10.7873/DATE.2014.023
55. G. Wang, J. Chen, M. Dai, E. Zheng, WAMF-FPI: a weight-adaptive multi-feature fusion network for UAV localization. Remote Sens. **15**(4), 910 (2023). https://doi.org/10.3390/rs15040910
56. C. Prasad, A review of self-heating effects in advanced CMOS technologies. IEEE Trans. Electron Devices **66**(11), 4546–4555 (2019). https://doi.org/10.1109/TED.2019.2943744
57. H. Jiang, H. Sagong, J. Kim, et al., Advanced self-heating model and methodology for layout proximity effect in FinFET technology, in *2020 IEEE International Reliability Physics Symposium (IRPS). Dallas, TX, USA*, (IEEE, 2020), pp. 1–5. https://doi.org/10.1109/IRPS45951.2020.9128322
58. https://e2e.ti.com/blogs_/archives/b/precisionhub/posts/how-to-calculate-the-effects-of-resistor-self-heating
59. R. Kishida, T. Asuke, J. Furuta, K. Kobayashil, Extracting BTI-induced degradation without temporal factors by using BTI-sensitive and BTI-insensitive ring oscillators, in *2019 IEEE 32nd International Conference on Microelectronic Test Structures (ICMTS). Kitakytushu, Japan*, (IEEE, 2019), pp. 24–27. https://doi.org/10.1109/ICMTS.2019.8730967
60. Y. Liu, X. Chen, Z.H. Zhao, et al., SiC MOSFET threshold-voltage instability under high temperature aging, in *2018 19th International Conference on Electronic Packaging Technology (ICEPT). Shanghai, China*, (IEEE, 2018), pp. 347–350. https://doi.org/10.1109/ICEPT.2018.8480781
61. R.J. Sivadasan, V. Shah, F.C. Huard, et al., NBTI aged cell rejuvenation with back biasing and resulting critical path reordering for digital circuits in 28 nm FDSOI, in *2018 Design, Automation & Test in Europe Conference & Exhibition (DATE). Dresden, Germany*, (IEEE, 2018), pp. 997–998. https://doi.org/10.23919/DATE.2018.8342154
62. D. Lorenz, M. Barke, U. Schlichtmann, Efficiently analyzing the impact of aging effects on large integrated circuits. Microelectron. Reliab. **52**(8), 1546–1552 (2012). https://doi.org/10.1016/j.microrel.2011.12.029

63. S. Mishra, P. Weckx, J.Y. Lin, et al., Fast & accurate methodology for aging incorporation in circuits using adaptive waveform splitting (AWS), in *2020 IEEE International Reliability Physics Symposium (IRPS). Dallas, TX, USA*, (IEEE, 2020), pp. 1–5. https://doi.org/10.1109/IRPS45951.2020.9129351
64. J. Diaz-Fortuny, J. Martin-Martinez, R. Rodriguez, et al., A noise and RTN-removal smart method for parameters extraction of CMOS aging compact models, in *2018 Joint International EUROSOI Workshop and International Conference on Ultimate Integration on Silicon (EUROSOI-ULIS). Granada, Spain*, (IEEE, 2018), pp. 1–4. https://doi.org/10.1109/ULIS.2018.8354740
65. A. Santana-Andreo, P. Martín-Lloret, E. Roca, R. Castro-Lopez, et al., Characterization and analysis of BTI and HCI effects in CMOS current mirrors, in *2022 18th International Conference on Synthesis, Modeling, Analysis and Simulation Methods and Applications to Circuit Design (SMACD). Villasimius, Italy*, (IEEE, 2022), pp. 1–4. https://doi.org/10.1109/SMACD55068.2022.9816212
66. D. Son, G.-J. Kim, J. Kim, et al., Effect of high temperature on recovery of hot carrier degradation of scaled nMOSFETs in DRAM, in *2021 IEEE International Reliability Physics Symposium (IRPS). Monterey, CA, USA*, (IEEE, 2021), pp. 1–4. https://doi.org/10.1109/IRPS46558.2021.9405153
67. A. Prakash, P. Jain, Optimization of leakage power dissipation in CMOS inverter using self biased and W/L scaling techniques, in *2023 2nd International Conference on Futuristic Technologies (INCOFT). Belagavi, Karnataka, India*, (IEEE, 2023), pp. 1–5. https://doi.org/10.1109/INCOFT60753.2023.10425379
68. J. Keane, T.-h. Kim, X. Wang, C.H. Kim, On-chip reliability monitors for measuring circuit degradation. Microelectron. Reliab. **50**(8), 1039–1053 (2010). https://doi.org/10.1016/j.microrel.2010.04.024
69. S. More, Aging degradation and countermeasures in deep-Submicrometer analog and mixed signal integrated circuits. PhD dissertation, Munich, 2011, 153p
70. C. Nunes, P.F. Butzen, A.I. Reis, R.P. Ribas, BTI, HCI and TDDB aging impact in flip-flops. Microelectron. Reliab. **53**(9–11), 1355–1359 (2013). https://doi.org/10.1016/j.microrel.2013.07.044
71. S. Peng, W. Jin, L. Chen, S. Tan, Data-driven fast electrostatics and TDDB aging analysis, in *Proceedings of the 2020 ACM/IEEE Workshop on Machine Learning for CAD (MLCAD '20)*, (Association for Computing Machinery, New York, NY, 2020), pp. 71–76. https://doi.org/10.1145/3380446.3430620
72. A. Daher, A. Haidar, A novel direct reference input Schmitt trigger design, in *DISP'19 International Conference on Digital Image & Signal Processing. Oxford University, United Kingdom*, (2019), pp. 1–4
73. https://eepower.com/technical-articles/analog-electronics-basics-an-overview-of-bicmos-operational-amplifiers/#
74. N. Czepl, D. Zupan, Noise behaviour of EMI improved folded cascode amplifier input stages including layout parasitics, in *2021 Austrochip Workshop on Microelectronics (Austrochip). Linz, Austria*, (2021), pp. 1–4. https://doi.org/10.1109/Austrochip53290.2021.9576855
75. V.S. Melikyan, S.A. Ghukasyan, S.S. Harutyunyan, et al., The aging influence minimization method for operational amplifiers with thin oxide transistors. Proc. RA NAS NPUA Ser. Tech. Sci. **75**(4), 527–535 (2022). https://doi.org/10.53297/0002306X-2022.v75.4-527
76. H.T. Kostanyan, H.V. Margaryan, V. Janpoladov, et al., The minimizaton method of transistor ageing influence on modern voltage references, in *2020 IEEE 40th International Conference on Electronics and Nanotechnology (ELNANO). Kyiv, Ukraine*, (IEEE, 2020), pp. 335–338. https://doi.org/10.1109/ELNANO50318.2020.9088844
77. H. Sahakyan, A. Hayrapetyan, S. Harutyunyan, et al., A low dropout voltage regulator with higher than Supply output voltage and load based frequency control system, in *2020 IEEE 40th International Conference on Electronics and Nanotechnology (ELNANO), Kyiv, Ukraine*, (IEEE, 2020), pp. 366–369. https://doi.org/10.1109/ELNANO50318.2020.9088754

78. A. Ballo, A.D. Grasso, G. Palumbo, A review of charge pump topologies for the power management of IoT nodes. Electronics **8**(5), 480 (2019). https://doi.org/10.3390/electronics8050480
79. J. Kim, P.K.T. Mok, C. Kim, A 0.15 V input energy harvesting charge pump with dynamic body biasing and adaptive dead-time for efficiency improvement. IEEE J. Solid State Circuits **50**(2), 414–425 (2015). https://doi.org/10.1109/JSSC.2014.2375824
80. M.H. Kamel, Z.K. Mahmoud, S.W. Elshaeer, et al., Comparative design of NMOS and PMOS capacitor-less low dropout voltage regulators (LDOs) suited for SoC applications, in *2019 36th National Radio Science Conference (NRSC)*, (IEEE, 2019), pp. 305–314. https://doi.org/10.1109/NRSC.2019.8734659
81. W.-C. Lai, S.-L. Jang, C.-Y. Huang, M.-H. Juang, Fully-integrated CMOS DC-DC boost converter, in *2019 IEEE Asia power and energy engineering conference (APEEC). Chengdu, China*, (IEEE, 2019), pp. 84–88. https://doi.org/10.1109/APEEC.2019.8720729
82. M. Chen, M. Chung, C. Yang, A low-PDP and low-area repeater using passive CTLE for on-chip interconnects, in *Symposium on VLSI Circuits (VLSI Circuits). Kyoto, Japan*, (IEEE, 2015), pp. 244–245. https://doi.org/10.1109/VLSIC.2015.7231273
83. V.S. Melikyan, A.S. Sahakyan, H.H. Dingchyan, et al., High PSRR and accuracy receiver active equalizer, in *IEEE 34th International Scientific Conference on Electronics and Nanotechnology (ELNANO). Kyiv, Ukraine*, (IEEE, 2014), pp. 194–197. https://doi.org/10.1109/ELNANO.2014.6873944
84. K. Zhang, A 4 Gb/s Adaptive CTLE based on data edge counting and 8B10B coding for DisplayPort, in *2022 7th International Conference on Integrated Circuits and Microsystems (ICICM). Xi'an, China*, (2022), pp. 453–458. https://doi.org/10.1109/ICICM56102.2022.10011261
85. http://tera.yonsei.ac.kr/class/2020_2_2/lecture/Design%20Exercise%208%20CTLE.pdf
86. V.S. Melikyan, A.V. Matevosyan, A.S. Petrosyan, A.A. Martirosyan, et al., High quality factor 5.0 Gbps CTLE circuit for SERDES serial links, in *2018 IEEE East-West Design & Test Symposium (EWDTS). Kazan, Russia*, (IEEE, 2018), pp. 1–5. https://doi.org/10.1109/EWDTS.2018.8524731
87. S.S. Harutyunyan, Aging protected two stage CTLE for high speed data recievers, in *2023 IEEE East-West Design & Test Symposium (EWDTS). Batumi, Georgia*, (IEEE, 2023), pp. 1–4. https://doi.org/10.1109/EWDTS59469.2023.10297111
88. D. Tzelepis, V. Psaras, E. Tsotsopoulou, et al., Voltage and current measuring technologies for high voltage direct current supergrids: a technology review identifying the options for protection, fault location and automation applications. IEEE Access **8**, 203398–203428 (2020). https://doi.org/10.1109/ACCESS.2020.3035905
89. V. Sahakyan, R. Soghomonyan, S. Harutyunyan, et al., Aging protected precision CMOS current reference with process, supply voltage and temperature compensation, in *2024 IEEE East-West Design & Test Symposium (EWDTS), Yerevan, Armenia*, (IEEE, 2024), pp. 01–04. https://doi.org/10.1109/EWDTS63723.2024.10873802
90. L. Prilenski, Bandgap reference design at the 14-nanometer FinFET node. Thesis, 2017
91. B. Razavi, *Design of Analog CMOS Integrated Circuits*, 2nd edn. (Mc Graw Hill India, 2017), 782p
92. *HSPICE Reference Manual* (Synopsys Inc., 2018), 855p
93. S.S. Harutyunyan, H.T. Kostanyan, M.T. Grigoryan, et al., A reliable pmos based charge pump architecture. Proc. RA NAS NPUA Ser. Tech. Sci. **LXXIII**(2), 181–187 (2020)
94. H. Omran, H. Alahmadi, K. Salama, Matching properties of Femtofarad and sub-Femtofarad MOM capacitors. IEEE Trans. Circ. Syst. I Regul. Pap. **63**, 1–10 (2016). https://doi.org/10.1109/TCSI.2016.2537824
95. W. Kraus, B. Stelzig, T. Tille, D. Schmitt-Landsiedel, MOS capacitances used in mixed-signal circuits and their operative range. Adv. Radio Sci. **1** (2003). https://doi.org/10.5194/ars-1-295-2003
96. M.T. Grigoryan, A.A. Atanesyan, G.H. Hakobyan, S.S. Harutyunyan, Two stage CTLE for high speed data receiving, in *2020 IEEE 40th International Conference on Electronics*

and Nanotechnology (ELNANO). Kyiv, Ukraine, (IEEE, 2020), pp. 374–377. https://doi.org/10.1109/ELNANO50318.2020.9088865
97. H.T. Kostanyan, S. Harutyunyan, G. Voskanyan, S. Ghukasyan, Transistor aging minimization method for multi output level converter design. Vestnik. Fiziko-Matematičeskie i Estestvennye Nauki. **1**, 89–100 (2022). https://doi.org/10.48200/1829-0450_pmn_2022_1_89
98. V.S. Melikyan, S.A. Ghukasyan, S.S. Harutyunyan, et al., A method For minimizing aging influence on rail-to-rail operational amplifier with thin oxide transistors. Известия вузов. Электроника. **3**, 45–52 (2023)
99. C. He, X. Dai, H. Xing, D. Chen, New layout strategies with improved matching performance. Analog Integr. Circ. Sig. Process **49**, 281–289 (2006). https://doi.org/10.1007/s10470-006-9705-1

Chapter 5
Development of Means to Improve the Efficiency of Information Transmission Systems Between Integrated Circuits

5.1 General Issues of Developing Means to Improve the Efficiency of Information Transmission Systems Between Integrated Circuits

5.1.1 Need to Develop Means to Improve the Efficiency of Information Transmission Systems Between Integrated Circuits

Integrated Circuits (ICs) are complex electronic devices composed of interconnected semiconductor elements, such as transistors, resistors, capacitors, and diodes. These components are manufactured on the same semiconductor platform, typically silicon, and are designed to perform various functions, such as signal processing, amplification, and data transmission. ICs play a key role in modern computing devices, machinery, aerospace, communication, and automation fields, providing high speed, energy efficiency, and cost-effectiveness [1–5].

Among the most critical components in ICs are the input-output (I/O) interfaces. They provide connectivity both between ICs and from ICs to the external environment. Efficient I/O interface design is crucial for high-speed data transmission, power efficiency, and ensuring signal integrity [6–8].

I/O interface consists of a transmitter, receiver, tuning, and biasing circuits (Fig. 5.1) [9]. The transmitter circuit is designed to convert a digital signal into an electrical signal and send it over a channel. The receiver circuit converts and recovers the electrical signal coming from the channel [10].

Data transmission in I/O interfaces can be implemented serially or in parallel. With parallel transmission, it is possible to increase the amount of transmitted data, as in this case, multiple transmitter and receiver circuits operate simultaneously. However, for long distances, serial data transmission is used because synchronizing a single signal is easier; consequently, the channel cost is lower, and it also occupies

V. Melikyan, K. Li, *Parametric Optimization Techniques of Integrated Circuits*,
https://doi.org/10.1007/978-3-032-20400-4_5

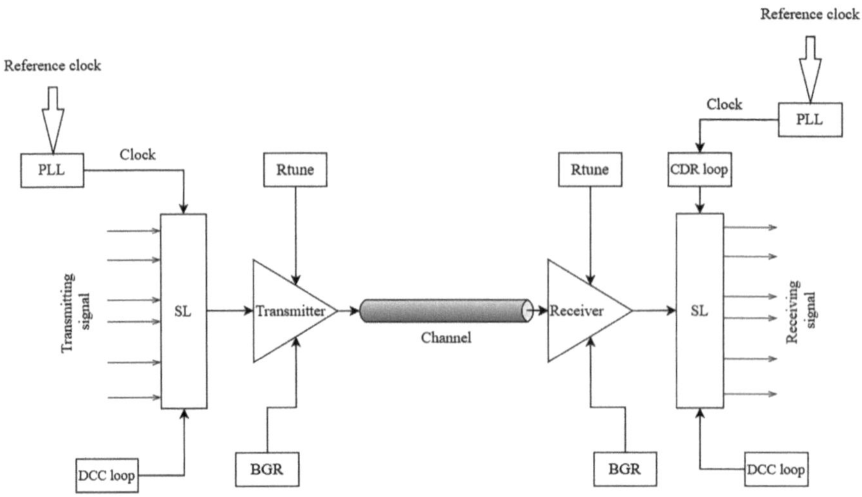

Fig. 5.1 Input/output interface architecture

a smaller area. To ensure the same speed of serial data transmission, a higher clock frequency is required compared to parallel transmission [11–13].

To increase noise immunity, serial data transmission is implemented using differential signaling. Non-differential signal transmission can lead to distortions that occur during signal propagation through the channel [9, 14].

To ensure efficient transmission of high-frequency signals, a Phase-locked loop (PLL) is required [15–17]. The reference clock signal, generated by a quartz oscillator, is applied to the PLL input and multiplied by an appropriate factor to produce the desired clock frequency. The resulting clock signal is provided to the Serializer (S) [18–20], which converts parallel data into a serial data stream at a specified frequency. The data package is sent to the receiver via the transmitter circuit. To read the received information, the necessary phase alignment is provided by the Clock and data recovery loop (CDR loop) [21–23], which adjusts the clock phase based on the received data signal phase. The serial data is deserialized at the receiver, and its original structure is restored via a specific circuit to reduce the signal frequency, as well as for parallel data processing and use [24].

Another essential support circuit for efficient information transmission in I/O systems is the Bandgap reference (BGR) [25–28], which provides the necessary stable supply and reference voltages. Its operation is based on the use of bipolar transistors and their properties. For the same current, the base-emitter junction voltage drop decreases with increasing temperature, meaning it has an inverse temperature dependence. By combining this characteristic with a positive temperature coefficient, a stable voltage characteristic independent of temperature is obtained [29–31].

Resistor tuning (Rtune) [32] is designed to prevent possible signal reflections in the channel. Signal reflections can lead to information distortions and inefficient transmission, which can result in erroneous data reading. Rtune matches the output impedance of the transmitter and the input impedance of the receiver to the characteristic impedance of the transmission channel [33, 34].

Efficient design of Duty Cycle Correction loops (DCC) [35] is an important task for efficient information transmission. The probability of erroneous reading during information reception can sharply increase when the duty cycle deviates. Thus, the durations of the logical "1" and "0" levels must be equal; a 50% duty cycle must be maintained [36, 37].

The main sub-circuit ensuring efficient information transmission from I/O interfaces to the channel is the transmitter circuit. It consists of a level shifter (LS), a control tri-state circuit, buffers, and output cascades, including current-limiting resistors and ESD protection diodes (Fig. 5.2) [38, 39].

Due to the low supply voltage inside ICs, there is a need for voltage level conversion circuits that interface between internal IC voltage levels and external voltage levels; therefore, to ensure reliable transmission of information, there is a need to develop efficient LS circuits [40–43].

The continuous scaling of transistor dimensions enables increasing the number of transistors used per unit area, which leads to increased functionality of ICs. However, in technologies smaller than 14 nm, secondary effects emerge that were not significant in older technologies or had a very small effect. This necessitates developing new efficiency improvement methods. It also complicates the design process [44].

Modern multifunctional ICs have become an integral part of human life. The continuous expansion of their application areas in systems such as medicine, autonomous systems, machinery, and transportation communication has led to new, stricter requirements for ensuring reliability in ICs. Inefficient information transmission in ICs can lead to erroneous signal reading and cause failures, which, in turn, can lead to undesirable consequences. For example, inefficient and defective

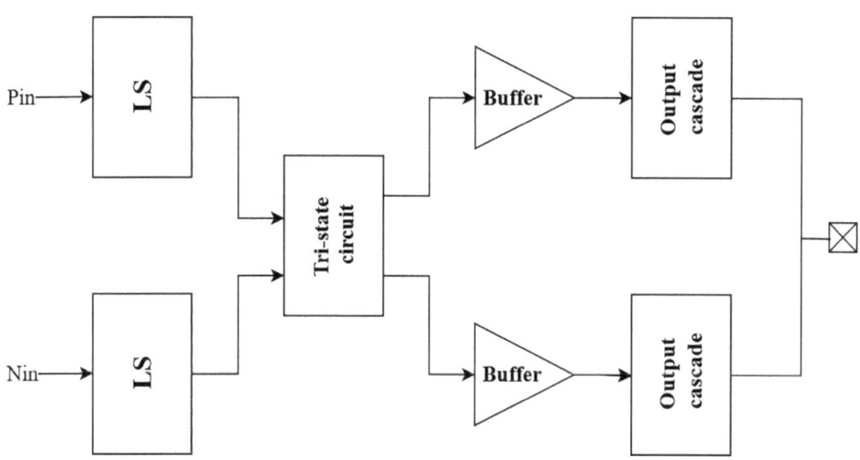

Fig. 5.2 Structure of a transmitter

IC processing in 2016 was the cause of the massive failure of the famous Samsung company's Galaxy Note 7 mobile phones. As a result, the phones overheated and caught fire, thus posing a danger to humans. Due to the failure, the corporation lost 5.3 billion dollars [45]. There are many such examples, and they are the reason for the increasingly stringent efficiency standards for modern ICs.

5.1.2 *Key Factors Determining the Degradation of Integrated Circuit Efficiency*

The main factors for degradation of IC efficiency are:

- Speed (performance) [46, 47],
- Noise [48, 49],
- Power Consumption [50–52],
- Reliability [53–63].

The continuous reduction of the channel length in CMOS transistors has led to an increase in information transmission speed. This means that new, higher standards are needed for the speed of information transmission through I/O interfaces. Therefore, the efficient design of high-speed I/O interfaces has become a critical challenge.

The information transmission speed is determined by the bit rate (Fig. 5.3) [46].

Current high-speed I/O interfaces provide a wide range of information transmission, reaching speeds of up to tens of Gb/s (Table 5.1) [47, 48].

Noise. Another factor causing degradation of I/O interface efficiency is noise. Noises are random signals that are not expected data and are described by a statistical approach; they can be divided into two groups [49]:

- External
- Internal.

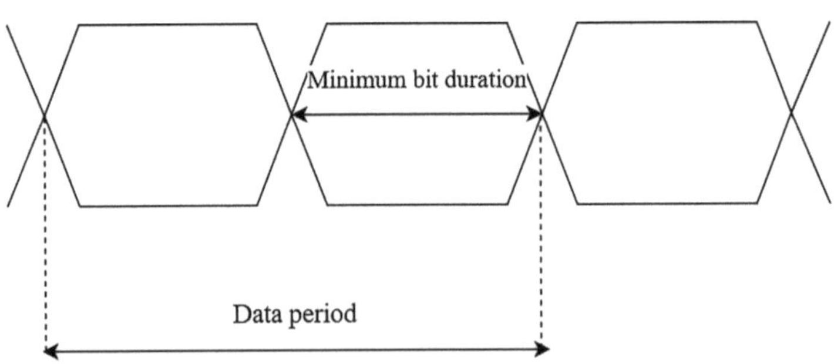

Fig. 5.3 Determination of information transmission speed

Table 5.1 Information transmission standards

I/O standards	Clock signal period (p/s)	Clock frequency (GHz)	Data transfer rate (Gbit/s)
USB4.2	25	40	80
PCEI 6.0	31.25	32	64
USB4.1	50	20	40
PCEI 5.0	62.5	16	32
USB 3.2	100	10	20

External noises are those generated by sources outside ICs. These can be capacitive couplings in the IC environment and the electromagnetic field. The impact of external noise on ICs can be reduced by amplifying signals and proper shielding and layout design.

Signal-to-noise (SNR) ratio can be improved by signal amplification [14].

$$\mathrm{SNR}_{\mathrm{dB}} = \frac{P_{\mathrm{sig}}}{P_{\mathrm{no}}}, \tag{5.1}$$

$$\mathrm{SNR}_{\mathrm{dB}} = 20\log\frac{P_{\mathrm{sig}}}{P_{\mathrm{no}}}, \tag{5.2}$$

where $\mathrm{SNR}_{\mathrm{dB}}$ is the signal-to-noise ratio expressed in decibels, and P_{sig} and P_{no} are the signal and noise powers, respectively.

Internal noise is the inherent noise generated by ICs or their sub-circuits. Inherent noise can appear in various terminals of CMOS transistors. There are many types of internal noise [49]:

- Thermal
- 1/f or Flicker (Pink)
- Shot
- Substrate
- Generation and Recombination.

Thermal noise is also called Johnson-Nyquist noise. This type of noise has a random nature and arises due to Brownian motion. Due to the random characteristic, the average value of noise is approximately zero. Therefore, the spectral power density of the noise is measured. The spectral power density of thermal noise for resistive elements (Fig. 5.4) [1, 49] is:

$$\overline{V}_n^2 = 4\mathrm{kTR}\Delta f, \tag{5.3}$$

where k is Boltzmann's constant, T is the temperature, R is the resistance, and Δf is the frequency change.

Like other noise sources, $1/f$ noise is not easily predictable. Among the reasons for its occurrence is the roughness of the oxide layer surface. At the boundary between the oxide layer and the die, there are crystal lattice defects at the

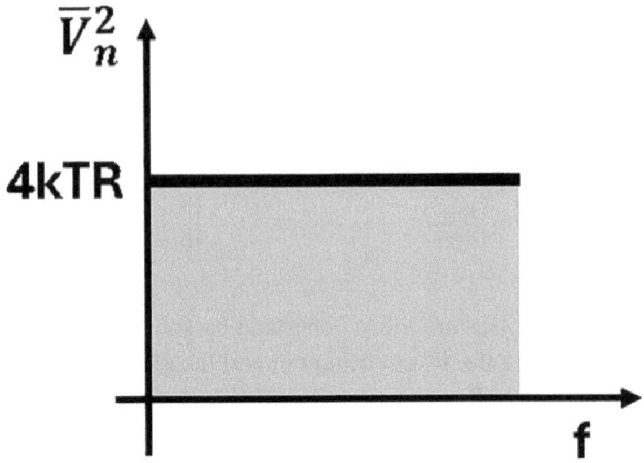

Fig. 5.4 Spectral power density of thermal noise

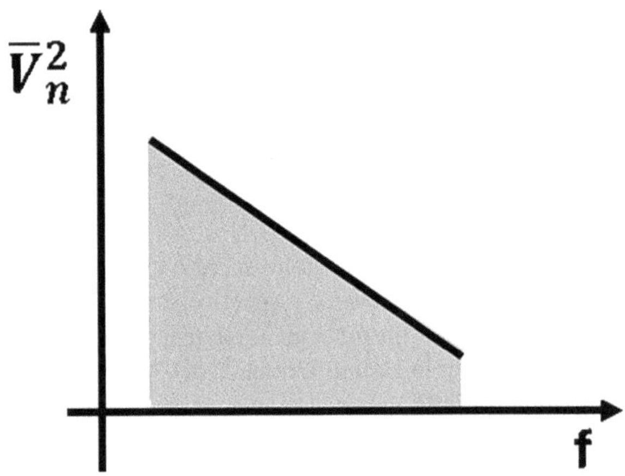

Fig. 5.5 Spectral power density of 1/*f* noise

oxide-semiconductor interface cause charge carriers to randomly enter and exit the channel, leading to current fluctuations. This noise is inversely proportional to frequency and transistor dimensions (Fig. 5.5) [1, 49], and the spectral power density of the noise is expressed by the following formula:

$$\overline{V}_n^2 = \frac{K}{WLC_{\text{ox}}}\frac{1}{f}, \tag{5.4}$$

where K is a process-dependent constant, C_{ox} is the gate capacitance per unit area, and f is the frequency.

Any constant current flowing through a P-N junction causes shot noise. Shot noise is mainly present in bipolar transistors and junction diodes. In CMOS transistors, it occurs in the channel and P-N junction regions [1, 49].

Capacitive couplings present in the transistor substrate do not have a significant effect at low frequencies, but as the frequency increases, the effects of substrate noise also increase, and small currents arise in the substrate (Fig. 5.6) [49].

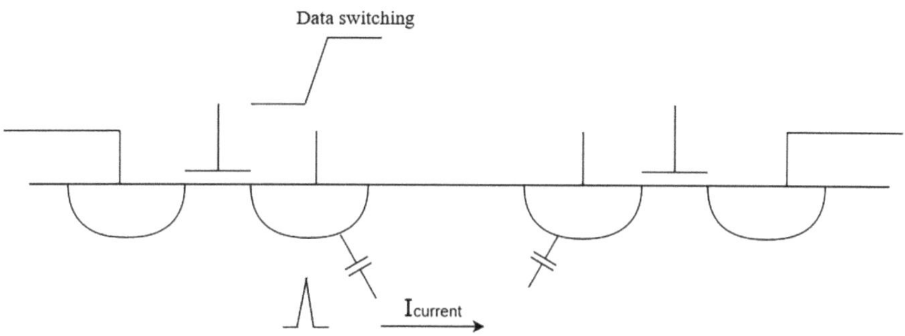

Fig. 5.6 Substrate noises in a CMOS transistor

In the channel of CMOS transistors, generation and recombination of charge carriers occur, which leads to changes in the currents flowing through the transistor channel, and noise is generated. The continuous reduction of the transistor channel leads to an increase in this type of noise.

However, a significant reduction of inherent noise during the design process is not possible; it is mainly determined by technological processes.

Power Consumption. Ensuring power efficiency in modern ICs has become one of the key requirements in modern IC design. Reducing power consumption has become a critical consideration in battery-powered and portable devices [50].

The power consumption of any circuit depends on the leakage currents of transistors. Research shows that 40% of the total power consumed is due to leakage currents, and the continuous scaling of transistors can increase this indicator [51].

The term Power Consumption is defined as the amount of energy dissipated from ICs in the form of heat. Power consumption can be divided into four main types [52]:

- Dynamic Power Consumption
- Static Power Consumption
- Short-Circuit Power Consumption
- Leakage Power Consumption.

The total power can be calculated by the following formula:

$$P_{\text{total}} = P_{\text{dynamic}} + P_{\text{static}} + P_{\text{leakage}} + P_{\text{short}} : \tag{5.5}$$

Dynamic power consumption is the amount of energy consumed during the charging and discharging of the circuit (Fig. 5.7) [52].

Dynamic power consumption can be of three main types:

- Switching power consumption
- Short-circuit power consumption
- Power consumption due to random transitions.

The dynamic power is calculated by the following formula [52]:

Fig. 5.7 Dynamic power consumption of a CMOS inverter

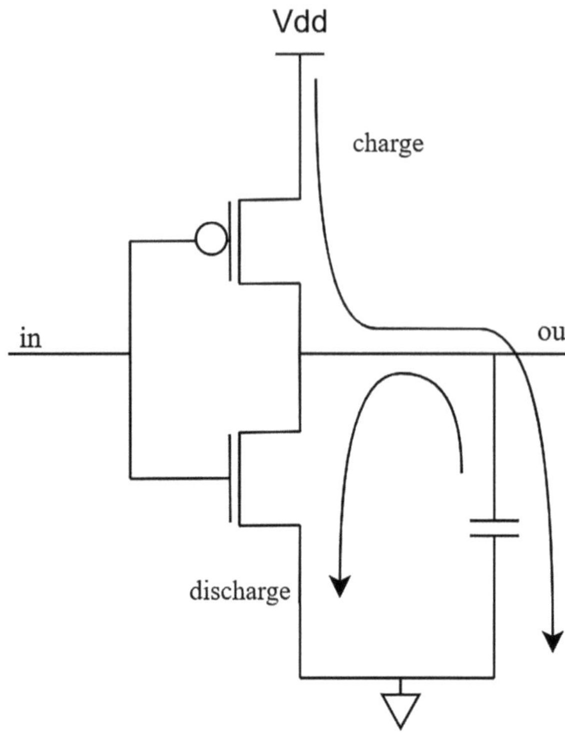

$$P_{\text{dynamic}} = K \cdot C_{\text{out}} \cdot Vdd \cdot 2f, \tag{5.6}$$

where Vdd is the supply voltage, C_{out} is the capacitance, K is the average number of positive clock edges, and f is the clock frequency.

The amount of power consumed in the circuit when it is in an idle state is called static power consumption. This loss mainly arises from subthreshold and reverse-biased diode leakage currents. The static power is calculated by the following formula [52]:

$$P_{\text{static}} = I_{\text{static}} \cdot Vdd, \tag{5.7}$$

where I_{static} is the static current, and Vdd is the supply voltage.

Static power loss is mainly due to the following currents (Fig. 5.8) [52]:

$I1$—Reverse-biased P-N junction current
$I2$—Subthreshold leakage current
$I3$—Oxide tunneling current
$I4$—Gate current
$I5$—Gate-induced drain leakage current
$I6$—Gate-induced bulk leakage current.

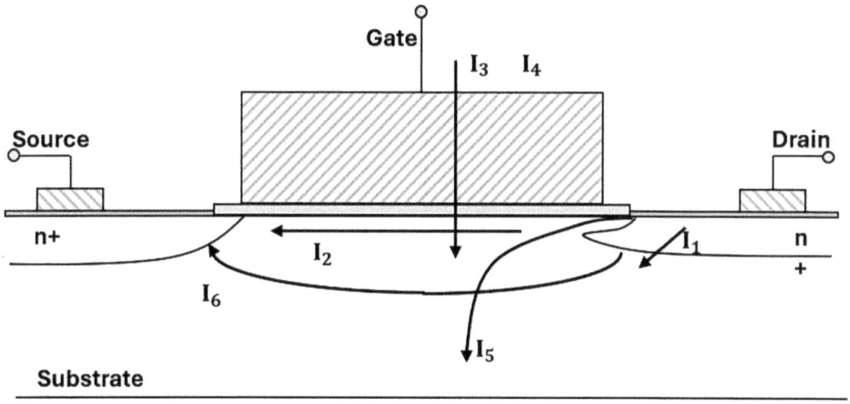

Fig. 5.8 Static currents in an NMOS transistor

Short-circuit power consumption occurs when the supply voltage is directly shorted to ground. This often happens when PMOS and NMOS transistors are simultaneously on (Fig. 5.9) [52]. This energy loss reduces the system's efficiency and reliability.

The short-circuit power is calculated by the following formula [52]:

$$P_{\text{short}} = I_{\text{short}} \cdot V\text{dd}, \tag{5.8}$$

where I_{short} is the short-circuit current, and Vdd is the supply voltage.

Leakage power loss is defined as the amount of power lost when the circuit is in standby mode. It mainly arises due to the following leakage currents [52]:

- Subthreshold leakage current
- Gate-induced drain leakage current
- Reverse-biased junction leakage current.

Reliability. The number of failures in ICs per unit time is called circuit reliability. It can be described by the failure rate graph (Fig. 5.10) [53–55].

As can be seen, the frequency graph consists of three parts:

- Early-life failures (Infant Mortality)
- Constant (Random) failures
- Saturation failures.

Early-life failures occur at the beginning of the IC's lifetime and are detected in a relatively short period; the frequency of these failures decreases over time and reaches a stable level.

Constant failures occur during the IC's lifetime. The frequency of these failures remains constant over time.

Saturation failures occur at the end of the IC's life, and they determine the circuit's lifespan. The frequency of saturation failures increases over time. The main

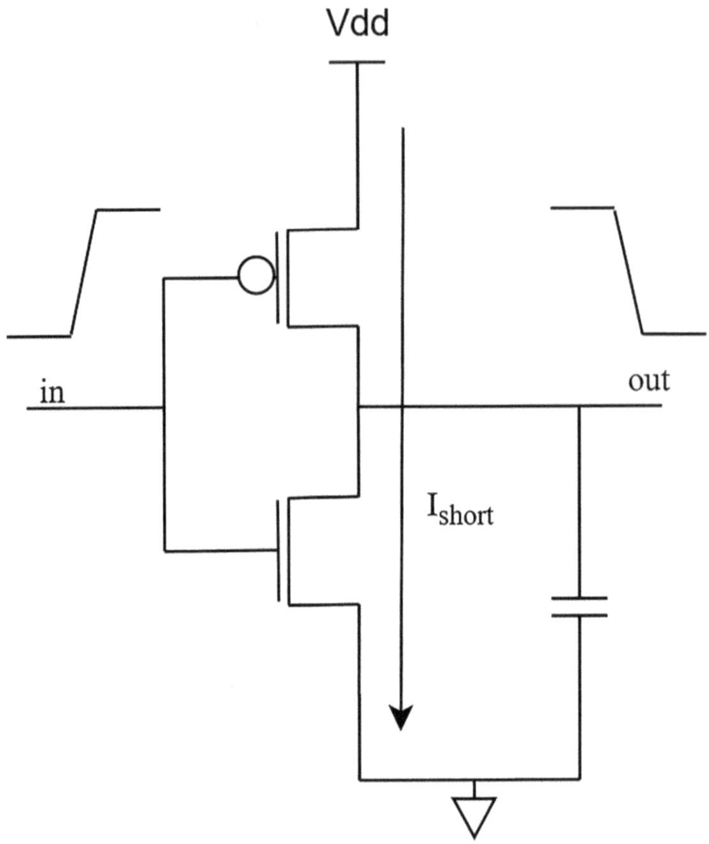

Fig. 5.9 Short-circuit power consumption of a CMOS inverter

cause of this type of failure is aging phenomena, the impact of which leads to degeneration of circuit parameters and failures [55].

Aging is the continuous degradation of transistor parameters over time, leading to a reduction in the operating life of ICs and a sharp decrease in reliability [56–62]. Among the main degradation mechanisms affecting the reliability of CMOS circuits are traps, which are defects in the crystalline structure of the oxide layer. Over time, charge carriers with high kinetic energy, passing through the oxide layer, accumulate in these traps, leading to the degradation of transistor parameters, particularly a shift in the threshold voltage (Fig. 5.11) [57, 58].

Mechanisms leading to transistor aging also include Bias Temperature Instability (BTI) [59–62], Hot Carrier Injection (HCI) [59–61], and electromigration.

The BTI phenomenon occurs at high temperatures of 100–150 °C and under the influence of a vertical electric field in the channel. It leads to the formation of oxide traps and the generation of interface states at the Si/SiO_2 surface. This phenomenon is of two types: negative BTI in PMOS transistors and positive BTI in NMOS transistors. With the scaling of the gate oxide layer thickness, the impact of BTI and

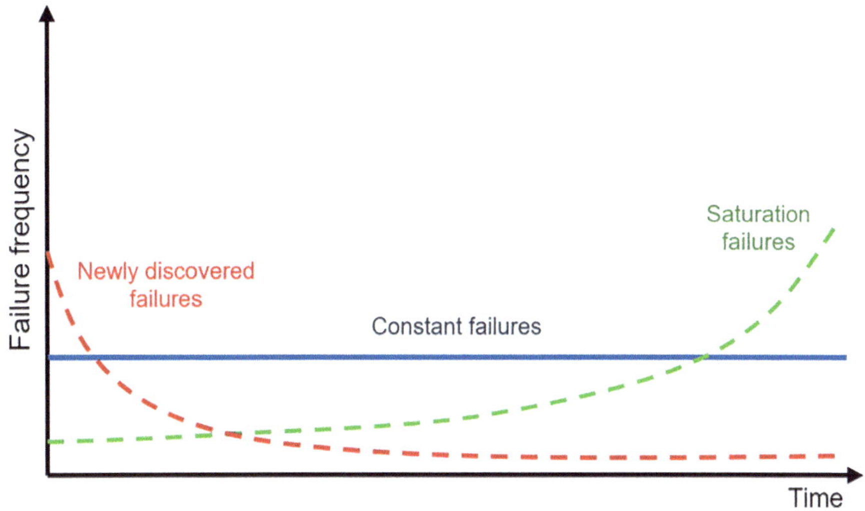

Fig. 5.10 Failure rate curves (bathtub curve)

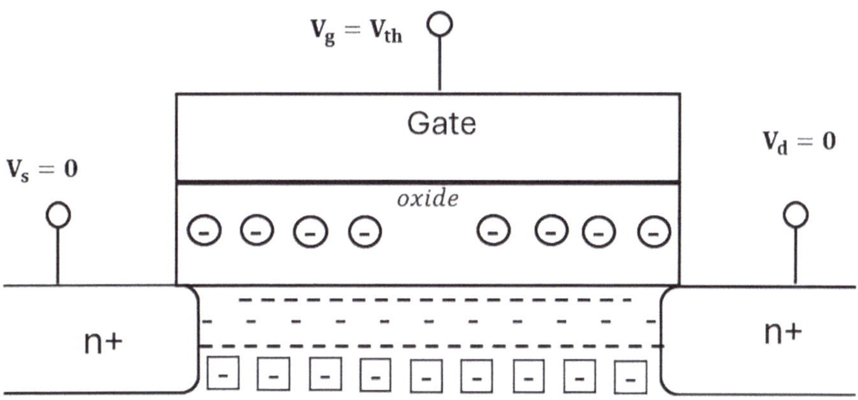

Fig. 5.11 Aged NMOS Transistor

other degradation mechanisms on the transistor threshold voltage also increases. In 45 nm and smaller CMOS technologies, materials with a higher dielectric constant (K) are used to overcome the increase in leakage currents. Despite all this, the BTI phenomenon remains one of the main reliability challenges [59–62].

The effect of BTI is a self-recovering process. When the transistor is in the off state ($V_{gs} = 0$), over time, the charge carriers that appeared in the oxide traps can return to the channel. However, this recovery is not complete, as the charge carriers in the oxide traps cannot completely return (Fig. 5.12) [59–62].

Thus, it can be said that in CMOS transistors, threshold voltage degradation occurs under the influence of an electric field, and recovery occurs in its absence.

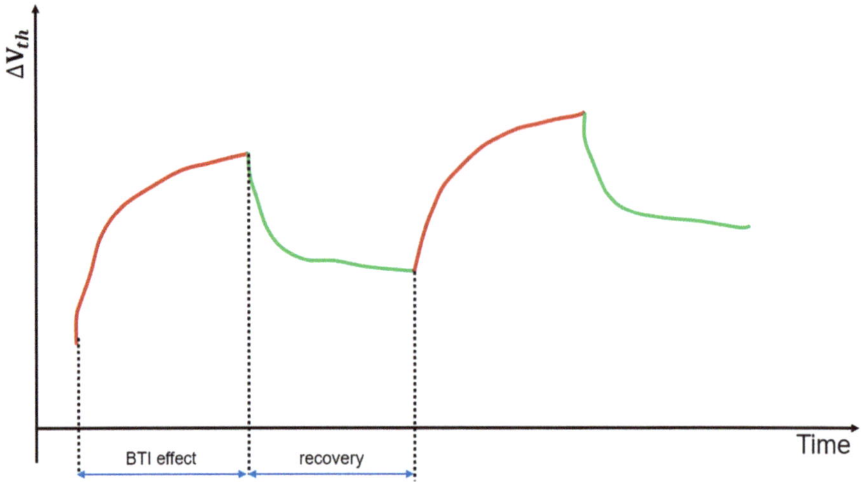

Fig. 5.12 Stages of BTI impact and recovery on a transistor

For NMOS and PMOS transistors, the change in V_{th} in cases of positive and negative BTI is modeled by the following formula [59–62]:

$$\Delta V_{th} = A\left(\frac{V_g}{t_{ox}}\right)^m e^{\frac{\Delta E}{kT}} L^{\alpha} W^{\beta} t^n, \tag{5.9}$$

Where V_g is the gate–source voltage, ΔE is the thermal energy, k is Boltzmann's constant, T is the temperature, L is the channel length, W is the channel width, t is the stress time, n is the time exponent, and A, m, α, β are experimentally determined constants.

The Hot Carrier Injection (HCI) phenomenon occurs under high drain–source voltage, under the influence of which the generated electric field, accelerating the charge carriers in the channel and endowing them with high kinetic energy (called hot), creates the possibility of overcoming the Si-SiO_2 potential barrier and penetrating the gate oxide. Some of the charge carriers that enter the oxide, reaching the gate, cause gate current, while others, due to defects in the oxide layer, become trapped, thereby increasing the number of existing traps.

Models for the change in transistor threshold voltage and passing current have been developed to assess the impact of the HCI phenomenon [59–61]:

$$\frac{\Delta I_d}{I_d} = x \cdot D_{HCI}, \tag{5.10}$$

$$\Delta V_{th} = y \cdot D_{HCI} \tag{5.11}$$

where D_{HCI} is the general model of HCI degradation, x and y are the separation components of the effects, whose numerical values lie in the 0–1 range, and D_{HCI} is determined by the following expression [59–61]:

$$D_{\text{HCI}} = B \cdot V_{\text{ds}}^{p} \cdot e^{\frac{\Delta E}{kT}} \cdot L^{\delta} \cdot t^{q}, \tag{5.12}$$

where V_{ds} is the transistor's drain–source voltage, T is the temperature, t is the duration of BTI impact, L is the channel length, k is Boltzmann's constant, and ΔH is the activation energy. Components B, p, δ, and q are technology-dependent parameters.

The HCI phenomenon includes several injection mechanisms [59–61]:

- Channel Hot Electron (CHE) injection, which occurs when an electron, on its way from source to drain, does not collide with other particles and continuously gains sufficient kinetic energy to overcome the potential barrier of the insulating layer.
- Drain Avalanche Hot Carrier (DAHC) injection occurs when the transistor is in deep saturation. As a result of impact ionization occurring in the drain region, electron-hole pairs are generated, which move under the influence of the existing electric field in the channel, causing further ionization. Due to the collision, some electrons gain sufficient energy and overcome the potential barrier, ending up in the oxide layer, while the holes move toward the bulk region, increasing the substrate current.
- Secondary Generated Hot Electron (SGHE) injection arises as a result of further impact ionization caused by hot electrons generated due to DAHC.
- Substrate Hot Electron (SHE) injection mainly occurs when the voltage value applied to the transistor's bulk is different from zero. In this case, the channel charge carriers gain energy and are highly likely to overcome the oxide potential barrier.

This mechanism has a symmetrical distribution along the channel length, as it does not depend on the source and drain voltages.

Electromigration is the phenomenon of material transfer due to the influence of a high electric field. This phenomenon mainly occurs at the ends and edge sections of conductors. In the case of high currents, the charge carrier particles, interacting with the atoms of the conductor, displace them, creating "voids" and "hillocks". As a result of this phenomenon, not only can a change in the conductor's resistance occur, but even its rupture. This phenomenon is characterized by the Mean Time To Failure and is modeled by Black's law [63]:

$$\text{MTTF} = \frac{A}{J^{n}} \exp\left(\frac{E_a}{kT}\right), \tag{5.13}$$

where MTTF is the mean time to failure, A is a constant depending on the conductor's cross-sectional area, J is the current density, E_a is the activation energy, k is Boltzmann's constant, T is the temperature, and n is the scaling factor.

Thus, in accordance with the requirements for high speed and high sensitivity, the development of new methods to improve the efficiency of information transmission in analog ICs is highly relevant. During the design of these means, the impact of the main factors causing the degeneration of information transmission efficiency must be taken into account, as well as the main requirements for small area and low power consumption.

The Impact of Performance Degradation Factors on Analog and Mixed-Signal Circuits

Analog and mixed-signal circuits play a critical role in modern ICs. They are the most sensitive circuits in ICs and are responsible for the efficient reception, processing, and transmission of information.

The impact of aging effects is a significant threat to analog and mixed-signal circuits. Unlike digital circuits, where aging effects mainly affect logic gate delay, in analog and mixed-signal circuits, a parameter degradation can be observed. In all cases, effects negatively affect the reliability of ICs and can lead to inefficient information transmission and failures [64].

Voltage Level Shifter (LS). The development of energy-efficient systems and digital circuits is a major challenge in the IC design process. Supply voltage reduction can significantly reduce power consumption. However, low supply voltages limit the circuit's speed and affect the linearity and gain of sub-circuits. Therefore, to ensure energy efficiency, the multi-supply voltage approach is used, i.e., each section of the IC has its own local supply voltage. In this case, the presence of a voltage level conversion circuit is necessary [40–43].

LSs are used as intermediate and auxiliary circuits and are of great importance in information transmission systems. They perform low-to-high and high-to-low voltage level conversions, thus providing the connection between the internal and external domains of ICs. Current requirements include high speed, low power consumption, reliability, and small occupied areas.

Modern basic architecture (Fig. 5.13) [40–43, 65] does not allow LS construction without the use of transistors with thick gate oxide, because the high (VDDH) and low (VDD) supply voltages can have significant voltage differences. Additionally, to reduce production costs, not all technologies have transistors with thick gate oxide or are designed to withstand the high voltage differences that arise between transistor pins, causing overvoltage stress. When analyzing the architecture of the basic low-to-high LS circuit, it is necessary to understand the weaknesses. The input signal is processed by the M3 and M4 transistors.

The input operates at VDD voltage level. and the VDDH voltage, which is the circuit's high supply level, forms the output signal. The logical "1" level turns on the M3 transistor, while M4 remains off. When M3 turns on, the gate of the M2 transistor is pulled to ground, causing it to turn on. In this case, the Voutp output pin reaches the logical "1" level, equal to VDD. On the other hand, Voutp approaches the logical "0" level because the M1 transistor turns off. It should be noted that when VDDH is applied to the input, and the source of the M3 transistor is connected

Fig. 5.13 Basic high-to-low LS architecture

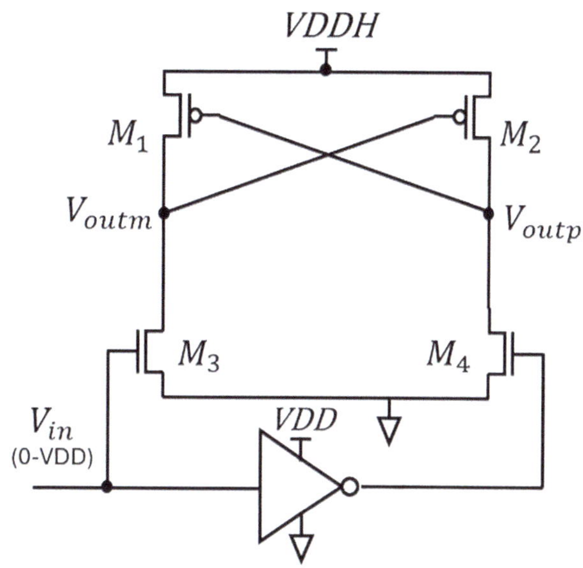

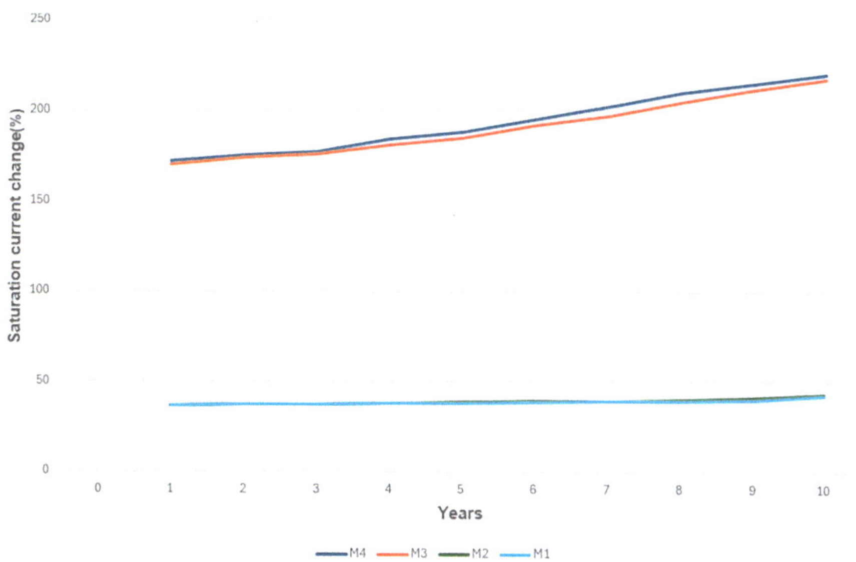

Fig. 5.14 Changes in transistor saturation currents over 10 years

to a low potential point, the gate–source voltage of M3 equals VDDH, creating unacceptable stress on the M3 transistor [40–43, 65].

The LS circuit was designed for the SAED 14 nm FinFET technology, and 10-year aging simulation was performed using Synopsys' HSPICE [66] tool, at VDDH = 1.2 V, VDD = 0.7 V, and 125 °C (Figs. 5.14 and 5.15) [67, 68], showing

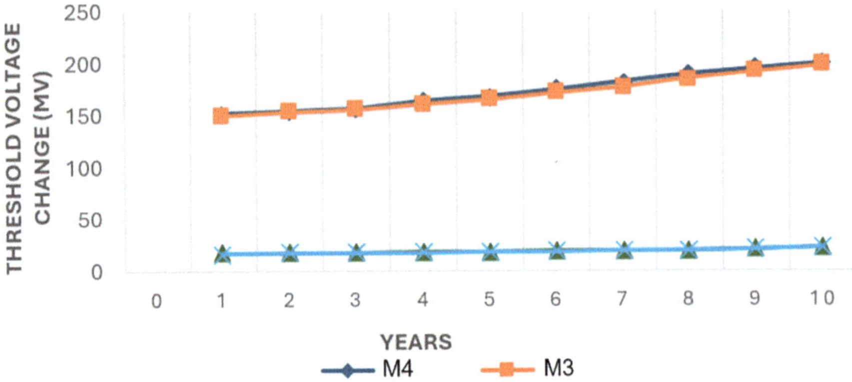

Fig. 5.15 Changes in transistor threshold voltages over 10 years

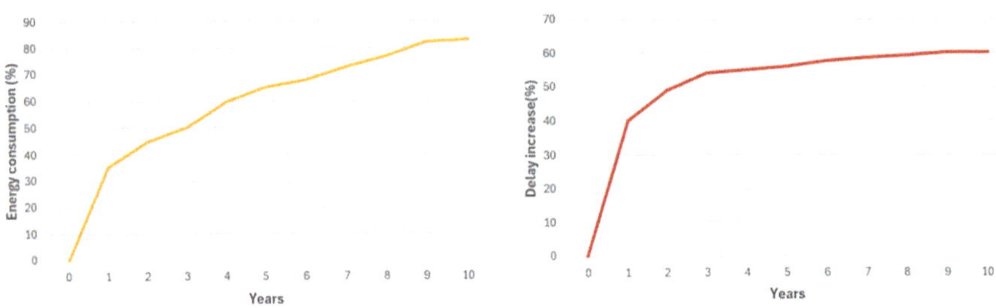

Fig. 5.16 Changes in the increase of power consumption and delay time of basic LS circuit over 10 years

maximum degradation of saturation current and threshold voltage), near (approximately 60% and 220 mV, respectively).

Under 10-year operating life conditions, the circuit's power consumption and delay time (Fig. 5.16) [67, 68] increased by approximately 80% and 60%, respectively.

A similar problem will also arise during the design of low-to-high LSs, as it has an architecture similar to the high-to-low LS (Fig. 5.17) [67, 68], and the same problems mentioned earlier arise.

Thus, due to aging effects, the parameters of the basic LS circuits degrade, which over time leads to an increase in circuit power consumption and a decrease in speed, thus reducing LS circuit efficiency and transmission system performance.

Bandgap Reference (BGR). In some analog circuits, for example, voltage regulators, digital-to-analog and analog-to-digital converters, it is necessary to have a highly accurate reference voltage source. In other words, to have an efficient information transmission system, a highly accurate, energy-efficient, and reliable BGR circuit is needed, which will be independent of the technological process, supply voltage, and temperature changes. It is also necessary to ensure applicability in the

Fig. 5.17 Low-to-high LS architecture

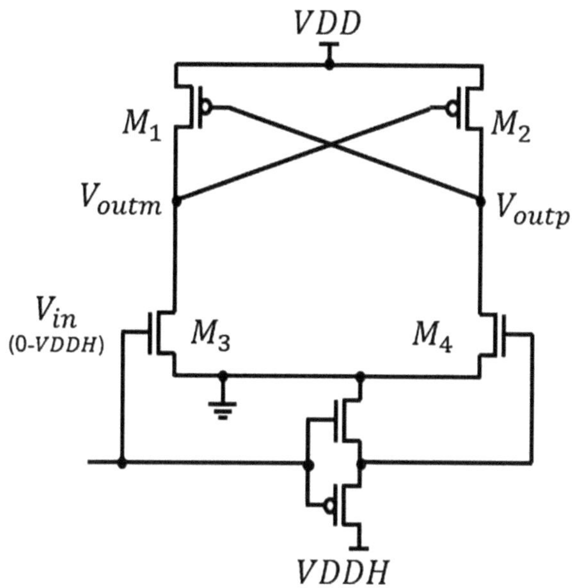

latest technologies, where, due to reduced production costs, there are no transistors with thick gate oxide.

The idea underlying the operation of the BGR circuit involves combining two voltages: one with a positive temperature coefficient (the voltage difference between the base and emitter of two bipolar transistors) and the other with a negative temperature coefficient (base-emitter voltage, V_{be}). This combination provides a stable reference voltage [1, 14, 69].

$$V_{bias} = V_{be} + KV_t, \tag{5.14}$$

Where V_{be} is the base-emitter voltage of the bipolar transistor, K is a constant value, and V_t is the thermal voltage.

As mentioned earlier, modern technologies do not have transistors with thick gate oxide, so to meet modern standards, the basic BGR circuit (Fig. 5.18) [69] was designed using only transistors with thin gate oxide.

The circuit uses an operational amplifier to equalize the voltages at points A and B. The output reference voltage of the OpAmp is connected via feedback to the bases of the Q1 and Q2 bipolar transistors.

A basic OpAmp with folded cascode was designed (Fig. 5.19) [69]. These OpAmps are widely used in BGR circuits because they provide high output gain.

In addition to the main circuit, the OpAmp includes additional transistors that allow the system to enter off mode, thus ensuring the energy efficiency of the RVS circuit. These transistors are indicated by dotted lines: M41, M42, M43, M44, and M54.

Fig. 5.18 Architecture of the basic BGR circuit

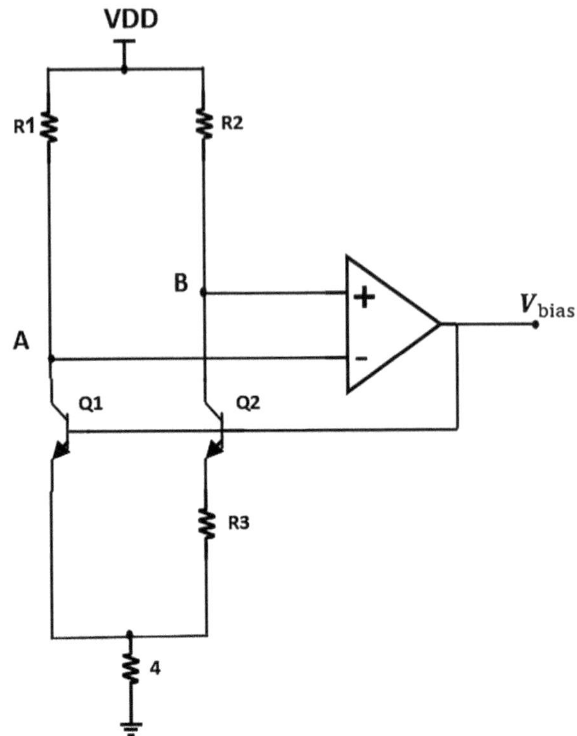

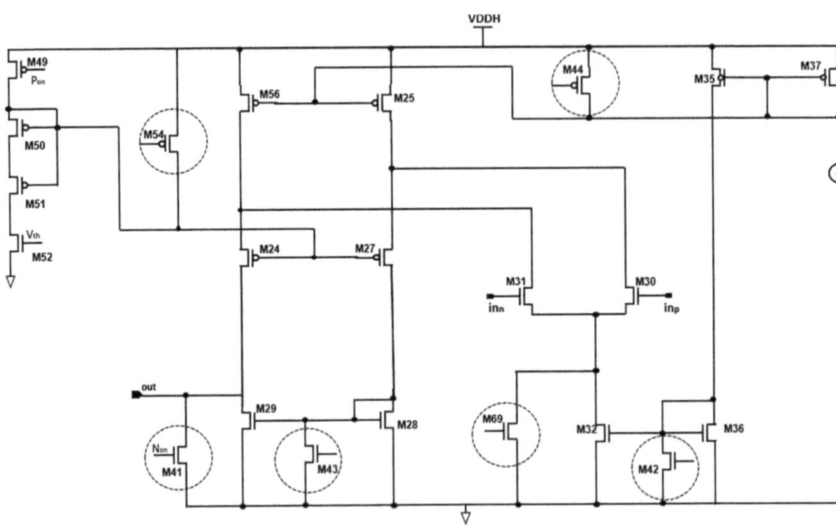

Fig. 5.19 Circuit of the basic folded cascode OpAmp

The OpAmp circuit was designed for the SAED 14 nm FinFET technological process, and amplitude-frequency and phase-frequency HSPICE [66] simulations were performed for different PVT conditions (Fig. 5.20) [69].

As a result of studying the circuit, it becomes clear that:

- The M41 transistor makes the OpAmp output voltage 0 V in off mode and sets the same value on the drains of the M29 and M24 transistors.
- The M43 transistor will make 0 V the gate of the M29 transistor, the drain, and the drain of the M27 transistor.
- The M42 transistor makes 0 V the gate of the M32 transistor, the gate and drain of the M36 transistor, as well as the drain of M35.
- The M44 transistor will connect the gates of the M56, M25, and M35 transistors, as well as the gate and drain of M37, to the high potential VDD value.
- The M54 transistor will connect gates of the M24, M27, and M51 transistors, the drain of M49, as well as the source and gate of M50, to the high potential VDD value.

A 10-year aging HSPICE [66] simulation was performed for the basic RVS circuit under conditions of VDD = 1.2 V voltage and 125 °C temperature, from which the maximum degradation values of the circuit's main parameters are visible (Table 5.2) [69].

Summarizing the simulation results, it becomes clear that the reference voltage and current values have deteriorated by 33.3% and 31.5%, respectively.

Thus, the main parameters of the basic BGR, designed using only transistors with thin gate oxide, do not meet modern reliability standards and will lead to circuit failure. Such parametric degradations will also negatively affect the efficiency of the entire information transmission system, since the BGR is the most important sub-circuit of the system.

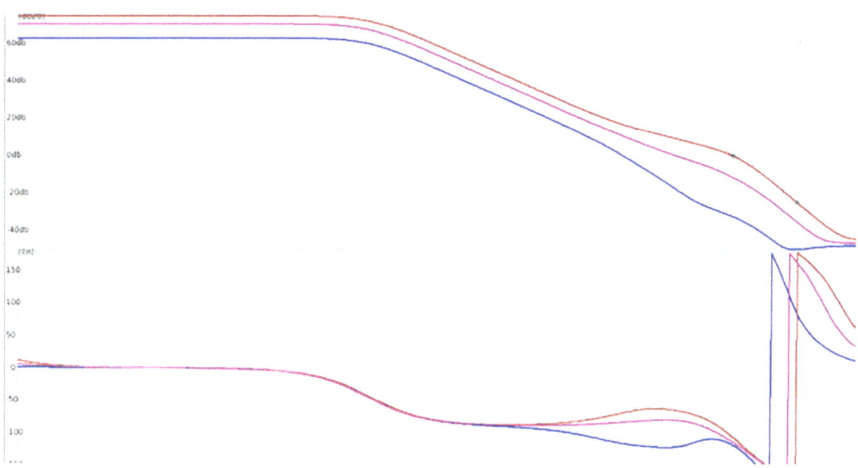

Fig. 5.20 Amplitude–frequency and phase–frequency characteristics of the OpAmp

Table 5.2 Values of the main parameters of the BGR circuit

Parameter	Initial	After 10 years of aging
Gain (dB)	67,3	45,7
V_{bias} (mV)	800±8	620±35
I_{ref} (mA)	0,25	0,2

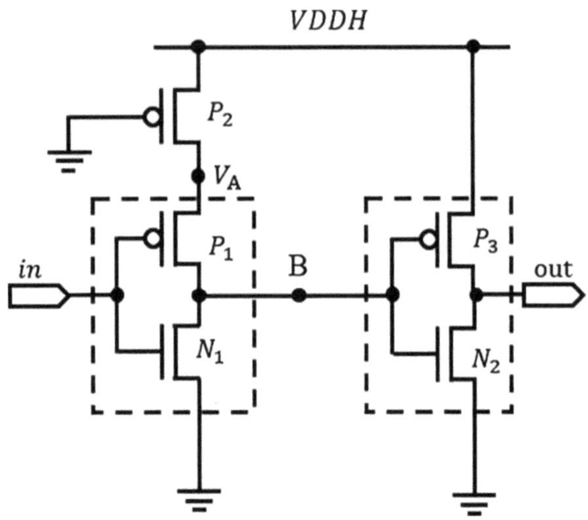

Fig. 5.21 Two-stage low-to-high LS circuit

5.1.3 *Existing Solutions and Key Issues to Reduce the Impact of Efficiency Degradation Factors*

The main issues discussed previously can lead to parameter degradations in the mentioned sub-circuits, which in turn can cause failures and lead to inefficient information transmission. To avoid all this, it is necessary to develop new methods that will reduce the impact of efficiency degradation factors. There are many developed methods to solve these problems, each of which has advantages and disadvantages.

This subchapter will present some of the known approaches aimed at reducing efficiency degradation factors and discuss their problems and shortcomings.

Existing Means of Increasing the Efficiency of a Level Shifter (LS)

The above showed the problems in basic LS circuits caused by efficiency degradation phenomena, which led to a decrease in the efficiency of transmission systems.

A two-stage low-to-high voltage level conversion method aimed at increasing efficiency in LS circuits has been considered. The method is based on a step-up structure where each inverter of the repeater circuit is powered by different voltages (Fig. 5.21) [70, 71].

The output inverter, composed of the P3 and N2 transistor pair, is powered by VDDH voltage, while the primary inverter, composed of the P1 and N1 pair, is powered by a voltage lower than VDDH. The P2 transistor is always on and is used to reduce the supply voltage level applied to P1, thus avoiding overvoltages between the pins of the P1 transistor. When the input voltage switches to 0, node B is charged through the P1 and P2 transistors. Thus, V_d becomes VDDH $-R_2I_{sd}$, where R~2~ is the equivalent resistance of the connected P2 transistor, which reduces the source voltage of the P1 transistor, and I_{sd} is the current flowing from source to drain of that transistor. It is important to note that the CMOS transistors used in the work satisfy the following condition: $|V_{thp}|$>VDDH – VDD, i.e., the P1 transistor is off when a VDD voltage level is applied to its input. Failure to satisfy this condition can result in static current between the supply voltage and ground when VDD voltage is applied to the input. The main difference of this method, compared to the basic low-to-high LS circuit, is that the number of transistors used is reduced, and there is no feedback structure here [70, 71].

The proposed two-stage LS circuit can operate as a power management circuit if the 0 voltage value applied to the gate of the P2 PMOS transistor is replaced by a control signal (Fig. 5.22) [70, 71]. The proposed LS circuit can cut the connection between the previous and next circuits, thus serving as a power management circuit. This is quite efficient in low power consumption designs, such as multi-supply systems.

The proposed two-stage LS circuit was compared with the basic low-to-high LS circuit (Table 5.3) [70, 71].

From the comparison results, the delay time and power values of the circuit designed using the proposed means, compared to the basic circuit, decreased by 13.26% and 21.18%, respectively. Thus, it becomes clear that the proposed means increase the efficiency of low-to-high LS circuits.

Considering the requirements of long-term operation standards for modern ICs, a multi-step 10-year aging simulation was performed for the existing circuit (Fig. 5.23) [70, 71].

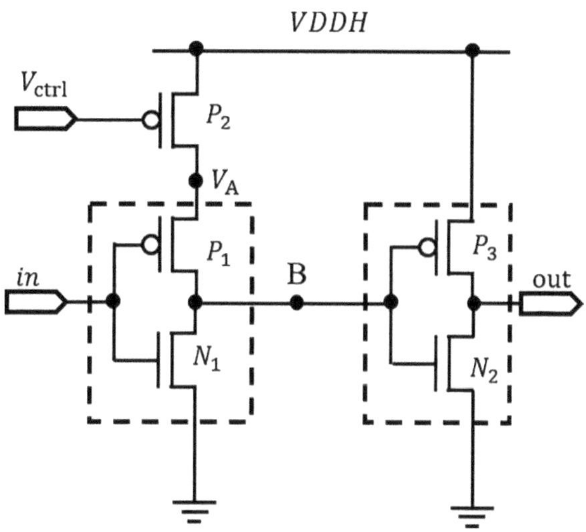

Fig. 5.22 Two-stage LS as a power management circuit

Table 5.3 Comparison of delay and power consumption of LS circuits

LS structure	Power (μW)	Delay time (ps)
Basic	8,88	119
Two-stage	7,84	98,2

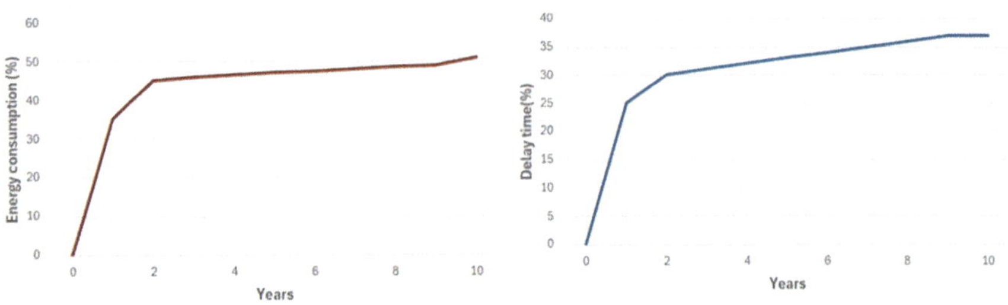

Fig. 5.23 Changes in the increase of power consumption and delay time of the two-stage LS circuit over 10 years

From the simulation results, it becomes clear that the power consumption and delay values deteriorated by 51% and 37%, respectively, over 10 years.

Thus, with the application of the existing method, power consumption and delay time were improved in the LS, and the impact of aging phenomena was mitigated. Despite this circumstance, the impact of aging phenomena on the circuit's main parameters is quite high, which can lead to further problems. The main disadvantage of the proposed method is the reduction of the LS's functional operation, since in this case there is a single output instead of basic differential outputs. This will lead to a sharp reduction in the scope of the circuit's application. Therefore, considering the above, there is a need to develop new methods.

Existing Means of Increasing the Efficiency of Reference Voltage Sources Based on Embedded Cascode Operational Amplifiers

When studying the factors of efficiency degradation in RVS circuits previously, it was noted that the main problems arise in the OpAmp, which is the main part of the RVS circuit, and replacing it with transistors with thin gate oxide. In this case, the impact of aging phenomena increased, leading to significant degradation of the circuit's parameters in off mode.

A known approach to reduce aging phenomena in a folded cascode OpAmp in off mode has been considered. Significant degradation of the parameters of the input transistors of the folded cascode OpAmp was observed in off mode.

To solve the problem, T1 and T2 CMOS switches were added for cutting and independent control of the OpAmp's input transistors. P4, P5, P6, N7, and N8 transistors were also added to reduce the impact of aging phenomena (Fig. 5.24) [72].

As mentioned earlier, the use of thick oxide transistors is not possible in modern technologies, which increases the impact of aging phenomena on the circuit and leads to efficiency degradation. A BGR circuit based on a folded cascode OpAmp was designed using the existing method, using only transistors with thin gate oxide (Fig. 5.25) [69, 72], for which a 10-year aging simulation was performed.

The simulation results were compared with the basic BGR circuit based on the folded cascode OpAmp (Table 5.4) [69, 72]. The results show that the main parameters of the BGR circuit using the existing method have significantly improved compared to the circuit using the basic method.

As a result of 10 years of aging using the existing BGR method, the gain coefficient deteriorated by 20.49%, the reference voltage value by 17.64%, and the reference current value by 13.63%.

Thus, with the application of the existing method, efficiency was increased in the BGR by mitigating the impact of aging phenomena on the main parameters of the BGR, but this method is not applicable in the latest technologies where there are no thick oxide transistors. Since the parametric degradations after 10 years of aging exceed the allowable 10% deviation and will lead to BGR circuit failure after approximately 6 years, there is a need to develop new methods that will increase the reliability of the BGR and ensure its applicability in the latest technologies.

Existing Methods for Managing Low Power Consumption in ICs

Power efficiency is one of the important requirements for efficient information transmission in ICs. There are several sub-circuits connected in the system, each performing a specific function. However, it may happen that some sub-circuits are temporarily inactive. If it is possible to turn off the supply voltage of such

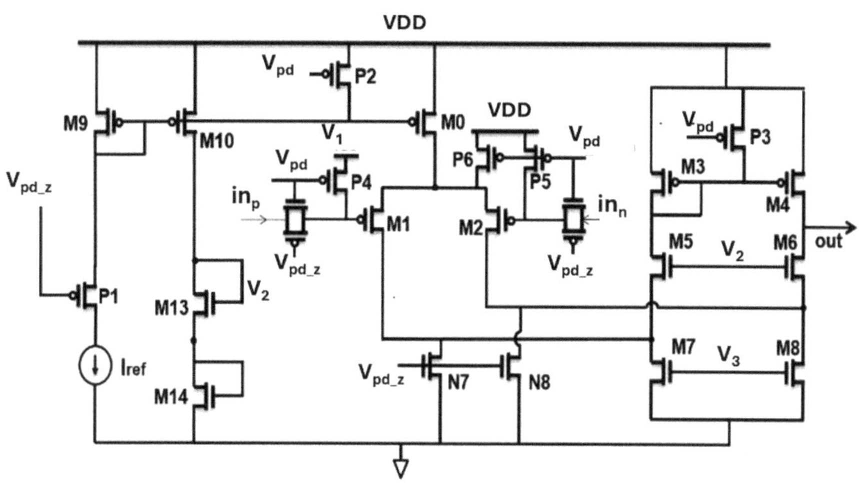

Fig. 5.24 Existing folded cascode OpAmp circuit for reducing aging phenomena

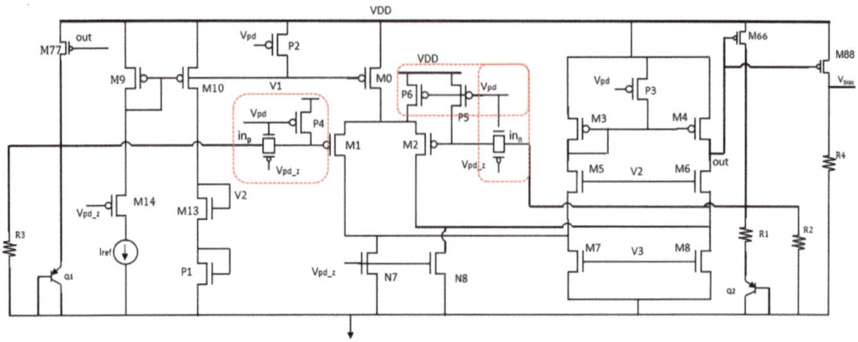

Fig. 5.25 BGR circuit based on folded cascode using the existing method

Table 5.4 Main parameters of the BGR circuit using existing and basic methods

	BGR using basic method		BGR using existing method	
Parameter	Initial	After 10 years aging	Initial	After 10 years aging
Gain (dB)	67,3	45,7	68,2	56,6
V_{bias} (mV)	800±8	620±35	800±8	680±20
I_{ref} (mA)	0,25	0,2	0,25	0,22

sub-circuits, then it will be possible to increase the energy efficiency of the entire system. Turning off or reducing the supply voltage will also have a positive effect on the system's reliability, as in that case they will be spared from overvoltage. There are many developed means to ensure this. The method of implementing low power consumption and standby mode is widely used in analog and digital circuits. It allows reducing the sub-circuit's power consumption to a minimum, while maintaining the necessary minimum supply level to bring the sub-circuit to operational state in the shortest possible time if necessary (5.14).

Many ICs have special input-output pins intended for standby mode control (Fig. 5.26) [73].

The advantage of having a pin for low power consumption control is that it requires a short switching time, is quite simple from an implementation and control perspective, and can be controlled by an external sub-circuit, and provides high energy efficiency, as it does not have additional current consumption in idle mode. However, this method has significant disadvantages: the presence of an external pin is necessary, which is a big problem for ICs with limited pins and can cause serious obstacles during physical design. An additional external logic presence is necessary, the inaccurate implementation of which will lead to functional failure of the entire IC. It should also be noted that it is quite sensitive to external noises. Therefore, the application of the method in ICs is quite limited.

However, there are other means that do not have external pins intended for standby mode control. Instead, they use built-in circuits with continuous time monitoring for standby mode detection and control (Fig. 5.27) [74, 76, 85].

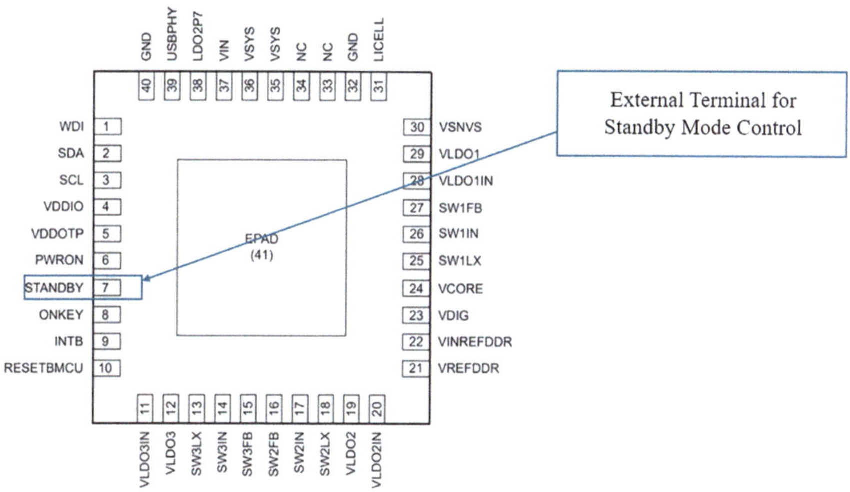

Fig. 5.26 External pin for standby mode control of PF1510 PMIC IC

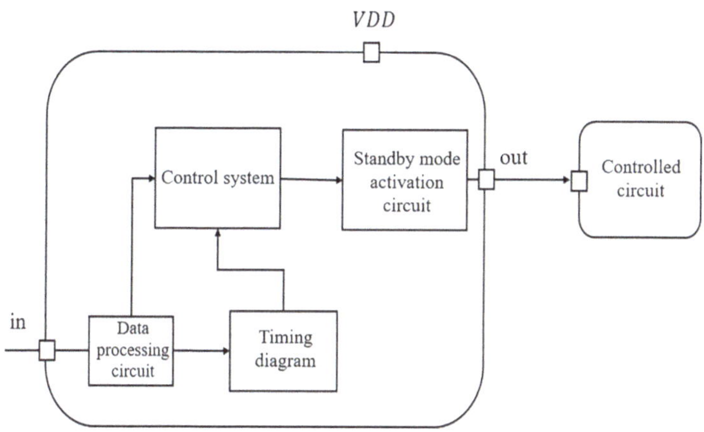

Fig. 5.27 Built-in system for standby mode control

Unlike the previous method, in this case, there is no need to use an additional external pin. The built-in structure requires the use of finite state machine logic, which constantly consumes static power. The input data is processed and controlled through information processing and time monitoring circuits for precise time control and subsequent determination of the standby mode state. The standby mode state is determined through the main control system, and the desired system's low power consumption mode is controlled through the standby mode activation circuit. With the application of this method, a certain time is required for mode switching [76]:

- 3 ms absence of input data is required to confirm the system's standby mode, and up to 10 ms time to switch the controlled system to low power consumption mode.
- Transition from standby mode to operational mode is done via a resume signal, which can have a duration from 1 ms to 15 ms, and approximately 10 ms of time is required to resume operational mode.

The application of the method provides great functionality in low power consumption modes and allows dynamically controlling power consumption. Among the main disadvantages of the method is the complex structure, due to which switching from operational mode to low power consumption mode requires a long time and consumes static power.

Thus, considering the advantages and disadvantages of both existing methods, there is a need to design a new built-in circuit for standby mode control that will not have static power consumption and will have a smaller switching time.

5.1.4 Proposed Principles to Increase the Efficiency of Information Transmission Systems Between Integrated Circuits

Considering the shortcomings of the existing methods for increasing the efficiency of information transmission systems between ICs discussed above, which do not sufficiently reduce the phenomena of efficiency degradation, there is a need to develop new means. Therefore, the following methods have been proposed:

1. Method to increase efficiency through the structure of a multi-output voltage level converter:

As a result of the simulation performed for the basic LS circuit above, it became clear that under the influence of aging phenomena, during a 10-year operational life, an 80% and 60% degradation in power consumption and delay time are observed, respectively, a significant part of which occurred during the first few years. The method considered above allowed reduced degradations, but in that case, the functionality of the LS decreased.

As a result of applying the proposed method, the functionality of the LS has been increased by providing six output voltage levels instead of one output in the existing method and two differential outputs in the basic circuit. The parametric degradations of the LS because of aging phenomena have also been reduced at the expense of an increase in area, which is due to the addition of extra transistors.

2. Method to increase the efficiency of the reference voltage source with thin oxide layer transistors:

Aimed at solving the problems discussed above, a method has been proposed that will allow avoiding failures due to aging phenomena in RVS circuits using only transistors with thin gate oxide. As a result of applying the method, it has been possible to ensure a 10-year failure-free operational life for the RVS for the latest technological processes at the expense of an increase in the area occupied on the semiconductor chip.

3. Method to implement low power mode control with differential information monitoring:

Considering the disadvantages of the two existing methods for low power consumption mode control presented above, an approach has been developed that will allow, without static power consumption, thanks to the monitoring of a differential signal, to switch the supply voltage of the controlled system in a short time compared to the existing method.

Conclusions

1. Along with the continuous scaling of modern technological processes, reducing the impact of efficiency degradation phenomena has become one of the most important fundamental issues in IC design. The impact of such phenomena can lead to parametric degradation of the sub-circuits of transmission systems, even up to failure. Therefore, the development of means to increase the efficiency of information transmission systems between ICs is a current problem.
2. The existing methods aimed at increasing the efficiency in information transmission systems between integrated circuits are not sufficient to overcome the current challenges, and in some cases, they are even absent, which leads to the need to develop new methods.
3. Approaches have been proposed to increase the efficiency of information transmission systems between integrated circuits, which, at the expense of an increase in power consumption, occupied area, and simulation duration within acceptable limits, ensure the necessary operating conditions.

5.2 Proposed Solutions to Increase the Efficiency of Information Transmission Systems Between Integrated Circuits

The previous chapter discussed and showed the existing approaches and their shortcomings for increasing the efficiency of the most important sub-circuits that are part of information transmission systems between ICs. The continuous growth of application areas of modern ICs has led to stricter requirements placed on them. Therefore, to meet the requirements, it is necessary to solve the problems that have arisen in the existing methods or develop approaches that will consider the phenomena of efficiency degradation in information transmission systems. Below are the proposed solutions for increasing the efficiency of information transmission systems between ICs, which correspond to current requirements and solve the shortcomings present in existing methods.

5.2.1 Method to Increase Efficiency Through the Structure of a Multi-Output Voltage Level Converter

As noted above, the efficiency problems in the circuit using the existing low-to-high LS method become more apparent in long-term operating conditions. Therefore, a method aimed at increasing efficiency in LS circuits has been developed, which will ensure the long-term operation of LSs and meet the current standards for efficiency [67, 68, 77].

From previous chapter, it became clear that in the basic low-to-high LS, significant degradations of transistor parameters were observed because of aging phenomena, which led to a decrease in the circuit's efficiency. To solve the problem, the known approach of cascoding was applied in the basic low-to-high LS circuit (Fig. 5.28) [68]. Using this method, it is possible to avoid overvoltage in transistors, as well as stress conditions.

A constant N_{bias} voltage value is applied to the gates of the M5 and M6 transistors, thus limiting the drain and source voltage levels, because, for voltages higher

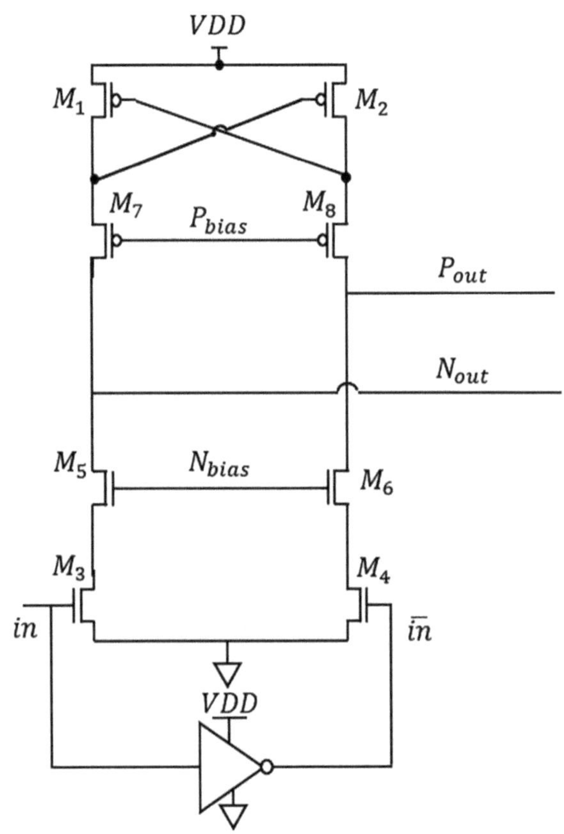

Fig. 5.28 LS circuit using the cascoding method

than N_{bias}, the transistors will be in the off state. However, the correct choice of constant voltage for these transistors allows obtaining the desired result [67, 68, 77].

The operating principle of the circuit using the cascoding method is as follows. The input signal is controlled by the M3 and M4 transistors. The input signal is formed by the VDD level, which is the circuit's low supply voltage, and the output signal is the high VDDH. When the input signal receives a logical "1" level, the M3 transistor turns on, and M4 remains off. Thanks to the M5 and M6 transistors, the switching of the input transistors is ensured without overvoltage problems. The output N_{out} and P_{out} signals will provide a stable voltage depending on the input voltage level. Thus, for an input VDD = 0.7 V low voltage level, a VDDH = 1.2 V high voltage level will be obtained at the output [40–43, 68].

The circuit was designed for the SAED 14 nm FinFET technological process, using only transistors with thin gate oxide, which ensures the circuit's application also in modern technologies where there are no thick oxide transistors.

A 10-year multi-step aging simulation was performed for the low-to-high LS circuit using the HSPICE [66] tool, under conditions of VDDH = 1.2 V, VDD = 0.7 V voltages and 125 °C temperature (Fig. 5.29) [68].

From the simulation results, in the LS using the cascoding method, the maximum degradation of the transistor saturation current is 10% instead of 60% for the basic circuit, and the maximum change in threshold voltage is 50 mV instead of 220 mV (Fig. 5.30) [68].

To increase efficiency, a method has been developed that allows having a reliable multi-output low-to-high LS circuit (Fig. 5.31) [68], thus providing greater functionality and efficiency.

The final cascade circuits, which were added to the cascoded low-to-high LS, allow obtaining additional output voltages. Their presence also ensures the reliability of subsequent circuits, because in this case it will be possible to apply a voltage value to the transistor pin that allows avoiding overvoltages [77, 78].

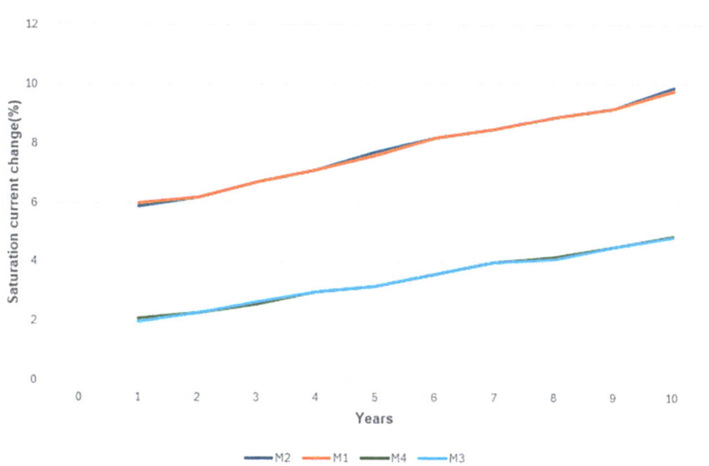

Fig. 5.29 Changes in saturation currents of the cascoded low-to-high LS transistors over 10 years

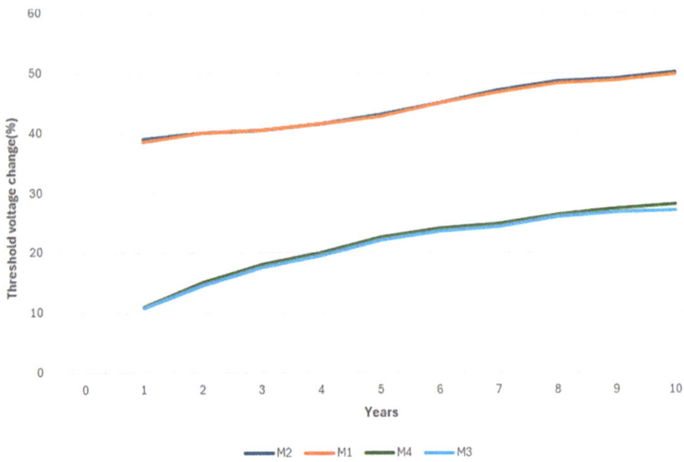

Fig. 5.30 Changes in threshold voltages of the cascoded low-to-high LS transistors over 10 years

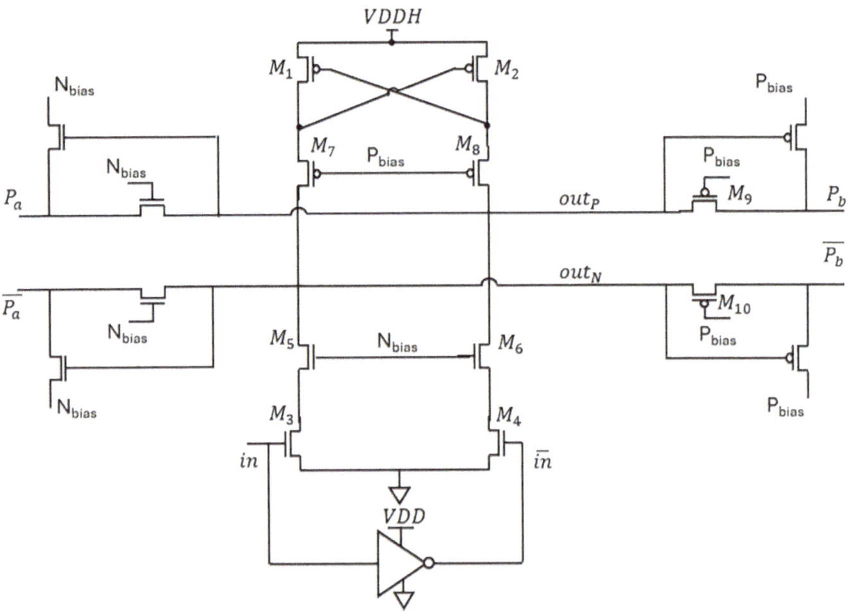

Fig. 5.31 Multi-output low-to-high LS circuit

The output voltages of P_b can have 0.4 V and 1.2 V voltage values, and the output voltages of P_a can have 0 V and 0.6 V (Fig. 5.32) [68]. This allows using PMOS and NMOS transistors with thin gate oxide as switches in the added cascades. For example, when the P_b output is 0.4 V, and the source of the M9 transistor is connected to a 1.2 V voltage, the PMOS can turn on without overvoltage problems.

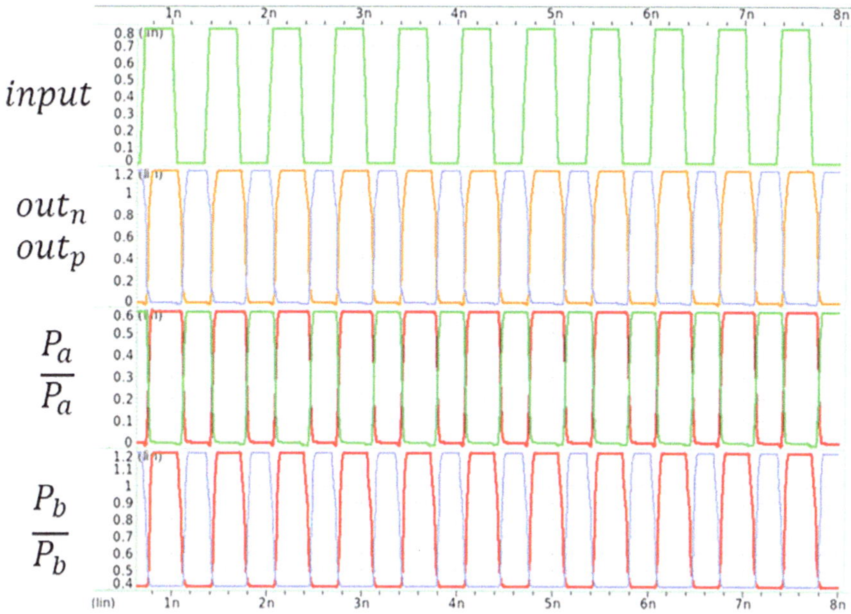

Fig. 5.32 Results of timing simulation of the multi-output low-to-high LS circuit

Using the proposed method, a 10-year aging simulation was performed for the multi-output LS circuit using the HSPICE [66] software tool under conditions of degradations VDDH = 1.2 V, VDD = 0.8 V voltages and 125 °C temperature (Fig. 5.33) [68].

From the obtained results, the maximum degradation of rising and falling edges of multi-output LS using the proposed method after 10 years of aging is 9.14% for the rising edge and 10% for the falling edge, which fully meets the current efficiency standards.

For an accurate assessment of efficiency, the changes in power consumption and delay time of multi-output LS circuit under 10-year operational life conditions were measured and compared with existing and basic LS circuits (Figs. 5.34 and 5.35) [67, 68, 71].

The change in power consumption of the proposed multi-output LS over 10 years was 17% instead of 51% and 83% for the existing and basic circuits, respectively.

The change in the increase of delay time was 12% for the proposed two-stage LS instead of 37% and 60% for the existing and basic circuits, respectively.

With the application of the proposed method, the physical design of the circuit increased by approximately twice, compared to the basic one (Figs. 5.36 and 5.37).

Thus, bccause of applying the proposed method, it was possible to increase the efficiency of the low-to-high LS by adding multiple outputs, which allow having additional output voltages. Their presence will also ensure the reliability of subsequent circuits.

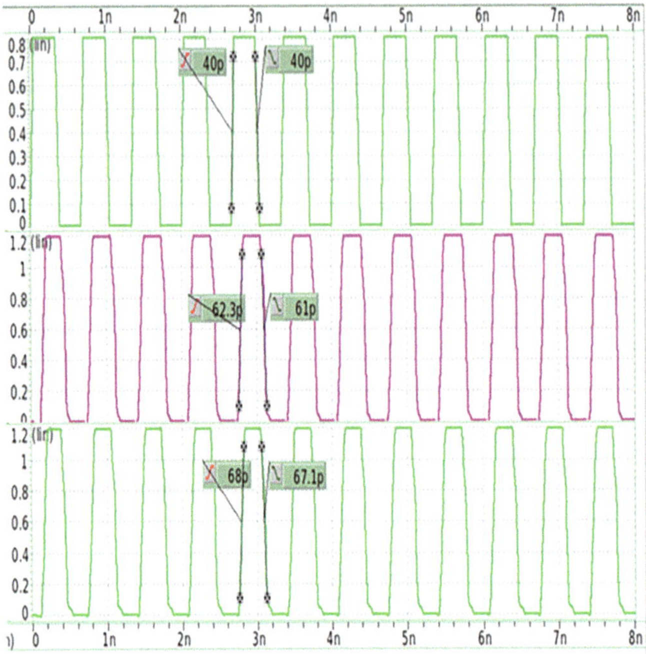

Fig. 5.33 Worst-case changes in rising and falling edges of multi-output LS using the proposed method over 10 years

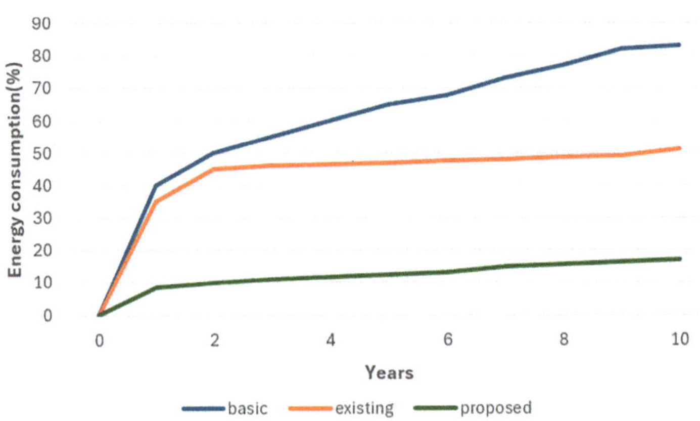

Fig. 5.34 Change in power consumption of the proposed LS circuit over 10 years

The degradation values of the increase in power consumption and delay time have also been reduced compared to the existing method by 3 and 3.08 times, respectively, and compared to the basic circuit by 4.8 and 5 times, at the expense of an approximately two-fold increase in the physical design area. It should also be noted that with the application of this method, the circuits were designed using only

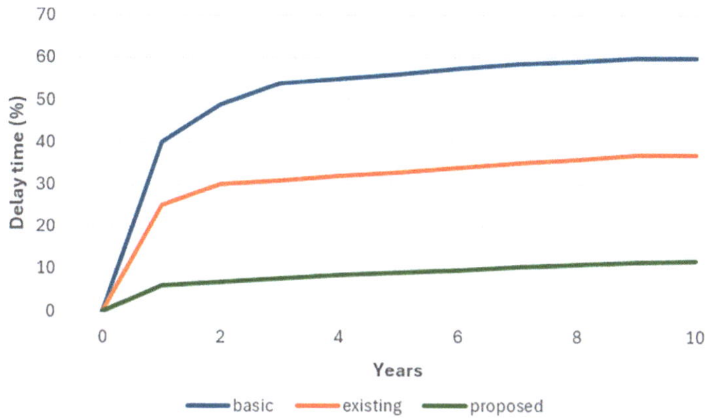

Fig. 5.35 Change in the increase of delay time of the proposed LS circuit over 10 years

transistors with thin gate oxide, which allows the application of the method in the latest technologies.

5.2.2 Method to Increase Efficiency of Bandgap Reference with Thin Oxide Layer Transistors

As noted above, the design of the BGR using the existing method, considering the requirements of modern technological processes, was done using only transistors with thin gate oxide. However, from the results of the 10-year simulation, it becomes clear that because of aging phenomena, after approximately 6 years, the BGR circuit will fail. Therefore, a method aimed at increasing efficiency in BGR circuits has been developed, using only transistors with thin gate oxide, which ensure the long-term operation of BRs and meet current reliability standards.

It was mentioned above that in the folded cascode OpAmp used in the basic BGR circuit, there are transistors between the pins of which overvoltages occur in off mode, leading to parametric degradation.

Aimed at solving the problem, a folded cascode OpAmp circuit constructed only with thin gate oxide transistors has been proposed, using CMOS switches and additional transistors (Fig. 5.38) [69]. The goal of the proposed method is to set corresponding 0 V or VDDH voltage values to all NMOS and CMOS transistor pins in off mode, which will allow avoiding overvoltage between the transistor pins.

As a result of the proposed modifications in the folded cascode OpAmp:

- The drain of the M35 transistor will have a VDDH value in off mode, thanks to the M58 transistor, instead of the previous 0 V value. The CMOS switch composed of M57 and M59 transistors breaks the connection between the drains of

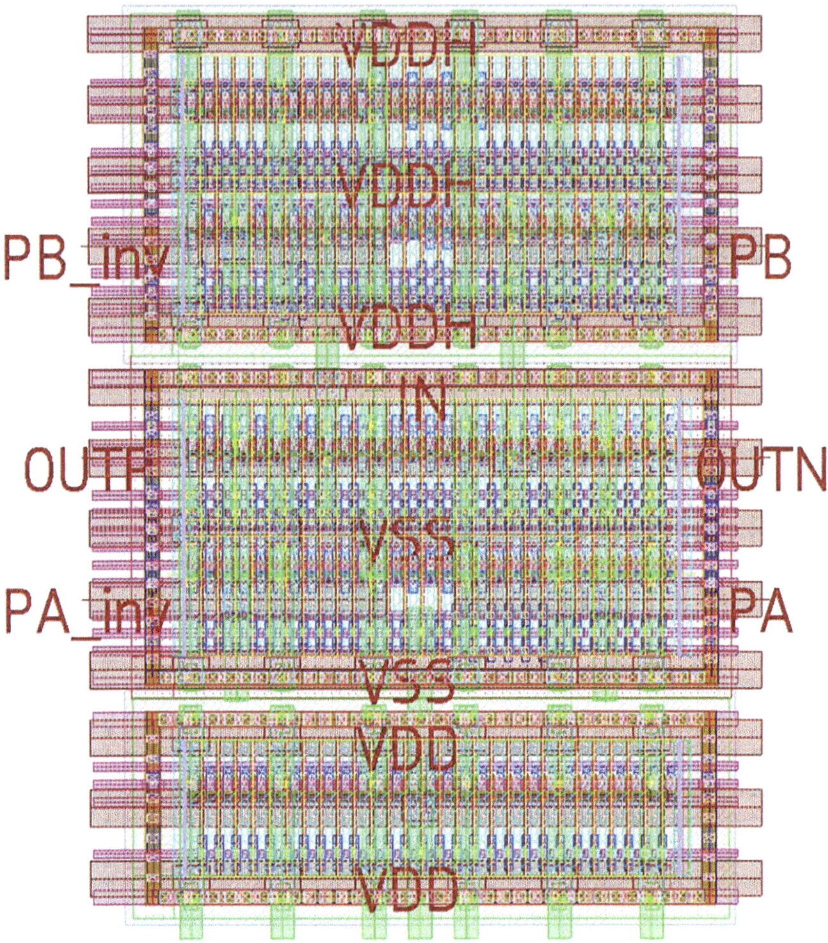

Fig. 5.36 Physical design of two-stage LS

M35 and M36 transistors in off mode, thus preventing a short circuit between VDDH and ground [69, 79–81].

- The M71 and M73 transistors make the drain and source values of the M27 transistor $V_{\text{sup.high}}$. As a result, the potential difference between all pins of the M27 transistor becomes 0 V, because the gate is also given a VDDH voltage value through the M54 transistor. To avoid a short circuit, the CMOS switch composed of M66 and M67 breaks the connection between the sources of the M27 and M28 transistors.
- The M70 and M72 transistors, to prevent overvoltage between the drain and source of the M24 transistor, connect them to VDDH. The M64 and M65 CMOS switch breaks the connection between the drains of M24 and M29 transistors to avoid a short circuit between VDDH and ground.

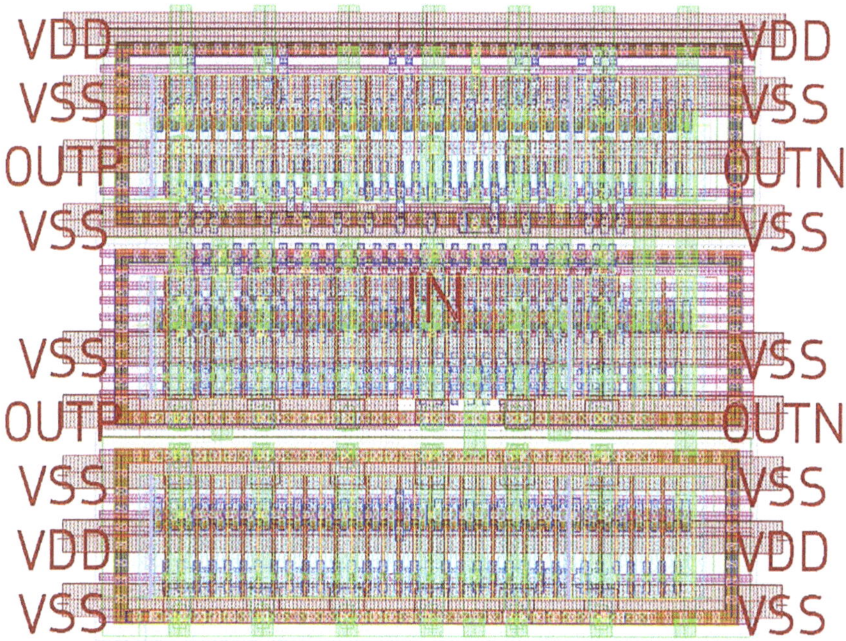

Fig. 5.37 Physical design of basic LS

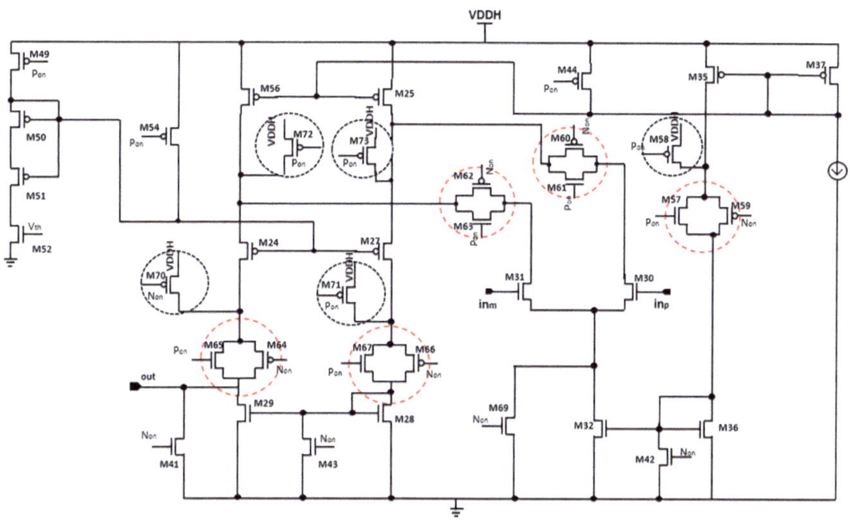

Fig. 5.38 Folded cascode OpAmp circuit using the proposed method

- The M72 and M73 transistors connect the drains of the M56 and M25 transistors to VDDH. The CMOS switches composed of M62, M63 and M60, M61 transistors break the connections of the drains of the M31 and M30 transistors from the drains of the M56 and M25 transistors, respectively.
- The M69 transistor grounds the drain of M32 in off mode.

A 10-year aging simulation was performed for the folded cascode OpAmp using the proposed method with the HSPICE software tool under conditions of VDDH = 1.2 V, and 125 °C temperature (Table 5.5) [69].

Thus, because of applying the proposed method in the folded cascode circuit, the M27, M24, M56, M25, M32, M30, and M31 transistors were protected from overvoltages. After the changes, the maximum threshold voltage degradation was 27 mV instead of the previous 245 mV, and the maximum drain current degradation changed from 23.9% to 3.9% [69].

An RVS circuit based on a folded cascode OpAmp using the proposed method, composed only of thin gate oxide transistors, was designed (Fig. 5.39) [69], for which a 10-year aging simulation was performed, and the obtained results were compared with the existing BGR circuit based on the folded cascode OpAmp (Table 5.6) [69, 72].

The results show that as a result of 10 years of aging of the BGR circuit using the proposed method, the gain coefficient deteriorated by 7.32% compared to 20.49% for the BGR using the existing method, the reference voltage value by 12.67% instead of 17.64% for the existing method, and the reference current value by 8.69% instead of 13.63% for the existing method.

By mitigating the parametric degradations of the BGR circuit using the proposed method, the operational life span was extended from 6 years for the BGR using the existing method (Fig. 5.40) to 8 years (Fig. 5.41). A 10% deviation from the initial value of the reference voltage and current was taken as the circuit failure limit [82].

Thus, the reliability of the BGR increased by 33.3%, which testifies to the effectiveness of the proposed method. However, the circuit using the proposed method does not meet the 10-year reliability standards, either.

In the BGR (Fig. 5.42) [69], the impact of aging phenomena on the M6, M7, and M8 transistors was considered as forming an important part of the circuit. The M6 and M7 transistor pair equalizes the input voltages of the OpAmp, providing a temperature-independent voltage. The M8 transistor mirrors the current flowing through the M6 and M7 transistors and, through a voltage divider, provides different

Table 5.5 Maximum deviations of I_d and V_{th} of transistors as a result of 10-year aging of the depicted CP circuit

	Without method application		With method application	
Transistor name	ΔV_{th} (mV)	ΔI_d (%)	ΔV_{th} (mV)	ΔI_d (%)
M35	197	19,9	0,9	1,4
M31, M30	200	20,7	1,1	1,9
M25, M56	243	23,7	25	3,5
M27, M24	245	23,9	27	3,9

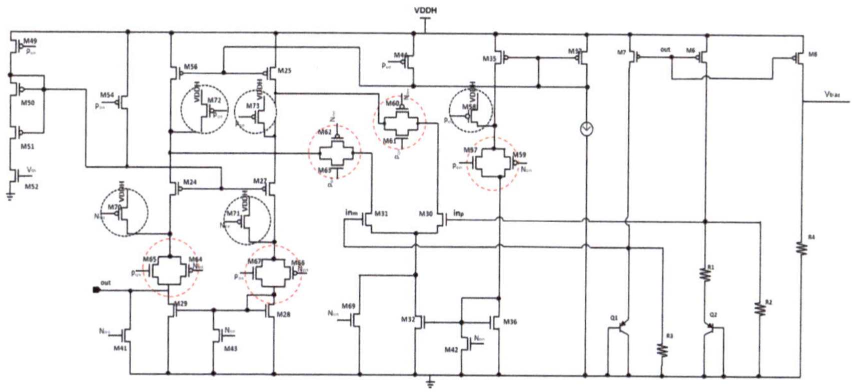

Fig. 5.39 BGR circuit based on folded cascode using the proposed method

Table 5.6 Main parameters of the folded cascode-based BGR using proposed and existing methods

	BGR using existing method		BGR using proposed method	
Parameter	Initial	After 10 years aging	Initial	After 10 years aging
Gain (dB)	68,2	56,6	67,4	62,8
V_{bias} (mV)	800 ±8	680 ±20	800 ±8	710 ±15
I_{ref} (mA)	0,25	0,22	0,25	0,23

reference voltage values. By reducing the impact of aging phenomena on the M6, M7, and M8 transistors, it will be possible to mitigate the degradation of the change in the circuit's reference voltage value.

The aging of bipolar transistors used in the BGR is mainly due to the BTI phenomenon. However, in this case, the bases of the bipolar transistors are connected to the collector, which prevents reverse biasing. Despite this, possible degradation should not be ruled out. If the resistors undergo degradation, then the changes in their parameters will be proportional, so the aging of the resistors will not affect the reference voltage value. Nevertheless, to be fully protected from aging, it is necessary to reduce the probability of parameter degradation of resistors [69, 72, 79–81].

- The sources of the M6, M7, and M8 transistors are connected to a high supply voltage value, and the drain voltage changes in operational mode, but in off mode it will have a 0 V voltage value. The output of the OpAmp, which is 0 in off mode, is connected to the gates of these transistors, because of which the transistors remain in the on state. This causes overvoltage between the sources and gates of transistors.
- The bases and collectors of Q1 and Q2 bipolar transistors are connected to a low voltage point. The emitter of Q1 is connected to the negative input of the OpAmp, and the emitter of Q2 is connected to a resistor, the other pin of which is connected to the positive input of the OpAmp. In off mode, a non-constant, "floating" voltage will appear at the OpAmp inputs, which are conditioned by bipolar transistors.

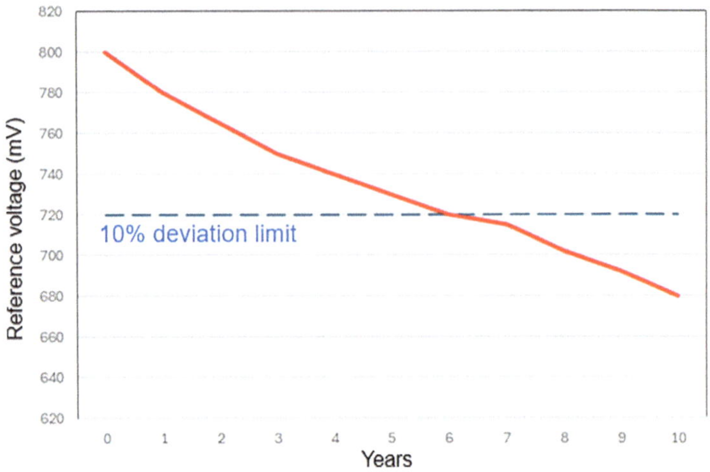

Fig. 5.40 Change in reference voltage of the folded cascode-based BGR using the existing method

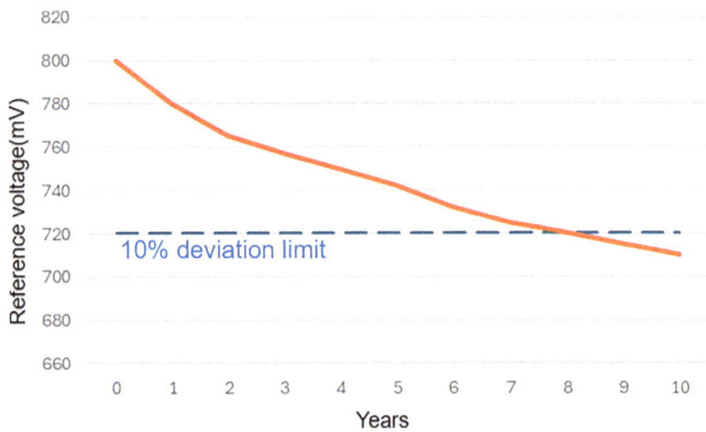

Fig. 5.41 Change in reference voltage of the folded cascode-based BGR using the proposed method

- The R4 resistor is connected between a low potential point and the drain of M8 transistor, which in off mode will also have a "floating" voltage value. This can cause a small leakage current, which is undesirable for minimizing the effects of aging phenomena.

Thus, to avoid overvoltages in off mode, the proposed method was also applied in case of the above-mentioned devices (Fig. 5.43). Through the added CMOS switches and additional transistors, corresponding 0 V or VDDH voltage values are set to all necessary pins in off mode.

- To turn off the M6, M7, and M8 transistors in off mode, their gates are given a VDDH voltage value through the M19 transistor. Similarly, through the M10,

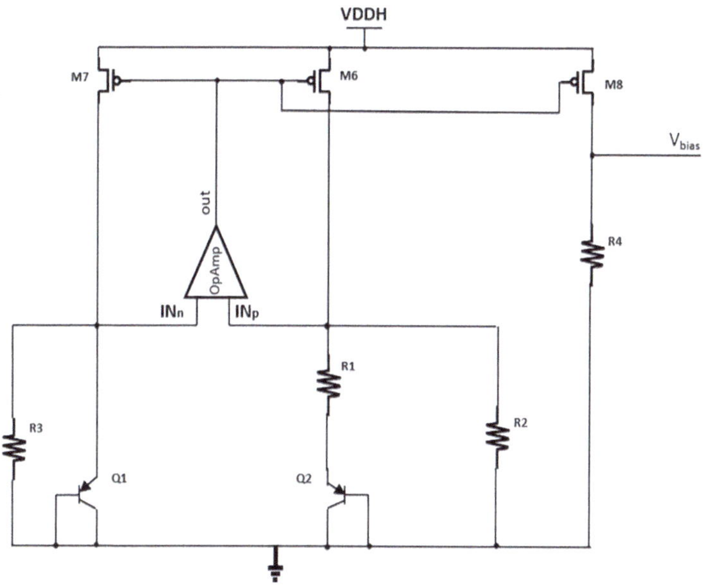

Fig. 5.42 Simplified circuit of folded cascode OpAmp-based BGR

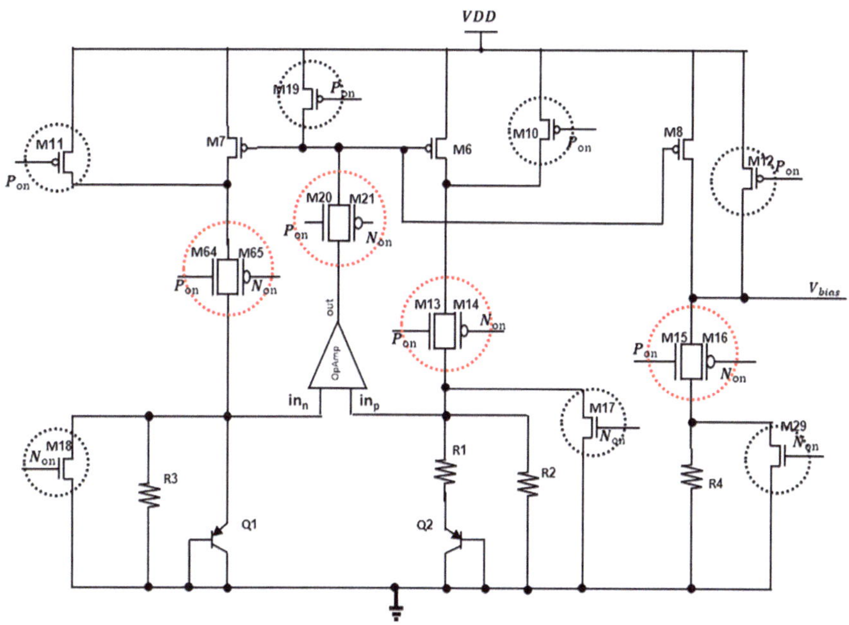

Fig. 5.43 BGR circuit using the proposed method

M11, and M12 transistors, the drains of the M6, M7, and M8 transistors receive a VDDH voltage value.

- The M18 transistor gives 0 V to the emitter of the Q1 bipolar transistor in off mode. The M64-M65 CMOS switch breaks the connection between the drain of the M7 transistor and the emitter of Q1 and the resistor R3, preventing a short circuit between VDDH and ground.
- In off mode, through the M17 transistor, the pin of the R1 resistor will be cut off to a 0 V voltage value. The M13-M14 CMOS switch breaks the connection between the drain of the M6 transistor and the R1, R2 resistors, preventing a short circuit between VDDH and ground.
- In off mode, through the M29 transistor, the voltage drop across the R4 resistor will be 0 V. The M15-M16 CMOS switch breaks the connection between the drain of the M8 transistor and the R4 resistor, preventing a short circuit.
- The M20-M21 CMOS switch breaks the connection between the OpAmp output and the gates of the M6, M7, M8 transistors to avoid a short circuit. Through the M17 and M18 transistors, the gates of the M30 and M31 input transistors of the OpAmp will assume a 0 V voltage value in off mode. And the M13-M14 and M64-M65 CMOS switches will break the connection between the OpAmp inputs and the M6, M7 transistors.

An BGR circuit using the proposed method, composed only of thin gate oxide transistors, was designed, for which a 10-year aging simulation was performed (Table 5.7 and Fig. 5.44).

Using the proposed method, the reference voltage value of the BGR circuit deteriorated by 6.66% because of 10 years of aging, and the reference current value changed by 4.16%, which do not exceed the 10% deviation limit, so no circuit failure will be recorded during the 10-year operational period.

Using the proposed method, the physical design of the BGR circuit increased by 11.2% compared to the basic one (Figs. 5.45 and 5.46).

Thus, because of applying the proposed method, it was possible to increase the efficiency of the BGR circuit using transistors with thin oxide gate by increasing reliability and making it more than 10 years, instead of 6 years for the existing method, but at the expense of an 11.2% increase in the physical design area.

Table 5.7 Main parameters of BGR using proposed methods

Parameter	Initial	After 10 years of aging
Gain (dB)	67,4	62,8
V_{bias} (mV)	800±8	750±12
I_{ref} (mA)	0,25	0,24

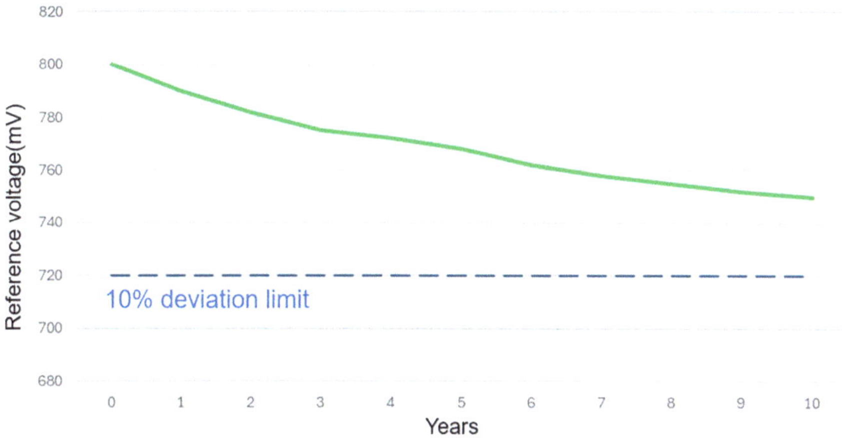

Fig. 5.44 Change in reference voltage of BGR using the proposed method over 10 years

Fig. 5.45 Physical design of the BGR using the basic method

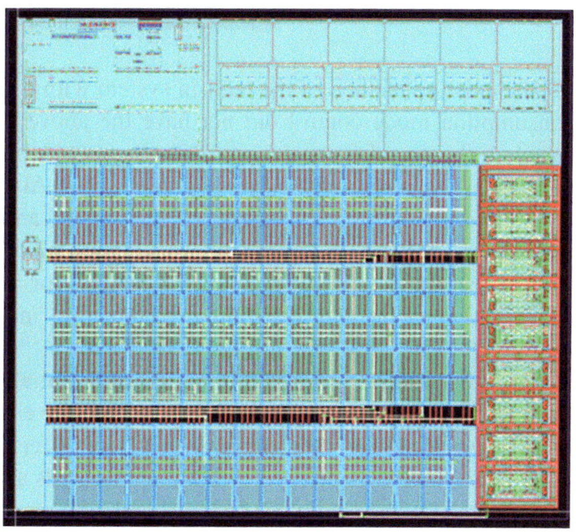

5.2.3 *Method to Implement Low Power Mode Control with Differential Information Monitoring*

As noted above, there are existing methods for low power consumption mode control in ICs to increase power efficiency: using an external pin and with a built-in system with time monitoring.

The existing method using an external pin does not have great applicability, because external logic is needed for control, as well as an external pin for the method's implementation, which is a big problem for ICs with limited pins.

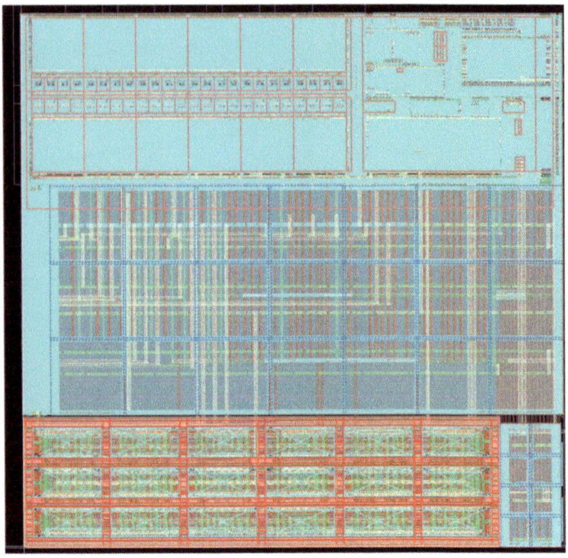

Fig. 5.46 Physical design of the BGR using the proposed method

The existing method of the built-in system with time monitoring for low power consumption mode control did not have the above problems, but the method is based on the use of finite state machine logic, which has a complex structure and constantly consumes static power [73–76, 83, 84]. Also, 3 ms absence of input data is required to confirm the system's standby mode, and up to 10 ms time to switch the controlled system to low power consumption mode.

To solve the mentioned problems, a method for built-in control of low power consumption mode with differential information monitoring has been proposed (Fig. 5.47) [85].

The proposed method is as follows: the differential lines of information are used to connect or disconnect circuits. It does not require either a constant supply voltage or an external control standby pin. Instead, the differential lines of information are used, which in the circuit serve as both input signals and power sources. Because of this, the need for a standby mode control pin and monitoring circuits is eliminated, and it also does not have constant power consumption.

Currently, to switch the power supply of an unnecessary sub-circuit, the presence of another sub-circuit in the system is necessary. This means that a system with low power consumption is needed, which will constantly monitor whether a given sub-circuit does not need power supply now. For an easy solution to the arisen problem, it is necessary to embed the proposed monitoring system into the circuit. Thus, the sub-circuit can be activated only if there is input information intended for processing. This means that the transmitted information itself can activate or deactivate the power supply.

During information transmission over differential lines, one of the signals always has a high voltage level, otherwise both are 0. If these are used as parallel inputs in a diode voltage divider, it will be possible to obtain the desired constant voltage at

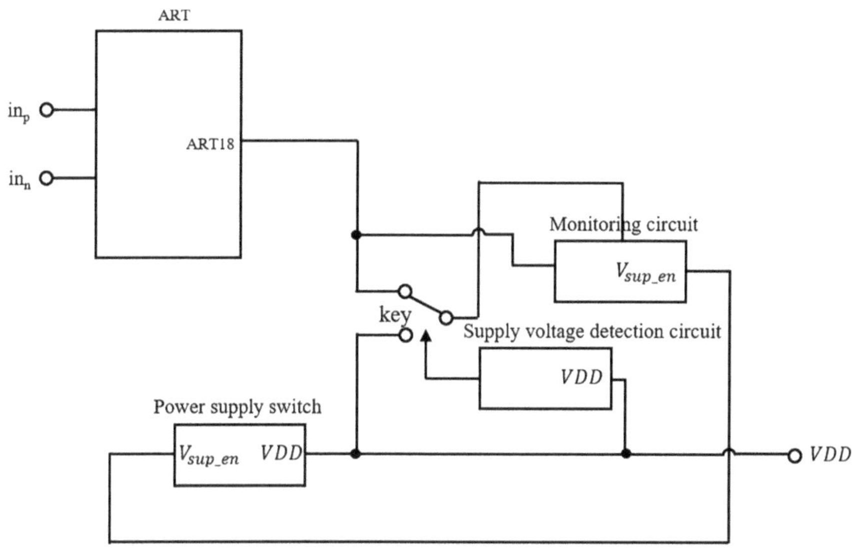

Fig. 5.47 Circuit using the method of built-in control of low power consumption mode

the output when information is transmitted. The divider's output is used as a control signal for low power consumption mode and as the supply voltage for the monitoring sub-circuit. Through the monitoring sub-circuit, in the absence of information, the supply VDD voltage is disconnected from the controlled system, ensuring its low power consumption state. After a pause of arbitrary duration, because of resuming information transmission, the former voltage switches the output of the monitoring sub-circuit, which connects the supply VDD voltage through a supply voltage switching circuit. The supply voltage of the built-in system will be switched from Art18 to VDD when the latter crosses the threshold limit of the supply voltage detection circuit.

The ART sub-circuit is a voltage divider composed of transistors connected in series with diodes (Fig. 5.48) [85]. in_p and in_n are the differential data inputs, and Art18 is the circuit's output. In the absence of information on the differential line, the in_p and in_n inputs will receive a 0 V voltage value, otherwise they will have a 3.3 V voltage. Through the transistors with diode connection, the 3.3 V voltage value at the Art18 output will decrease and have a 1.8 V voltage value.

Two transition CMOS gates are used to set two different threshold voltages for the circuit. When VDD is equal to zero, the circuit's output is also zero, because the input of the first inverter is also 0. When the value of VDD increases from zero, it passes through the CMOS gate with the higher threshold value, i.e., the one controlled by V_1. When VDD exceeds the set voltage threshold, the circuit's output starts to follow the VDD voltage value, and that CMOS gate turns off, the other one turns on instead, changing the threshold value. When the VDD voltage value starts to decrease from its set value toward zero, the previously described reverse actions occur, but now with a different threshold value (Fig. 5.49) [85].

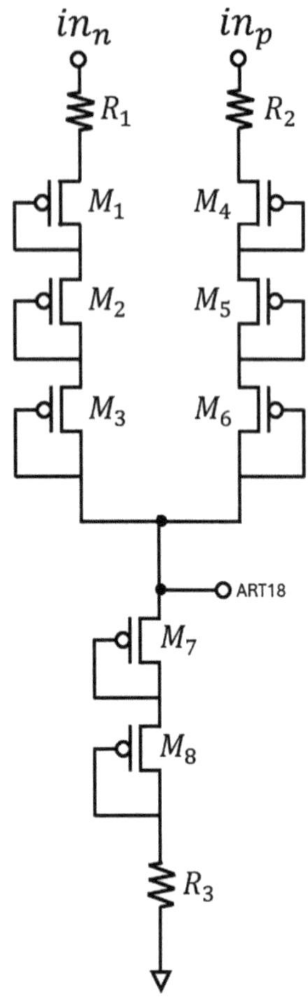

Fig. 5.48 ART data divider circuit

An HSPICE simulation was performed for the supply voltage detection circuit for all PVT cases. As a result, a maximum of 0.87 V and a minimum of 0.71 V were obtained for the low-to-high switching threshold, and for the high-to-low switching threshold, 0.69 V and 0.52 V, respectively (Table 5.8).

The difference between the switching threshold voltages ensures the noise immunity of the circuit (Fig. 5.50) [85].

In the absence of differential information, in_n and in_p, as well as VDD and Art18, will have a 0 V voltage value. In the presence of information on the differential line, in_n or in_p will have a high voltage value. A voltage close to VDD will be formed at the output of the ART circuit, which activates the monitoring circuit, which in turn will connect the other sub-circuits to the power source. When the voltage of the supply voltage switching node reaches the nominal value of 0.8 V, the output of the supply voltage detection circuit turns on and switches the power supply of the monitoring circuit from Art18 to the VDD supply voltage via a switch. When VDD reaches approximately 80% of the final value, the output of the supply voltage

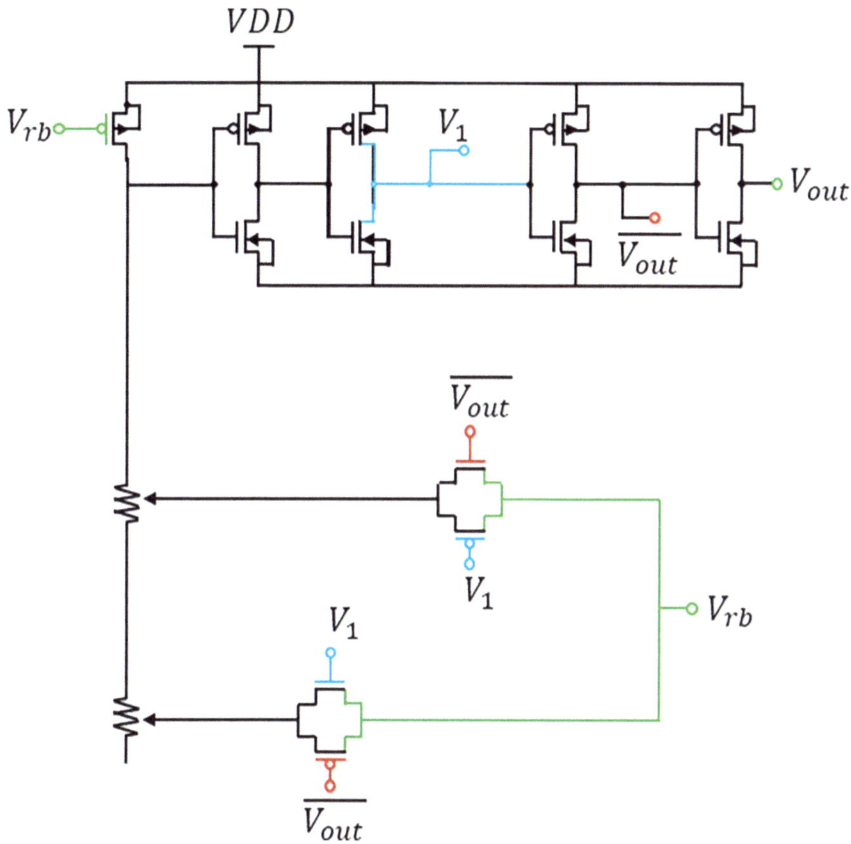

Fig. 5.49 Supply voltage detection circuit

Table 5.8 Change in switching threshold of the supply voltage detection circuit

	Minimum	Typical	Maximum
V_{th} High-to-Low (V)	0,52	0,58	0,69
V_{th} Low-to-High(V)	0,71	0,79	0,87

detection circuit is high enough to be able to completely switch the power supply of the monitoring circuit to VDD. The monitoring circuit, now powered by VDD, continues to monitor the information lines, and when information is absent again on the differential line, the power supply is disconnected again from the supply line via the supply voltage switch.

Timing simulation of the circuit using the proposed method was performed for all PVT cases (Fig. 5.51) [85].

The results of the performed timing simulation show that up to 100 μs time is required for switching from standby mode to operational mode and vice versa.

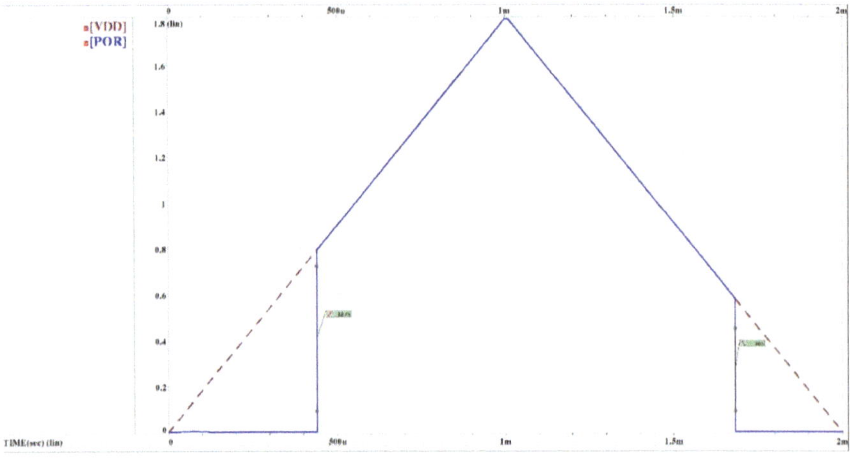

Fig. 7 Voltage detection circuit output and supply voltages

Fig. 5.50 Result of timing simulation of the supply voltage detection circuit

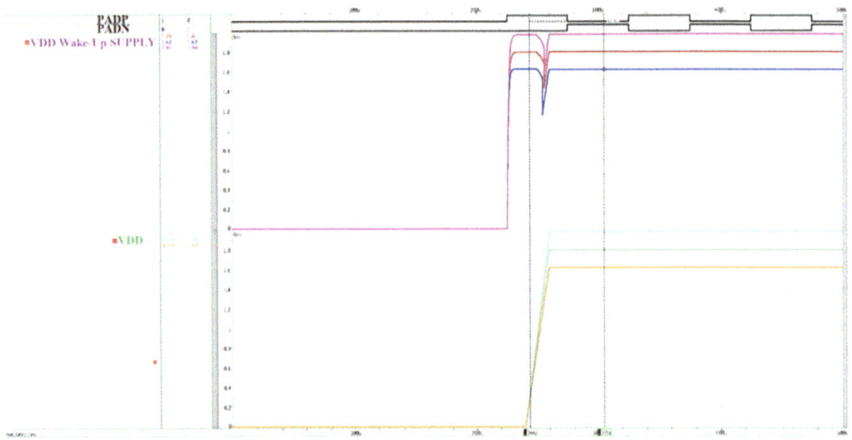

Fig. 5.51 Results of timing simulation of the circuit using the proposed method

Thus, with the application of the proposed method, the built-in system for standby mode control allows, because of differential information monitoring, to activate the supply voltage of the controlled circuit within a period of up to 100 μs, instead of up to 3 ms for the existing built-in method. However, in the case of applying the method, the activation process requires a temporary current expenditure from the differential lines of the information-sending circuit.

Conclusions

- A multi-output voltage level converter circuit has been developed, in which efficiency has been increased: the circuit's functionality has been increased, the impact of aging phenomena has been reduced, ensuring a reduction in the degradation of the increase in power consumption and delay from 83% and 60% to 17% and 12%, respectively, at the expense of an approximately two-fold increase in the physical design area.

- A reference voltage source circuit using only thin gate oxide transistors has been proposed, where the parametric degradations due to aging phenomena have been reduced, ensuring <10% deviation of the output reference voltage and current values during a 10-year operational life, at the expense of an 11.2% increase in area.
- An approach for built-in control of low power consumption mode with differential information monitoring has been created, ensuring activation of the low power consumption mode within up to 100 μs, instead of 3 ms, at the expense of a temporary current expenditure from the differential lines of the information-sending circuit.

References

1. B. Razavi, *Design of Analog CMOS Integrated Circuits*, 2nd edn. (Mc Graw Hill India, 2017), 782p
2. K. Zhang, Circuit design in nano-scale CMOS technologies, in *2018 International Symposium on VLSI Design, Automation and Test (VLSI-DAT). Hsinchu, Taiwan*, (IEEE, 2018), p. 1. https://doi.org/10.1109/VLSI-DAT.2018.8373276
3. G. Yang, H. Jung, J. Lim, et al., Standard cell design optimization with advanced MOL technology in 3 nm GAA process, in *2022 IEEE Symposium on VLSI Technology and Circuits (VLSI Technology and Circuits). Honolulu, HI, USA*, (IEEE, 2022), pp. 363–364. https://doi.org/10.1109/VLSITechnologyandCir46769.2022.9830450
4. R.K. Bassett, *To the Digital Age: Research Labs, Start-up Companies, and the Rise of MOS Technology* (Johns Hopkins University Press, 2002), pp. 53–54. ISBN 978-0-8018-6809-2
5. G.E. Moore, Cramming more components onto integrated circuits. IEEE Solid-State Circ. Soc. Newsl. **11**(3), 33–35 (2006). https://doi.org/10.1109/N-SSC.2006.4785860
6. M. Alian, Y. Yuan, J. Zhang, et al., Data direct I/O characterization for future I/O system exploration, in *2020 IEEE International Symposium on Performance Analysis of Systems and Software (ISPASS), Boston, MA, USA*, (IEEE, 2020), pp. 160–169. https://doi.org/10.1109/ISPASS48437.2020.00031
7. M. Jayasimha, M. Moorthy, Power supply noise coupling mitigation across high-speed serial links, in *2018 IEEE Electrical Design of Advanced Packaging and Systems Symposium (EDAPS), Chandigarh, India*, (IEEE, 2018), pp. 1–3. https://doi.org/10.1109/EDAPS.2018.8680880
8. B. Pandey, A.S. Tully, et al., Novel technique for manufacturing & in-system testing of large scale SoC using functional protocol based high-speed I/O, in *2022 IEEE 40th VLSI Test Symposium (VTS), San Diego, CA, USA*, (IEEE, 2022), pp. 1–7. https://doi.org/10.1109/VTS52500.2021.9794234
9. M. Moorthy, M. Jayasimha, Power delivery network design strategy for next generation high speed IO, in *2018 IEEE Electrical Design of Advanced Packaging and Systems Symposium (EDAPS), Chandigarh, India*, (IEEE, 2018), pp. 1–3. https://doi.org/10.1109/EDAPS.2018.8680860
10. V. Melikyan, A. Sahakyan, A. Shishmanyan, et al., Multi-rate Clock-Data recovery solution in high-speed serial links, in *2015 IEEE 35th International Conference on Electronics*

and Nanotechnology (ELNANO), Kyiv, UKraine, (IEEE, 2015), pp. 242–244. https://doi.org/10.1109/ELNANO.2015.7146883
11. C.K. Tseng, C.D. Tseng, Y.T. Hwang, Efficient DSP equalization algorithm and architecture designs for high speed serial links, in *2024 International Conference on Consumer Electronics - Taiwan (ICCE-Taiwan), Taichung, Taiwan*, (2024), pp. 621–622. https://doi.org/10.1109/ICCE-Taiwan62264.2024.10674041
12. H. Ko, High-speed serial link trend and technical challenge, in *2022 IEEE Asian Solid-State Circuits Conference (A-SSCC), Taipei, Taiwan*, (IEEE, 2022), pp. 1–3. https://doi.org/10.1109/A-SSCC56115.2022.9980758
13. H. Zhu, Q. Sun, X. Li, Simulation of 112 Gbps full-link interconnect system, in *2019 Cross Strait Quad-Regional Radio Science and Wireless Technology Conference (CSQRWC), Taiyuan, China*, (IEEE, 2019), pp. 1–3. https://doi.org/10.1109/CSQRWC.2019.8799127
14. H. Park et al., Crosstalk-included PAM-4 worst eye diagram estimation method for high-speed serial links, in *2021 IEEE 30th Conference on Electrical Performance of Electronic Packaging and Systems (EPEPS), Austin, TX, USA*, (IEEE, 2021), pp. 1–3. https://doi.org/10.1109/EPEPS51341.2021.9609114
15. R. Jacob Baker, *CMOS Circuit Design, Layout and Simulation*, 2nd edn. (Wiley, 2008), 1045p
16. R. Magerramov, V. Zaitsev, Research parameters of a PLL system based on active and passive low-pass filter in 0.25-um CMOS technology, in *2021 IEEE Conference of Russian Young Researchers in Electrical and Electronic Engineering (ElConRus), St. Petersburg, Moscow, Russia*, (IEEE, 2021), pp. 2587–2589. https://doi.org/10.1109/ElConRus51938.2021.9396668
17. J. Zhong, X. Yang, et al., A 0.016 mm^2 active area 4 GHz fully ring-oscillator-based cascaded fractional-N PLL with burst-mode sampling. IEEE Trans. Circ. Syst. II Express Briefs **70**(10), 3792–3796 (2023). https://doi.org/10.1109/TCSII.2023.3288120
18. R. Seetharaman, A. Menaka, K. Anandan, Analysis of serializer for high speed transmission, in *2022 IEEE International Conference on Nanoelectronics, Nanophotonics, Nanomaterials, Nanobioscience & Nanotechnology (5NANO), Kottayam, India*, (IEEE, 2022), pp. 1–4. https://doi.org/10.1109/5NANO53044.2022.9828996
19. A. Yadav, M. Patnala, M. Rizkalla, High performance Gnrfet based serializer, in *2019 IEEE National Aerospace and Electronics Conference (NAECON), Dayton, OH, USA*, (IEEE, 2019), pp. 458–464. https://doi.org/10.1109/NAECON46414.2019.9057943
20. Y.U. Jeong, J.H. Chae, et al., A low-power and low-noise 20:1 serializer with two calibration loops in 55-nm CMOS, in *2019 IEEE/ACM International Symposium on Low Power Electronics and Design (ISLPED), Lausanne, Switzerland*, (IEEE, 2019), pp. 1–6. https://doi.org/10.1109/ISLPED.2019.8824879
21. M. You et al., A 4×25 Gb/s de-serializer with baud-rate sampling CDR and standing-wave clock distribution for NIC optical interconnects, in *2021 IEEE International Conference on Integrated Circuits, Technologies and Applications (ICTA), Zhuhai, China*, (IEEE, 2021), pp. 253–254. https://doi.org/10.1109/ICTA53157.2021.9661911
22. K.A. Clark, Z. Liu, Modeling the performance of the clock phase caching approach to clock and data recovery. Lightwave Technol. **40**(6), 1681–1689 (2022). https://doi.org/10.1109/JLT.2021.3130955
23. N.H. Tho, H.J. Lee, T.J. An, J.K. Kang, A 0.32–2.7 Gb/s reference-less continuous-rate clock and data recovery circuit with unrestricted and fast frequency acquisition. IEEE Trans Circuits Syst II Express Briefs **68**(7), 2347–2351 (2021). https://doi.org/10.1109/TCSII.2021.3053581
24. S. Roh, M.C. Choi, D.K. Jeong, A maximum eye tracking clock-and-data recovery scheme with golden section search (GSS) algorithm in 28-nm CMOS, in *2021 18th International SoC Design Conference (ISOCC), Jeju Island, Republic of Korea*, (IEEE, 2021), pp. 47–48. https://doi.org/10.1109/ISOCC53507.2021.9613919
25. U.B. Joy, A. Chakraborty, J. Kandpal, M.S. Hussain, Design of a bandgap voltage reference circuit in 90 nm CMOS process, in *2024 15th International conference on computing communication and networking technologies (ICCCNT), Kamand, India*, (IEEE, 2024), pp. 1–5. https://doi.org/10.1109/ICCCNT61001.2024.10724041

26. J.A. Michaelsen, *Reference Circuits* (Department of Informatics, University of Oslo, Spring 2012), pp. 1–47
27. A. Lemonou, I. Sidiropoulos, et al., Curvature compensated and noise optimized bandgap reference circuit, in *2024 Panhellenic Conference on Electronics & Telecommunications (PACET), Thessaloniki, Greece*, (2024), pp. 1–4. https://doi.org/10.1109/PACET60398.2024.10497029
28. B. Razavi, The design of a low-voltage bandgap reference. IEEE Solid-State Circuits Mag. (summer 2021)
29. Y. Yuan et al., A high-order curvature compensated bandgap reference with 1.6 ppm/°C temperature coefficient, in *2024 4th International Conference on Electronics, Circuits and Information Engineering (ECIE), Hangzhou, China*, (2024), pp. 353–357. https://doi.org/10.1109/ECIE61885.2024.10626568
30. Akshaya R., Siva S.Y. Design of an improved bandgap reference in 180 nm CMOS process technology. 2017 2nd IEEE International Conference on Recent Trends in Electronics, Information & Communication Technology (RTEICT), Bangalore, India, IEEE, 2017, p. 521–524, doi:https://doi.org/10.1109/RTEICT.2017.8256651
31. H.T. Kostanyan, H.T. Kostanyan, et al., 5 V wide supply voltage bandgap reference for automotive applications, in *2019 IEEE 39th International Conference on Electronics and Nanotechnology (ELNANO), Kyiv, Ukraine*, (IEEE, 2019), pp. 229–232
32. B. Razavi, *RF Microelectronics*, 2nd edn. (Prentice Hall, 2012)
33. Texas Instruments Application Report: "LVDS Owner's Manual"
34. Y. Ding, Z. Shen, W. Shao, Y. Cui, A tunable electrical impedance matching system for dual-frequency transducers, in *2024 17th International Congress on Image and Signal Processing, BioMedical Engineering and Informatics (CISP-BMEI), Shanghai, China*, (2024), pp. 1–6. https://doi.org/10.1109/CISP-BMEI64163.2024.10906148
35. X. Han, Y. Wang, et al., A reference-frequency doubler with integrated duty-cycle corrector, in *2024 IEEE 10th International Symposium on Microwave, Antenna, Propagation and EMC Technologies for Wireless Communications (MAPE), Guangzhou, China*, (IEEE, 2024), pp. 1–4. https://doi.org/10.1109/MAPE62875.2024.10813950
36. R. Raghavendra, K. Singh, M.G. Agrawal, A clock duty cycle correction circuitry for ultra-wide frequency range using nested loops, in *2024 IEEE Nordic Circuits and Systems Conference (NorCAS), Lund, Sweden*, (IEEE, 2024), pp. 1–7. https://doi.org/10.1109/NorCAS64408.2024.10752455
37. M.S. Vazgen, A.A. Arman, et al., Duty-cycle correction circuit for high speed interfaces, in *2019 IEEE 39th International Conference on Electronics and Nanotechnology (ELNANO), Kyiv, Ukraine*, (IEEE, 2019), pp. 42–45. https://doi.org/10.1109/ELNANO.2019.8783779
38. W. Jung, J. Lee, A 8.4 Gb/s low power transmitter with 1.66 pJ/b using 40:1 serializer for DisplayPort interface, in *2020 International SoC Design Conference (ISOCC), Yeosu, Korea (South)*, (IEEE, 2020), pp. 41–42. https://doi.org/10.1109/ISOCC50952.2020.9333028
39. T. Nguyen-Viet, Q.C. Bui, et al., A power-efficient transmitter design for 3D-stacked memories in 28-nm CMOS technology, in *2024 Tenth International Conference on Communications and Electronics (ICCE), Danang, Vietnam*, (IEEE, 2024), pp. 153–156. https://doi.org/10.1109/ICCE62051.2024.10634605
40. S.B. Chavan, A.P. Chavan, Ultra-low power, area efficient and high-speed voltage level shifter based on wilson current mirror, in *2021 IEEE Mysore Sub Section International Conference (MysuruCon), Hassan, India*, (IEEE, 2021), pp. 108–113. https://doi.org/10.1109/MysuruCon52639.2021.9641657
41. A. Kapoor, A. Thapar, C.S. Jha, C.I. Kumar, A high performance and low power subthreshold voltage level shifter design, in *2024 37th International Conference on VLSI Design and 2024 23rd International Conference on Embedded Systems (VLSID), Kolkata, India*, (IEEE, 2024), pp. 623–627. https://doi.org/10.1109/VLSID60093.2024.00110
42. D. Zhao et al., A voltage level shifter with fast level translation speed, in *2022 IEEE 5th International Electrical and Energy Conference (CIEEC), Nangjing, China*, (IEEE, 2022), pp. 1333–1336. https://doi.org/10.1109/CIEEC54735.2022.9846289

43. A. Yuan, H. Zhao, et al., An ultra-low leakage and wide-range voltage level shifter for low-power digital CMOS VLSIs. IEEE Trans. Circuits Syst. II Express Briefs **71**(3), 1406–1410 (2024). https://doi.org/10.1109/TCSII.2023.3323016
44. A. Mocuta, P. Weckx, P. Demuynck, et al., Enabling CMOS scaling towards 3 nm and beyond, in *2018 IEEE Symposium on VLSI Technology. Honolulu, HI, USA*, (IEEE, 2018), pp. 147–148. https://doi.org/10.1109/VLSIT.2018.8510683
45. https://blog.qasource.com/top-epic-technology-failure-examples-that-shook-the-world
46. https://www.usb.org/usb4
47. https://pcisig.com/pci-express-6.0-specification
48. Archanaa M., Balamurugan K. Analysis of thermal noise and noise reduction in CMOS device. 2014 International Conference on Green Computing Communication and Electrical Engineering (ICGCCEE), Coimbatore, India, IEEE, 2014, p. 1–5, doi:https://doi.org/10.1109/ICGCCEE.2014.6922246
49. B. Razavi, *RF Microelectronics*, 2nd edn. (Prentice Hall, 2012), pp. 35–58
50. S.K. Dash, A. Bakshi, J., et al., Study of low power techniques in analog circuit and its application to design second generation current conveyors and voltage amplifiers, in *2023 International Conference on Recent Advances in Electrical, Electronics, Ubiquitous Communication, and Computational Intelligence (RAEEUCCI), Chennai, India*, (2023), pp. 1–6. https://doi.org/10.1109/RAEEUCCI57140.2023.10134520
51. N. Arumugam, M.P. Shakthi, S. Subramanian, SAPON approach: a new technique for low power VLSI design, in *2021 IEEE 2nd International Conference on Applied Electromagnetics, Signal Processing, & Communication (AESPC), Bhubaneswar, India*, (2021), pp. 1–6. https://doi.org/10.1109/AESPC52704.2021.9708489
52. V.S.R. Manchala, G.R. Murthy, Low-power and low-leakage design techniques in CMOS technology, in *2021 5th International Conference on Trends in Electronics and Informatics (ICOEI), Tirunelveli, India*, (2021), pp. 273–278. https://doi.org/10.1109/ICOEI51242.2021.9452866
53. E. Zio, Reliability optimization. IEEE Trans. Reliab. **73**(1), 50 (2024). https://doi.org/10.1109/TR.2024.3356822
54. D.W. Coit, History of system reliability optimization. IEEE Trans. Reliab. **73**(1), 49 (2024)
55. N. Hellwege, Aging-aware design methods for reliable analog integrated circuits using operating point-dependent degradation. Ph.D., Bremen, 2015, 146p
56. L. Ding, Reliability analysis and optimization of computer communication network based on policy optimization, in *2022 2nd International Conference on Networking, Communications and Information Technology (NetCIT), Manchester, United Kingdom*, (2022), pp. 458–462. https://doi.org/10.1109/NetCIT57419.2022.00114
57. X. Zhang, Z. Zhang, et al., Efficient aging-aware standard cell library characterization based on sensitivity analysis. IEEE Trans. Circuits Syst. II Express Briefs **70**(2), 721–725 (2023). https://doi.org/10.1109/TCSII.2022.3212123
58. A. Santana-Andreo, P. Martín-Lloret, et al., Characterization and analysis of BTI and HCI effects in CMOS current mirrors, in *2022 18th International Conference on Synthesis, Modeling, Analysis and Simulation Methods and Applications to Circuit Design (SMACD), Villasimius, Italy*, (2022), pp. 1–4. https://doi.org/10.1109/SMACD55068.2022.9816212
59. M. Ullmann, K. Jech, Puschkarsky, et al., Impact of mixed negative bias temperature instability and hot carrier stress on MOSFET characteristics. Part I: experimental. IEEE Trans. Electron Devices **66**(1), 232–240 (2019). https://doi.org/10.1109/TED.2018.2873419
60. M. Ullmann, S.T. Jech, et al., The impact of mixed negative bias temperature instability and hot carrier stress on single oxide defects, in *2017 IEEE International Reliability Physics Symposium (IRPS). Monterey, CA, USA*, (IEEE, 2017), pp. XT-10.1–XT-10.6. https://doi.org/10.1109/IRPS.2017.7936424
61. M. Jech, B. Ullmann, G. Rzepa, et al., Impact of mixed negative bias temperature instability and hot carrier stress on MOSFET characteristics. Part II: theory. IEEE Trans. Electron Devices **66**(1), 241–248 (2019). https://doi.org/10.1109/TED.2018.2873421

62. K. Son, H. Hong, Shim, et al., New insight into negative bias temperature instability degradation during self-heating in nanoscale bulk FinFETs. IEEE Electron Device Lett. **40**(9), 1354–1357 (2019). https://doi.org/10.1109/LED.2019.2930077
63. N. Pande, Z. Zhou, M.H. Lin, et al., Characterizing electromigration effects in a 16 nm FinFET process using a circuit based test vehicle, in *2019 IEEE International Electron Devices Meeting (IEDM)*, (IEEE, 2019), pp. 5.3.1–5.3.4. https://doi.org/10.1109/IEDM19573.2019.8993548
64. V.S. Melikyan, S.A. Ghukasyan, S.S. Harutyunyan, et al., The aging influence minimization method for operational amplifiers with thin oxide transistors. Proc. Repub. Armen. Natl. Acad. Sci. Natl. Polytech. Univ. Armen. Ser. Tech. Sci. **75**(4), 527–538 (2022). https://doi.org/10.53297/0002306X-2022.v75.4-527
65. R.N. Ray, M.M. Tripathi, C.I. Kumar, High performance energy efficient CMOS voltage level shifter design, in *2022 2nd International Conference on Intelligent Technologies (CONIT), Hubli, India*, (2022), pp. 1–5. https://doi.org/10.1109/CONIT55038.2022.9847677
66. *PrimeSimTM HSPICE Reference Manual: Commands and Control Options* (Synopsys Inc., 2020), 823p
67. S.A. Ghukasyan, E.T. Papyan, R.A. Ivanyan, S.S. Mkrtchyan, G.A. Voskanyan, A minimization method for transistor aging influence in highto-low voltage level converter design. Proc. Repub. Armen. Natl. Acad. Sci. Natl. Polytech. Univ. Armen. Ser. Tech. Sci. **77**(3), 322–330 (2024). https://doi.org/10.53297/0002306X-2023.v77.3-322
68. H. Kostanyan, S. Harutyunyan, G. Voskanyan, S. Ghukasyan, Transistor aging minimization method for multi output level converter design. Вестник Российско-Армянского университета: Физико-математические и естественные науки. Ереван **1**, 87–98 (2022)
69. G.A. Voskanyan, The reliability improvement method for bandgap reference with thin oxide transistors. Proc. Repub. Armen. Natl. Acad. Sci. Natl. Polytech. Univ. Armen. Ser. Tech. Sci. **77**(3), 322–330 (2024). https://doi.org/10.53297/0002306X-2023.v77.3-339
70. A. Kapoor, C.S. Jha, et al., High performance CMOS voltage level shifters design for low voltage applications, in *2023 International Conference for Advancement in Technology (ICONAT), Goa, India*, (2023), pp. 74–81. https://doi.org/10.1109/ICONAT57137.2023.10079972
71. B. Halak, V. Tenentes, D. Rossi, The impact of BTI aging on the reliability of level shifters in nano-scale CMOS technology. Microelectron. Reliab. **67** (2016). https://doi.org/10.1016/j.microrel.2016.10.018
72. H.T. Kostanyan et al., The minimizaton method of transistor ageing influence on modern voltage references, in *2020 IEEE 40th International Conference on Electronics and Nanotechnology (ELNANO), Kyiv, Ukraine*, (IEEE, 2020), pp. 335–338. https://doi.org/10.1109/ELNANO50318.2020.9088844
73. PF1510, Power management integrated circuit (PMIC) for low power application processors. Rev. 3–7 April, 2020
74. A. Amaya, L.E. Rueda, G.E. Roa, A multi-level power-on reset for fine-grained power management, in *2018, 28th International Symposium on Power and Timing Modeling, Optimization and Simulation (PATMOS), Platja d'Aro*, (2018), pp. 129–132
75. O. Petrosyan, R. Soghomonyan, H. Sahakyan, V. Sahakyan, A. Galstyan, G. Voskanyan, A. Melikyan, Fully integrated resistance calibration method using reference frequency and MOSFET capacitance, in *2024 IEEE East-West Design & Test Symposium (EWDTS), Yerevan, Armenia*, (IEEE, 2024), pp. 1–4. https://doi.org/10.1109/EWDTS63723.2024.10873752
76. https://www.usb.org/document-library/usb-32-revision-11-june-2022
77. H. Gomtsyan et al., Reliable all PMOS voltage doublers for low-voltage applications, in *2024 IEEE East-West Design & Test Symposium (EWDTS), Yerevan, Armenia*, (2024), pp. 1–4. https://doi.org/10.1109/EWDTS63723.2024.10873617
78. S. Morita, S. Bian, M. Shintani, T. Sato, Efficient analysis for mitigation of workload dependent aging degradation. IEEE Trans. Computer-Aided Des. Integr. Circuits Syst. **41**(12), 5515–5525 (2022). https://doi.org/10.1109/TCAD.2022.3149856
79. V.S. Melikyan, S.A. Ghukasyan, S.S. Harutyunyan, G.A. Voskanyan, N.A. Asatryan, A method for minimizing aging influence on rail-to-rail operational amplifier with thin oxide

transistors. Proc. Univ. Electron. **28**(3), 351–359 (2023). https://doi.org/10.24151/1561-5405-2023-28-3-351-359
80. S. Chand, C. Li, et al., Automating requirements review in the automotive sector: a tailored AI approach, in *2024 IEEE 32nd International Requirements Engineering Conference (RE), Reykjavik, Iceland*, (IEEE, 2024), pp. 492–493. https://doi.org/10.1109/RE59067.2024.00059
81. R. Gopalakrishnan, D.S. Vidhya, et al., Reducing energy consumption by stand-by mode, in *2024 5th International Conference on Mobile Computing and Sustainable Informatics (ICMCSI), Lalitpur, Nepal*, (2024), pp. 842–845. https://doi.org/10.1109/ICMCSI61536.2024.00132
82. D.N. Huy, G. Chen, K. Niitsu, 22 nm CMOS pW standby power flip-flops with/without security using dynamic leakage suppression logic, in *2022 IEEE 13th Latin America Symposium on Circuits and System (LASCAS), Puerto Varas, Chile*, (IEEE, 2022), pp. 1–4. https://doi.org/10.1109/LASCAS53948.2022.9789041
83. H.H. Sahakyan, R.M. Soghomonyan, D.K. Gabrielyan, G.A. Voskanyan, Data and power supply monitoring and switching system for improved power efficiency and signal integrity. Bull. High Technol. **2**, 57–65 (2020)
84. https://doc.qt.io/qtforpython/
85. *Custom Compiler™ Environment User Guide* (Synopsys Inc., 2022), 776p

Chapter 6
Development of Means to Mitigate the Crosstalk in Integrated Circuits Using Machine Learning

6.1 General Issues of Developing Means to Mitigate the Crosstalk in Integrated Circuits Using Machine Learning

6.1.1 Need and Importance of Developing Means for Detecting and Mitigating the Crosstalk in Integrated Circuits Using Machine Learning

The current development of integrated circuits (ICs) has led to a reduction in the size of their elements—down to nm [1–3], an increase in the density of elements—now more than 100 billion transistors [1–3], an increase in speed—synchronization signal frequency up to tens of GHz [3], and a decrease in the circuit's operating voltages—down to a few mV [4]. On the other hand, the current use of ICs in autonomous vehicles [5], medical equipment [6, 7], data processing systems, the military industry [6], and elsewhere, imposes stricter requirements on ICs: high circuit speed, uninterrupted data transmission, low power consumption, longevity, etc. Meeting such requirements has become a challenge for designers under the mentioned development conditions of ICs. This development leads to a reduction in signal switching time in ICs, an increase in the number of interacting elements, as well as an increase in the magnitude of the influence of elements on each other [8–10]. Therefore, among the existing challenges, failures caused by Interactions between elements (IBE) in the circuit stand out.

IBEs occur between "aggressor"–"victim" wires in the circuit [9]. The wire through which a signal change affects the voltage level of another wire is called the "aggressor," and the other wire is called the "victim" (Fig. 6.1) [10]. In ICs, IBEs can arise due to mutual inductance between them (Fig. 6.2a) [10–12] or mutual capacitance (Fig. 6.2b) [10, 11, 13].

V. Melikyan, K. Li, *Parametric Optimization Techniques of Integrated Circuits*,
https://doi.org/10.1007/978-3-032-20400-4_6

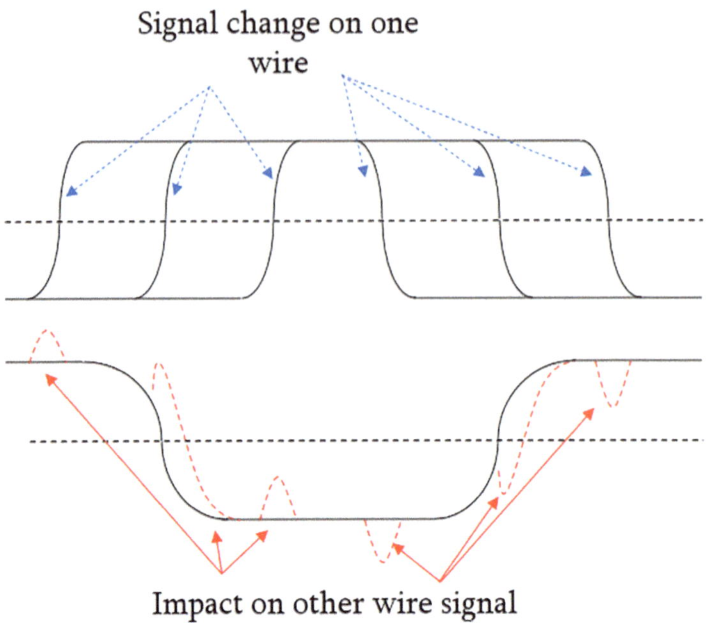

Fig. 6.1 Effect of signal change in one wire on another wire

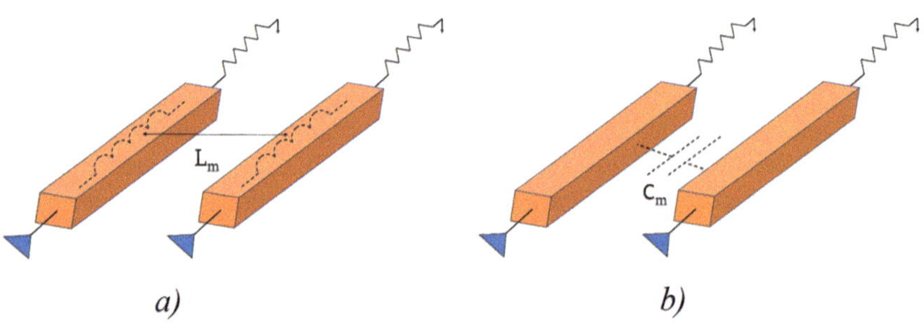

Fig. 6.2 Mutual inductance and capacitance between wires

The voltage change on the victim due to mutual inductance from the aggressor is described as follows [11, 12]:

$$V_{\text{vic},L_m} = L_{\text{m}} * \frac{dI_{\text{a}}}{dt}, \tag{6.1}$$

where L_{m} is the mutual inductance between the aggressor and the victim, and I_{a} is the current through the aggressor.

The voltage change on the victim due to mutual capacitance from the aggressor is described as follows [11, 13]:

$$I_{\text{vic},C_m} = C_m * \frac{dV_{\text{a}}}{dt}, \tag{6.2}$$

where C_m is the mutual capacitance between the aggressor and the victim, V_a is the voltage on the aggressor. As is known, the impedance of C_m is described as follows [13]:

$$X_{C_m} = \frac{1}{2\pi f C_m}, \tag{6.3}$$

where f is the switching frequency of the signal passing through the wire, C_m is the mutual capacitance between the wires. Depending on the physical parameters of the wires, the capacitance coefficient C_m is determined as follows [13]:

$$C_{m_f} = \frac{S}{d}, \tag{6.4}$$

where S is the overlap area of the two wires, and d is the distance between them. The consequences of IBE in ICs have been known for a long time [14, 15]. However, in modern ICs, it has become a major challenge for several reasons:

- The increase in the number of elements in ICs leads to a decrease in the distance between them. From (6.4), it follows that a decrease in the distance between wires increases the capacitance between them; therefore, decreasing the capacitive impedance (6.3). The latter means that a change in the voltage level on the aggressor will lead to a larger current on the victim (6.2).
- From (6.3) it follows that an increase in the speed of the IC reduces the capacitive impedance between its elements. Again, a change in the voltage level on the aggressor will lead to a larger current on the victim (6.2).
- A change in the voltage level on the aggressor can already lead to a change in the logical level of the victim's voltage because the circuit's operating voltages have decreased.
- Changes in the logical level on a wire can lead to undesirable switching in the logic circuits connected to it. Even if the output signal of that circuit has no effect on the function of the overall circuit, the additional switching still leads to an increase in the dynamic power consumption of the IC. As a consequence, the temperature may also increase due to self-heating as follows [16–20]:

$$T = T_a + P * R_{\text{th}}, \tag{6.5}$$

- where T_a is the temperature of the external environment surrounding the IC, P is the power consumption, and R_{th} is the thermal resistance.

- As a result of IBE, voltage value fluctuations may occur, which, without causing a change in the signal's logical level, still affect power consumption as follows:
  - Due to changes in the current passing through the wire, its power consumption changes, which in turn can lead to temperature changes due to self-heating [16–20].
  - Changes in the voltage value on the wire also lead to changes in the leakage current of the element connected to it; consequently, there are also changes in power consumption and temperature [16–20].

To mitigate changes in the circuit's power consumption due to IBE, it may be necessary to apply additional solutions, for example, low power design methods [21–23]. For instance, in [21], instead of an input resistor and comparator in a parallel analog-to-digital converter, inverters with different switching points are used, which allow reducing power consumption. If power consumption reduction is impossible, to mitigate the effects of self-heating, the approach proposed in [24–27] creates the possibility of reducing the temperature increase by increasing the size of transistors, and consequently, their area. An undesirable temperature increase can in turn lead to a number of other problems affecting the circuit's reliability, including its accelerated aging [28–31].

- Besides the above written, IBEs can cause changes in timing parameters of the signal on the victim wire, such as:

Change in Signal Switching and Delay Times. In the case of an aggressor and a victim switching in opposite directions, the victim's switching time will be larger (Fig. 6.3a) [32–34], and in the case of switching in the same direction, it will be smaller (Fig. 6.3b) [32–34]. Thus, the signal switching time, and consequently, the delay time, also changes.

The terms setup time and hold time are defined for sequential elements in digital ICs (Fig. 6.4) [35].

The time periods before and after the active edge of the synchronization signal of a sequential element in the circuit, during which its input signal must be unchanged, are called Setup time (ST) and Hold time (HT), respectively [32, 34]. Therefore, deviation from ST and HT means a change in the input signal during the setup and hold times, respectively. The relationship between ST and HT and the circuit's synchronization signal is expressed as follows [35]:

$$T_{\text{clk}-q} + T_{\text{pd}} + T_{\text{ST}} < T_{\text{clk}}, \tag{6.6}$$

where $T_{\text{clk-q}}$ is the delay from the clock signal to the information output of the sequential element, T_{pd} is the combinational circuit delay (Fig. 6.5) [35], T_{ST} is the ST, and T_{clk} is the clock signal frequency.

$$T_{\text{clk}-q} + T_{\text{pd}} > T_{\text{HT}}, \tag{6.7}$$

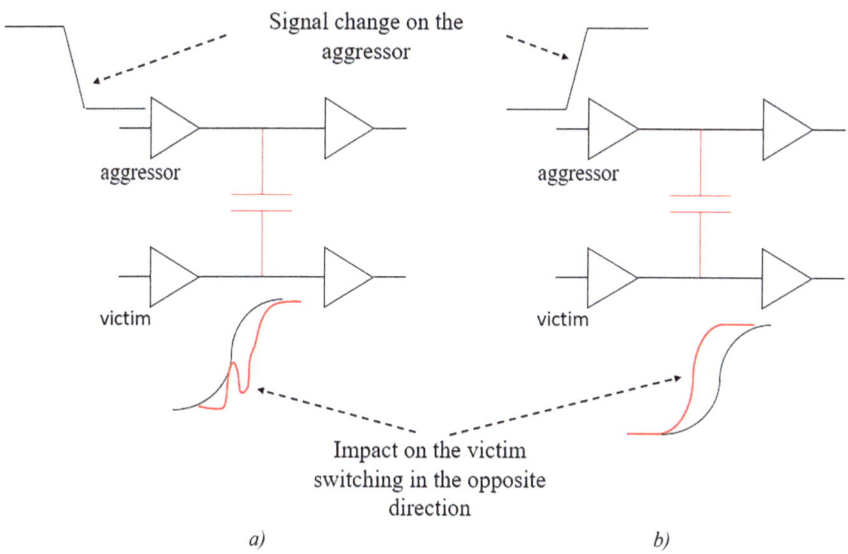

Fig. 6.3 Effect of aggressor change on the victim signal

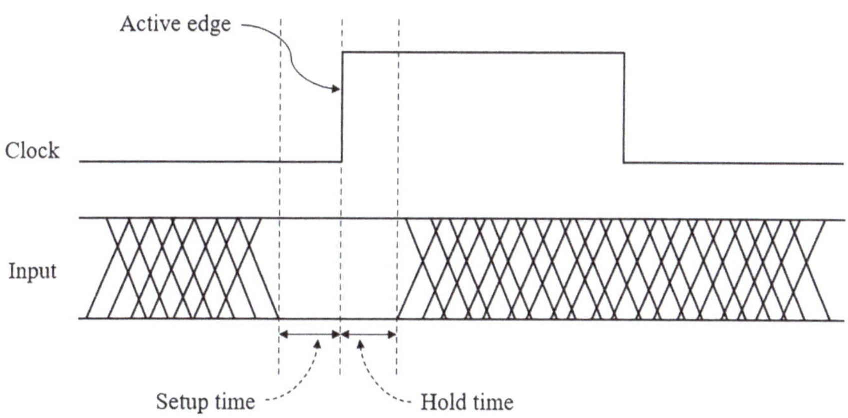

Fig. 6.4 Setup and hold times of a sequential element

where $T_{clk\text{-}q}$ is the delay from the clock signal to the information output of the sequential element, T_{pd} is the combinational circuit delay (Fig. 6.5) [35], and T_{HT} is the HT.

Change in Setup Time of Sequential Elements. As already mentioned, as a result of IBE, the victim signal's switching time can increase. Such a change can lead to a signal change within the ST (Fig. 6.6) [32–34].

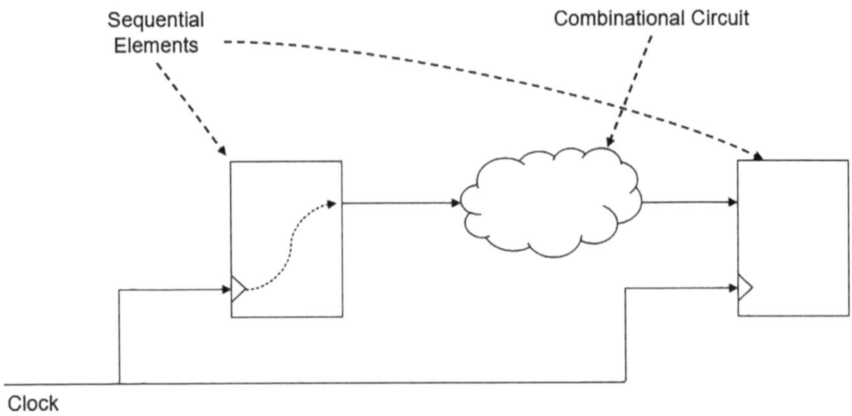

Fig. 6.5 Description of a circuit with sequential and combinational elements

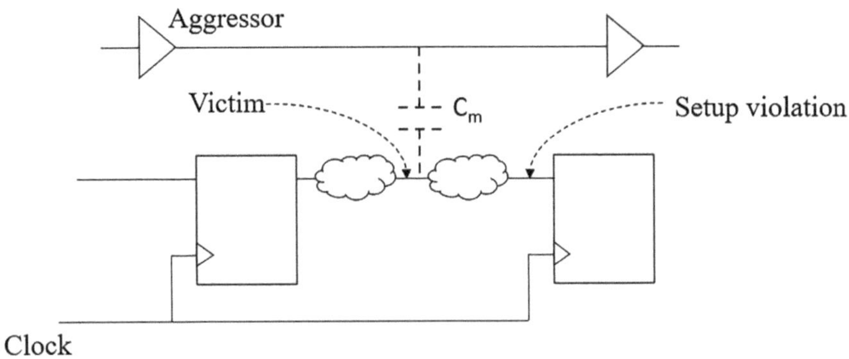

Fig. 6.6 Deviation from setup time due to IBE

Without IBE, the victim signal switches, preserving the ST. With IBE present, due to the change in switching time, the requirement for the signal to remain constant during the ST is violated (Fig. 6.7) [33].

From (6.6) it follows that there are several options to adjust the ST deviation. It is possible to reduce T_{pd}, but in that case, a deviation from HT will occur, or to increase T_{clk}, therefore decreasing the clock frequency.

Change in Hold Time of Sequential Elements. Usually in digital ICs, particular attention is paid to IBEs between Clock Tree (CT) nodes and other wires, because CT wires are distributed throughout the IC and work with the circuit's sequential elements (Fig. 6.8) [32–34]. Due to changes in the timing parameters of that signal, it reaches sequential elements at different times, leading to undesirable changes in circuit functions.

In an IC, another wire can be an aggressor for the synchronization signal. When the aggressor switches in the opposite direction to the clock signal change, its

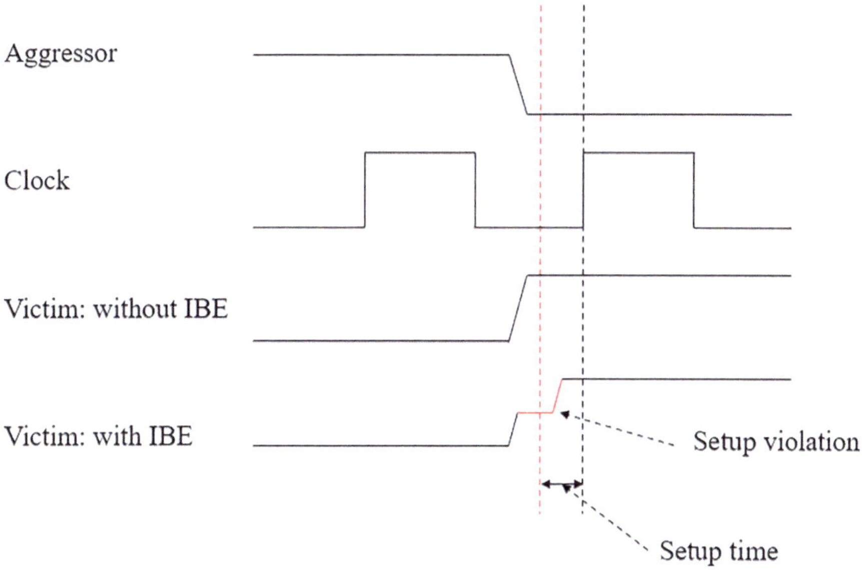

Fig. 6.7 Change in signal switching time due to IBE

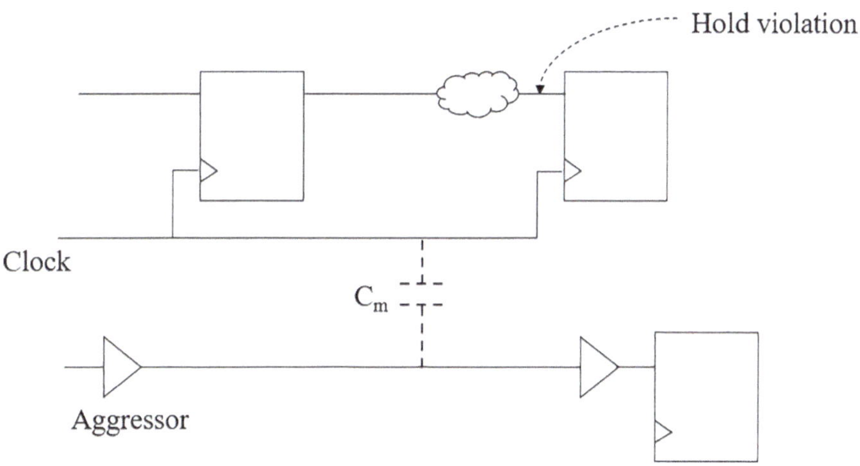

Fig. 6.8 Deviation from hold time due to IBE

switching time increases; therefore, the clock signal arrives at the sequential element inputs with a delay (Fig. 6.9) [33].

Thus, the HT is violated, and consequently, the IC will malfunction. From (6.7), it follows that the HT deviation can be adjusted by increasing T_{pd}, but in that case, a deviation from ST will occur.

IBEs are divided into several components. The effect of the aggressor on the victim wire at the end closest to it is called Near End Crosstalk (NEXT), and the

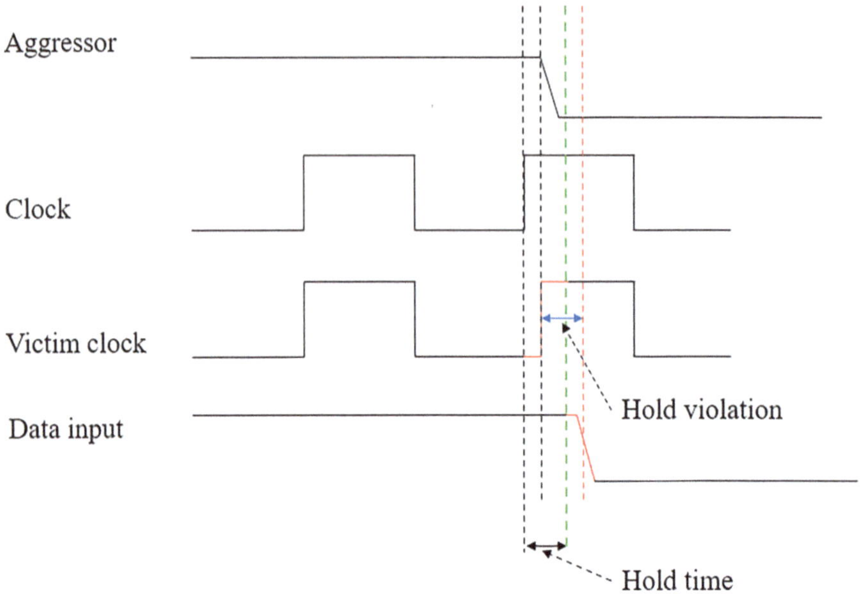

Fig. 6.9 Change in clock signal switching time due to IBE

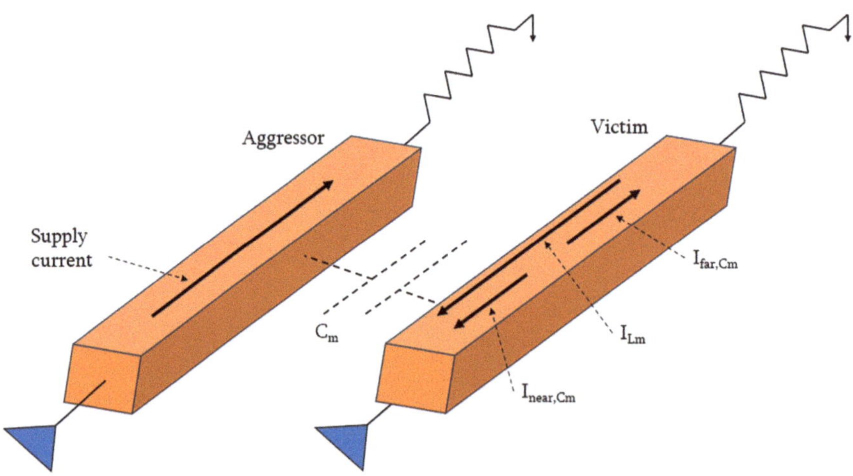

Fig. 6.10 NEXT and FEXT effects from aggressor to victim

effect at the farthest point from it is called Far End Crosstalk (FEXT) [10, 36–39]. The currents generated on the victim line will be described as follows (Fig. 6.10) [36, 37]:

$$I_{\text{near}} = I_{L_m} + I_{\text{near},C_m}, \tag{6.8}$$

$$I_{\text{far}} = I_{\text{far},L_m} - I_{L_m}, \tag{6.9}$$

where I_{near} is the current on the victim wire at the end near the aggressor, I_{L_m} is the current on the victim wire due to mutual inductance, I_{near,C_m} is the current on the victim wire at the end near the aggressor due to mutual capacitance, I_{far} is the current on the victim wire at the end far from the aggressor I_{far,L_m} is the current on the victim wire at the end far from the aggressor due to mutual inductance.

Therefore, for two similar wires, the NEXT coefficient will be expressed as follows [37, 38]:

$$\text{NEXT}_f = \frac{1}{4}\left(\frac{C_m}{C} + \frac{L_m}{\text{I}}\right), \tag{6.10}$$

For two similar wires, the FEXT coefficient will be expressed as follows [37, 38]:

$$\text{FEXT}_f = \left(\frac{C_m}{C} - \frac{L_m}{\text{I}}\right), \tag{6.11}$$

And the dependence of FEXT on the circuit's physical parameters and operating mode will be expressed as follows [37, 38]:

$$\text{FEXT} = \left(\frac{D_c}{T_a}\right) * \text{NEXT}_f, \tag{6.12}$$

where C_m is the mutual capacitance per unit length between the wires, L_m is the mutual inductance per unit length between the wires, C is the capacitance per unit length of one of the wires, L is the inductance per unit length of the victim or aggressor, D_c is the length of the overlapping section of the wires, and T_a is the switching time of the signal on the aggressor.

With the reduction in the size of elements in ICs and the increase in their density, the number of interconnects in the circuit has increased. In current ICs, accounting for the consequences of IBE using mathematical methods takes a time that is incompatible with the time requirements set by IC manufacturers. Besides, accounting for the consequences of IBE is complex and does not have a unique solution. If an attempt is made to solve such fundamental issues directly, it will require such machine and time resources that will not satisfy the time requirements for circuit design, verification, and production. Therefore, such fundamental issues must be solved by methods that work with the large number of elements in the circuit and require an acceptable time for design and production. It seems that the application of the currently widely developing methods of Artificial Intelligence (AI) will contribute to the effective solution of the same tasks without using inadmissible resources. The reason is that currently in the IC design and verification process, the involvement of Machine Learning (ML), which is a dominant part of AI, is increasing [40–42]. As a result, designers gain the ability to quickly model the circuit's behavior under different conditions, detect and prevent probable failures in the IC before its fabrication, and also predict them early. There already exist methods built on ML algorithms that allow predicting signal switching and delay times in the circuit with high accuracy (Table 6.1) [43].

ML also allows optimizing the circuit design flow and provides automation possibilities. Therefore, all this can be done without numerous iterations of the steps required for IC construction and testing, simultaneously reducing design time and costs. The above possibilities are also used for detecting, studying, and preventing IBE in ICs [44, 45]. All this is done thanks to ML algorithms, and the human factor is minimized as much as possible.

Thus, considering the role and involvement of ICs in modern systems, failures caused by IBE in the circuit can have irreversible consequences. Therefore, ensuring the reliability of the IC, from the perspective of detecting and preventing IBE in it, has been and remains an important fundamental issue. Consequently, there is a need to develop new methods that can detect IBE at the design stages of the circuit, allowing designers to consider them during the design process and build reliable circuits. To speed up the design and testing process as much as possible and exclude errors due to the human factor, it is preferable to use machine learning methods.

6.1.2 Existing Solutions and Key Issues for Detecting and Mitigating the Crosstalk in Integrated Circuits Using Machine Learning

From the fundamental issues presented in Sect. 6.1.1, it is clear that IBEs in circuits can cause failures in the entire IC. Until today, various scientific groups or individual specialists have developed methods and approaches for considering, solving, and verifying IBE, which have many different advantages over each other, but are not without drawbacks. To avoid these drawbacks, it is necessary to develop new approaches that will allow early detection and consideration of IBEs during the circuit design process. The proposed methods assume making changes to both

Table 6.1 Results of predicting switching and delay times using different ML methods

Method	Ratio of predicted to actual switching times	Ratio of predicted to actual delay times
Lasso regression	0,739	0,710
Neural network (with mean squared error)	0,861	0,891
Neural network (with custom error calculation)	0,856	0,882
Other software tool	0,909	0,947
Developed algorithm	0,974	0,971

physical and logical design of the circuit. However, with the current development of ICs, such solutions may already be insufficient to detect, consider, and eliminate the resulting failures.

Existing Approaches to Encoding Vibration Signals for IBE Mitigation
One of the common methods to avoid IBE in circuits is encoding the circuit's wires at the circuit-level design stage, especially in wires where several logical operations are implemented [46–49]. When adjacent wires switch simultaneously, signals on these lines affect each other, causing possible failures in a circuit. There are different types of encoding methods, but basically, the block diagram of encoding is similar (Fig. 6.11) [50].

In the first encoding variant, all logical "0" signals in the circuit are encoded as "00," and logical "1" signals as "10." Such encoding prevents changes in the logical states of signals in adjacent wires in opposite directions (Fig. 6.12) [50], without causing inadmissible switching. All such codes ultimately result in one complete data "word". If the data word width consists of N bits, and all data words are allowed, then the encoded word must have at least 2^N bits.

Another encoding variant is "selective" shielding [51, 52]. In case of ordinary shielding, it was assumed that high-frequency data buses are separated from each other by static signal wires. In case of an 8-bit "11,000,110" signal, the corresponding shielded signal will be a 16-bit "101,000,000,010,100" signal. That is, a "0" static signal is placed between every pair of signal wires. Instead, it is proposed to place the "0" signal only after logical "1" signals, thereby reducing not only the number of placed "0"s but also redundant "0"–"0" pairs, as well as the design process and the space occupied by the signal bus. With such encoding, the encoded version of the same 8-bit signal bus will be the "101000010100" 12-bit signal. Clearly, the encoding in this case depends on the number of logical "1"s in the data bus [52].

To consider another encoding variant, several terms are defined. An N-bit bus is considered, whose bits work with uniform data. At a given moment in time, the bus signal is represented as the current bus, and the bus at the next moment in time is termed the next bus. Encoding is performed using Hamming Distance (HD), which expresses the number of signals that have switched between logical high and low levels. There are two control signals: an inversion signal and a shift signal, which express whether the signal is inverted or shifted, respectively.

Bus Inversion (BI) Method. The BI method calculates the HD between the current bus and the next bus [53, 54]. If this distance is greater than half the number of *N* bits of the bus, then the inverted signal of the next bus is placed on the bus and the inversion signal is set to logical "1". Otherwise, when the HD is less than or equal

Fig. 6.11 Description of the wire encoding block diagram

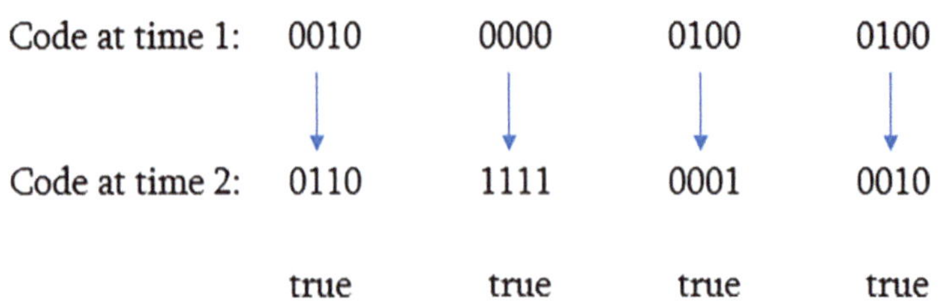

Fig. 6.12 Correct and incorrect signal switching as a result of wire encoding

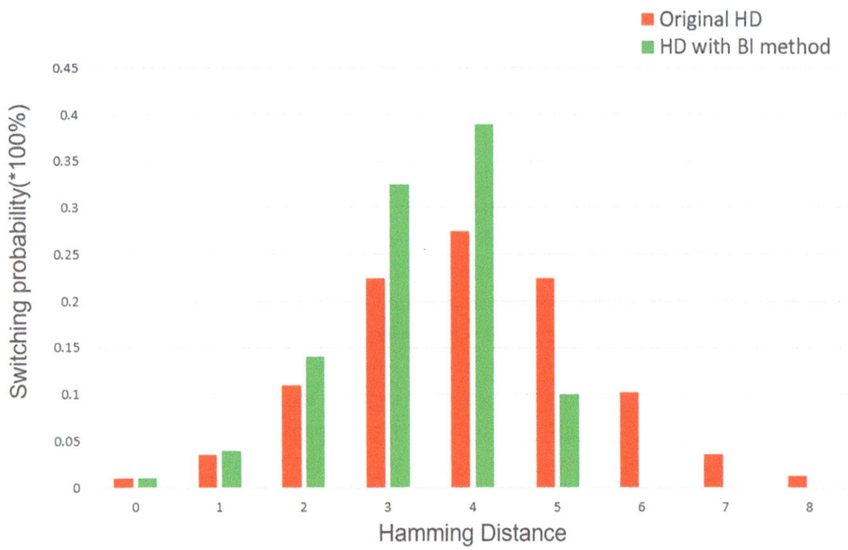

Fig. 6.13 Probability of signal inversion depending on HD for the BI method

to half the number of N bits of the bus, the next bus is placed on the bus, and the inversion signal is set to "0" [55, 56]. To implement the BI method, consider an N = 9-bit signal bus, when the current bus is "010000000" and the next bus is "000111111", where the first member of the bus is the inversion signal. In this case, the HD will be 7, which is greater than half the number of bits in the bus. Therefore, the inversion signal in this case will be set to "1", and the next bus to "111000000". For the formed signal, the HD will be 2, which is less than the initial value. Using this method, the probability of data switching will be highest for an HD value of 4, if the number of bits in the data bus is $N = 8$ (Fig. 6.13) [57].

A high probability of data switching means that there will be a minimal difference between the signal provided to the receiver and the actual signal. However, such a method does not ensure IBE reduction, nor power savings when the HD is less than the number of bits in the data bus.

Bus Shift (BS) Method. This method involves shifting the data bits to the right or left instead of inverting them [58], which will ensure power reduction. The bits of the next bus are shifted to the right or left, and the HD is calculated each time. Then, for the smallest HD value, the next bus is set, whose bit count is calculated as the base-2 logarithm of the number of bits in the current bus. To implement the BS method, consider an $N = 11$-bit signal bus when the current bus is "00010000000" and the next bus is "00001000000", where the first 3 bits of the bus are the shift signal bits (a 3-bit signal is needed for shifting an 8-bit signal) [57]. In this case, the HD will be 2. After method implementation, the next bus will be "00110000000" because the bits have been shifted once, and the shift signal has become "001". Thus, the HD becomes "1", which is less than its initial value. Therefore, the "00110000000" signal will be established on the next bus. The receiving logic, reading the first 3 bits of the next bus, will record that the signal has been shifted once, and therefore, at the beginning of working with the data, it will shift it once. Using such a method, the probability of data switching will be highest for an HD value of 3 if the number of bits in the data bus is $N = 8$ (Fig. 6.14) [57].

Bus Shift-Inverse Method (BSINV). The sequence of steps for such encoding is as follows: invert the next bus according to the BI method implementation, shift the next bus left according to the BS implementation to the left side, shift the next bus right according to the BS implementation to the right side, and calculate the smallest HD during the step execution [59]. For the implementation of this method, two additional bits are already needed to describe four possible cases: "00" for an unmodified signal, "11" for an inverted signal, "10" for a left-shifted signal, and "01" for a right-shifted signal.

The distribution of data switching probability using such a method is expressed in the form presented in Fig. 6.15 [57] for the case of a data bus with $N = 8$ bits:
To implement the BSINV method, consider an $N = 10$-bit signal bus, when the current bus is "0010000000" and the next bus is "0000111011", where the first 2 bits of the bus are the additional bits to describe the signal modification status. In this case, the HD will be 6. After performing the method implementation steps, the next bus will be in the following sequence: "1111000100" (BI implementation, HD = 4), "1001110110" (left BS implementation, HD = 7), "1111000100" (right BS implementation, HD = 5). In this case, the lowest HD value will be 4, for which the signal describing the modification status will be "11". " Therefore, the receiving logic will accept the next bus as the "1111000100" signal and perform the corresponding operation before working with the data bus.

Circular Shift-Inverse Method (CSINV). The design results according to the previous methods are summarized in this method, and their unification is proposed. If the previous method assumed shifting only 1 bit of data to the right or left, but for large-sized data this approach may not be so successful, then in this case it is more appropriate to perform data shifting for several bits at once [57]. The distribution of data switching probability using such a method is expressed as follows (Fig. 6.16) [57].

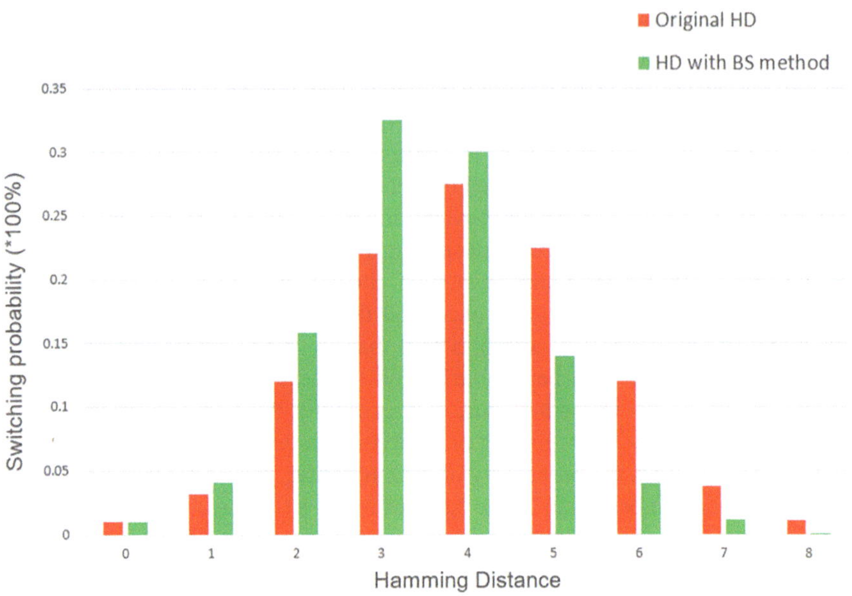

Fig. 6.14 Probability of signal inversion depending on HD for the BS method

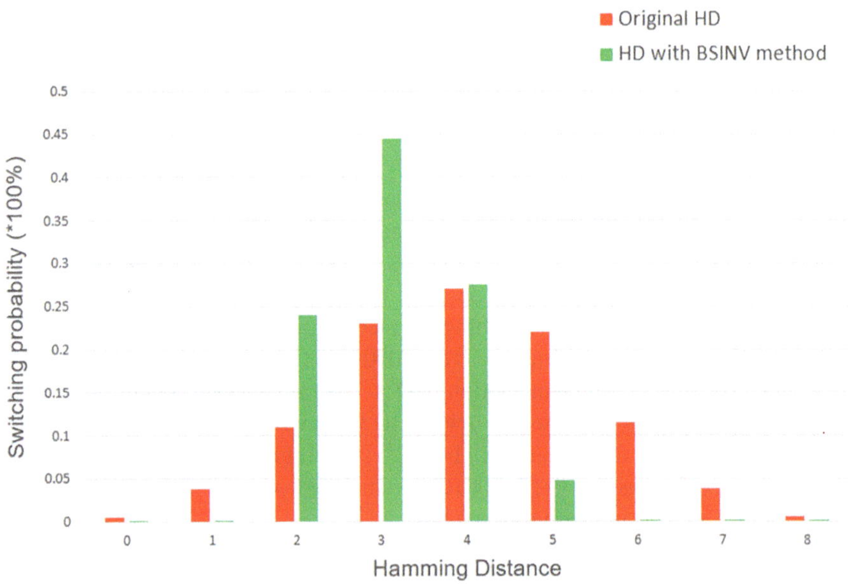

Fig. 6.15 Probability of signal inversion depending on HD for the BSINV method

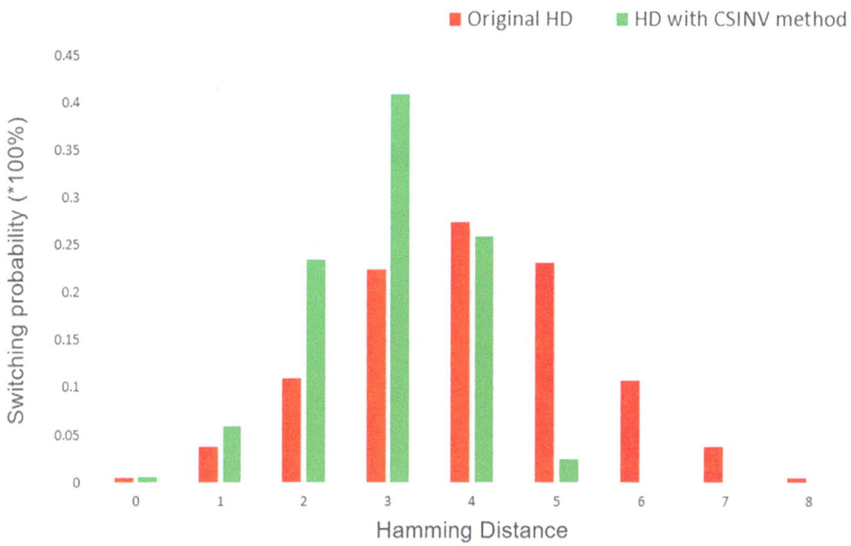

Fig. 6.16 Probability of signal inversion depending on HD for the CSINV method

The sequence of steps for implementing the CSINV method is as follows: perform BI transformation and calculate HD, circularly shift the next bus by the selected amount and calculate the HD each time, compare all values, and establish the corresponding signal on the next bus for the lowest HD. CSINV will choose the BI method only if the inversion signal is equal to "1", otherwise, it will consider the shift signal to calculate the number of shift steps for BS.

In this case, for obtaining the best results, a combination of wire encoding and other known solutions is performed, such as simultaneous bus encoding and the use of repeaters [60].

Thus, various methods of wire encoding are used to reduce IBEs in ICs. However, each encoding adds another logic to the circuit, and for such logics, changes must also be made in the circuits receiving those signals. The encoding variants also make changes to the buses under different conditions (for example, the BI method works when the HD is greater than half the number of bits in the bus, and in other cases does not make changes, not ensuring IBE reduction).

Current Method of Using Differential Signaling to Mitigate the Effects of IBE Another common way to reduce IBEs in circuits is the use of differential signals in the circuit [61]. This method requires having two wires carrying the signal simultaneously in the circuit, one of which will have the given signal, and the other will have the inverted version of the signal (Fig. 6.17) [62].

In a differential pair, the signal change must occur on both wires simultaneously. Since the receiving circuit considers the difference between these two signals as the input signal, any change that is common to both signals is effectively canceled even under significant external influence [63–66]. However, to achieve complete results

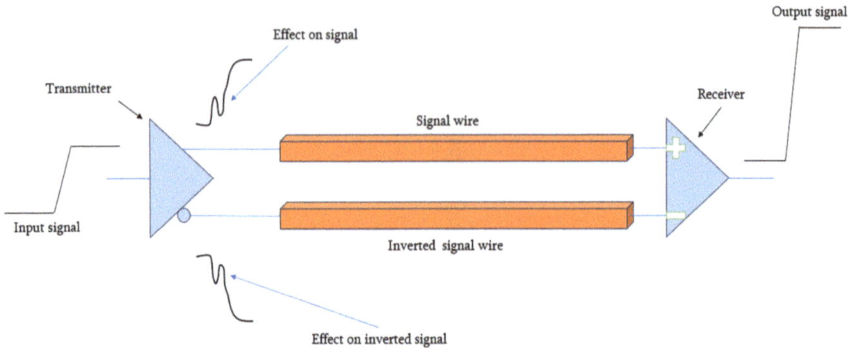

Fig. 6.17 Description of the output signal using differential signaling

with the application of this method, high symmetry of the signal-carrying wires is required, which is a challenge in the current development of ICs. Adding additional wires also implies changes in the IC, additional power consumption, and an increase in the circuit area. Besides all this, the number of elements that perform signal inversion and reception of two input signals also increases in the circuit.

Thus, for reducing IBE in circuits, the method of applying differential signals is proposed. The output signal of the receiving circuit will be without any changes. However, to achieve such a result, both designers and manufacturers require symmetry of the signal-carrying wires during design and production. Additional elements are added to the circuit, complicating the design and increasing the circuit area and element density.

Current Method of Applying IBE Mitigation Repeaters

Another common method for reducing IBE in ICs is the use of repeaters in signal buses [67–70]. In this approach, it is necessary to take into account various factors, including the physical arrangement of elements in the circuit, the characteristics of the transmission lines, and the specific requirements of the IC's application. In the case of the usual arrangement of wires in the circuit, the effect of the aggressor on the victim wire depends on the C_m capacitance between them (6.4). The victim wire propagating along the aggressor is divided into separate components using repeaters.

Thus, the overlap area between the wires is reduced, therefore the C_m capacitance (6.4), reducing the effect on each section of the wire (Fig. 6.18) [71].

Such a solution, besides reducing IBE, improves not only the timing parameters of the signals passing through the given bus (Fig. 6.19) [72], but is also used in the case of CT construction in the circuit [73].

They contribute to reducing the time difference of clock signal reception at the inputs of two sequential elements. To obtain such advantages using repeaters, designers have developed fast-working algorithms that allow most quickly locating the preferred locations for repeaters in the circuit [74]. However, the use of repeaters in ICs leads to an increase in their area, complication of the design process, an

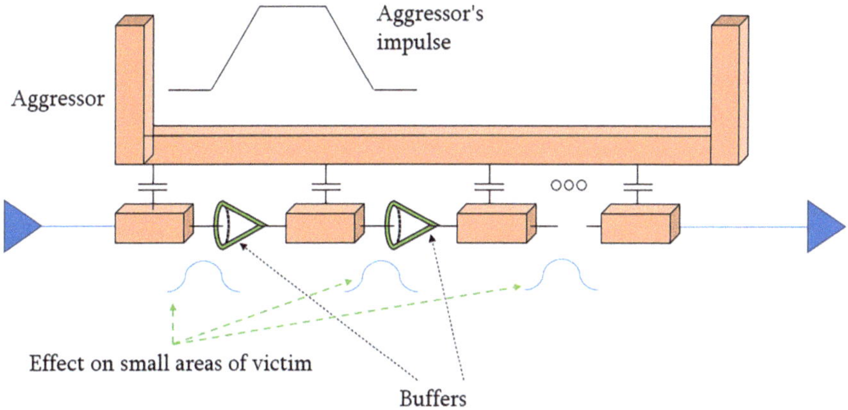

Fig. 6.18 Arrangement of wires in the circuit using repeaters

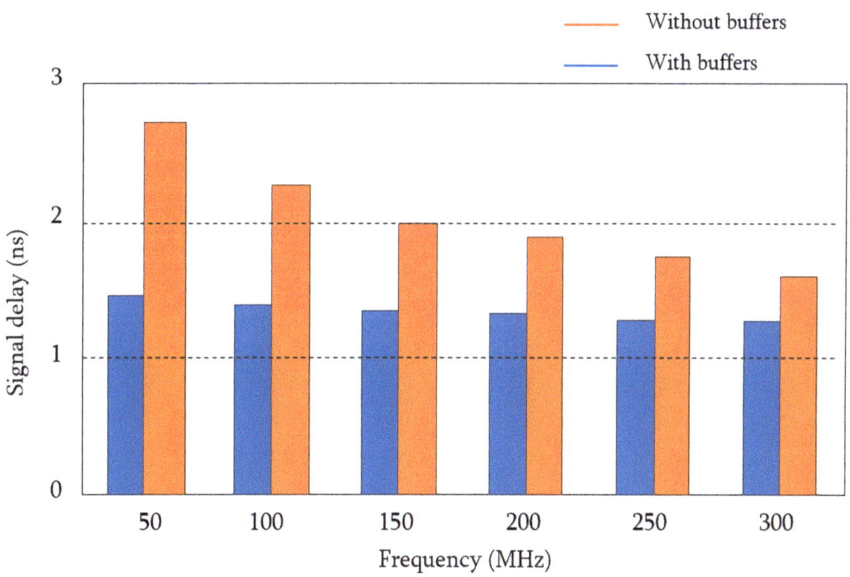

Fig. 6.19 Signal delay depending on frequency using repeaters

increase in the overall delay of the circuit due to the cumulative delay of the repeaters, resulting in a violation of timing requirements for the circuit, etc.

Thus, for reducing IBE in the circuit, the method of installing repeaters on the victim wire is proposed. With these, it will be possible to reduce IBE, but the design process becomes more complicated, and the installation of new clements in the circuit is more difficult along with the latest development of ICs.

Current Method of "Screening" for Mitigating the Consequences of IBE

As already mentioned, IBEs in ICs usually occur due to capacitive or inductive coupling between the aggressor and the victim. One common method to solve this problem is to increase the distance between high-frequency lines, which follows from (6.4). But this leads to an increase in "unused" space in the IC (Fig. 6.20) [75].

Therefore, such a solution is not optimal and preferable. To solve the problem, it is proposed to insert protective wires between information-carrying wires [76, 77], which is more commonly known by the term "screening". Such an approach also reduces electromagnetic interference on the circuit's wires [78]. In most cases, the protective layers between high-frequency lines are similar and connected to the circuit's "0" point. However, such a solution also does not provide results corresponding to the development of ICs. Therefore, it is proposed to place protective layers of different sizes between different pairs of wires (Fig. 6.21) [75]:

The application of screening reduces the effect of the aggressor on the victim line, and depending on the number of protective layers, it is smaller (Fig. 6.22a) [79]. The effect of the aggressor on the victim is reduced not only by increasing the number of protective layers but also by changing the sizes of those layers (Fig. 6.22b) [79].

The general view of the circuit after the insertion of protective layers will have the form presented in Fig. 6.23 [80].

The screening variant is quite common. Numerous solution studies have been conducted [81, 82], and there are different modifications of the solution to achieve the best result, as well as combinations with other solutions [83]. However, the application of this method is complex because changes must be made in the circuit to be able to integrate additional layers into it. And the components of the IC are located on a small area and quite close to each other, therefore, making such changes

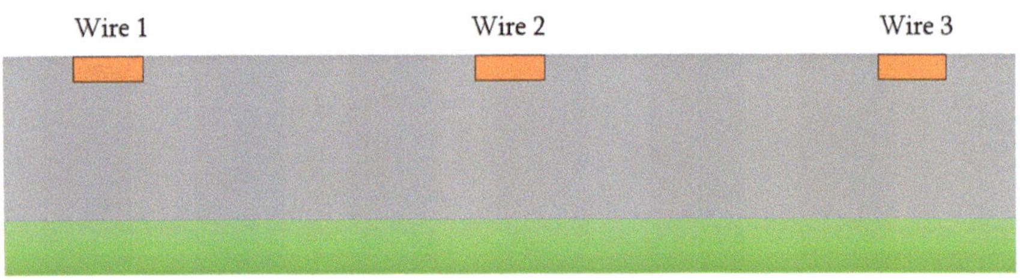

Fig. 6.20 Mutual arrangement of wires without "screening"

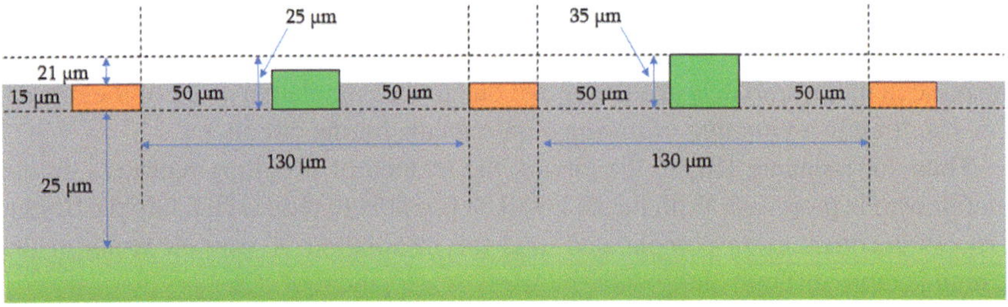

Fig. 6.21 Arrangement of wires with "screening" application using different protective layers from each other

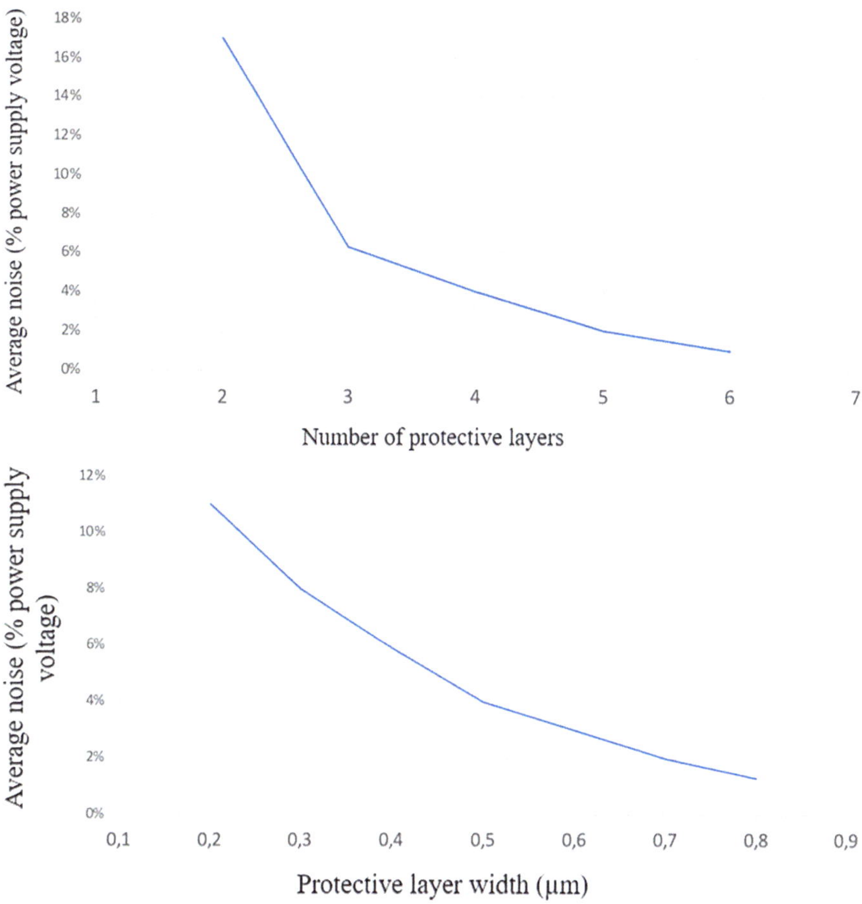

Fig. 6.22 Effect of aggressor on victim depending on the number and size of protective layers

is a challenge for designers. As a result of poor screening design, wire overlaps are also possible during the production process, which will lead to IC failures.

Thus, the application of screening is a common and effective method for reducing IBEs in ICs. However, it assumes that changes must be made in the circuit so that protective layers are placed between aggressors and victims. With the current development of ICs, the density of elements in it continues to increase, which complicates the application of the method.

Existing Methods for Improving CT to Mitigate the Effects of IBE

As already mentioned in Sect. 6.1.1, the highest frequency in an IC is possessed by the synchronization signals in it. Therefore, as can be seen from (6.3), they are the "main" aggressors of the circuit. And from (6.6) and (6.7) it can be seen that if synchronization signal is the victim of any signal, then timing parameter failures will occur. Therefore, it is very important that the IC's CT is built in such a way that

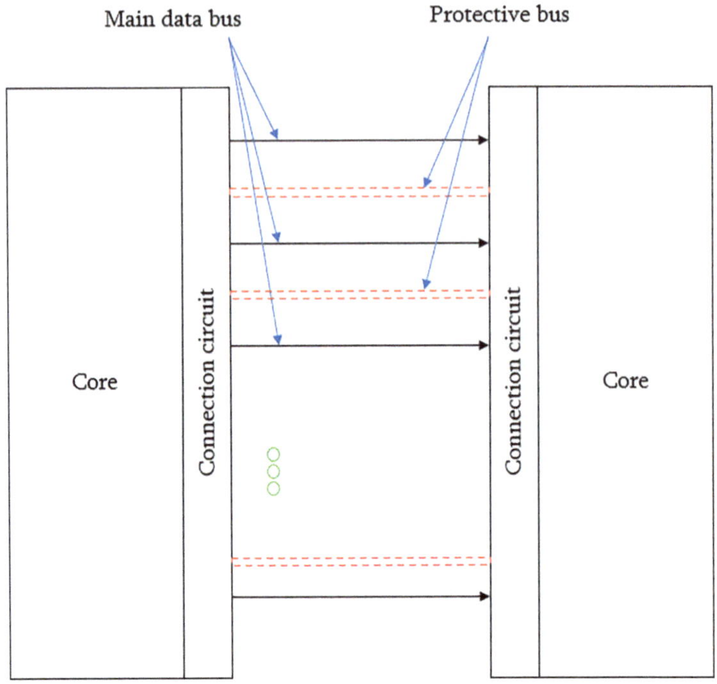

Fig. 6.23 General view of the circuit after the insertion of protective layers

there are minimal interactions between it and other signals. There are several methods for CT construction that allow avoiding the above circumstances. For IBE reduction, it is proposed to perform blocking of the existing synchronization signal (Fig. 6.24) [84, 85].

The blocked clock signal will also be determined by the corresponding state of the enable signal (Fig. 6.25) [84, 85].

In such an approach, one more logic is added for the clock signal-carrying wires, which allows activating or deactivating the clock signal before it reaches the circuit's sequential elements.

A method is also proposed when special design conditions are placed on certain wires of the CT, unlike the rest. For example, in [32], those clock-carrying wires of memory devices whose length is >100 μm (Fig. 6.26a) [32] are routed away from other wires. In the circuit considered in that work, the number of victims was 321; after applying the proposal, their number becomes 189 (Fig. 6.26b) [32].

The worst 10 IBE values were considered before the method application (Table 6.2) [32] and after the method application (Table 6.3) [32].

Other methods of CT construction have also been developed, which not only reduce the effect of the CT on other signals [86–88], but also achieve minimal skew between different clock signal edges [89, 90], as well as low power consumption [91]. However, changing the CT architecture leads to complication of the IC design, and changing the architecture in different parts of the circuit can lead to skew

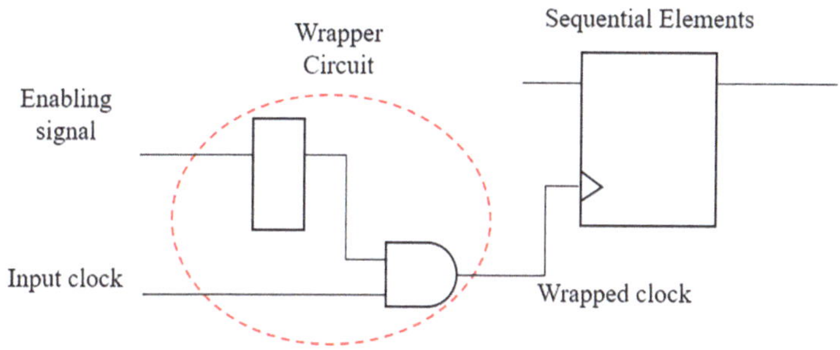

Fig. 6.24 Clock signal blocking circuit

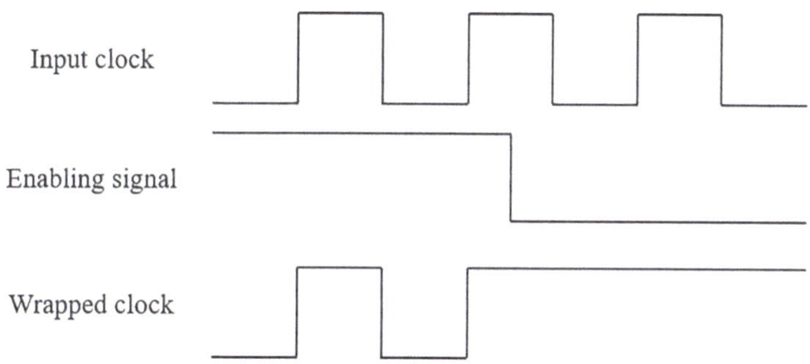

Fig. 6.25 View of the blocked clock signal

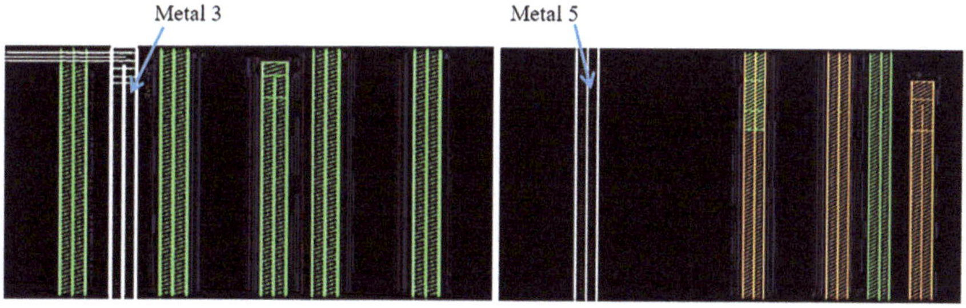

Fig. 6.26 View of the clock-carrying wire before proposal application

between different clock signals of the IC. The increase in clock-carrying wires proposed in different methods for reducing the consequences of IBEs leads to an increase in both area and power consumption.

Thus, it is proposed to make changes to the clock-carrying wires in the IC, which will allow reducing IBEs, but they can lead to changes in the circuit's parameters, which in turn lead to circuit failures.

Current Approach to Blocking the Routing of IBE Mitigation Wires

As it can be seen from (6.3) and (6.4), in the case of close routing of wires in an IC and their large overlap, the mutual capacitance coefficient between them is large.

Table 6.2 IBE before the clock-carrying wire arrangement proposal

Wire number	Wire length (μm)	Max delay due to IBE (ns)	Min delay due to IBE (ns)
1	303,2110	0,0000	−0,104,150
2	264,2910	0,073034	−0,073712
3	202,8640	0,047006	−0,046385
4	231,2360	0,033836	−0,036404
5	312,3380	0,023931	−0,036985
6	216,5460	0,010246	−0,034793
7	196,4480	0,019085	−0,031466
8	209,6950	0,027458	−0,029576
9	206,3540	0,000000	−0,029089
10	174,7880	0,017201	−0,028187

Table 6.3 IBE after the clock-carrying wire arrangement proposal

Wire number	Wire length (μm)	Max delay due to IBE (ns)	Min delay due to IBE (ns)
1	314,709	0	0
2	270,822	0,000073	−0,007326
3	200,454	0	0
4	235,765	0,032421	−0,036356
5	275,26	0	−0,012946
6	280,203	0	0
7	169,841	0,000968	−0,012191
8	202,702	0	−0,008138
9	227,48	0,008268	−0,021662
10	179,702	0,006678	−0,009295

This, in turn, leads to low capacitive resistance between them; consequently, there is an increase in interactions between them. To avoid such cases, it is proposed during the circuit design process, when the corresponding software tool creates a given wire in the circuit, to set certain blocking boundaries within which wire routing cannot be performed [92, 93]. Such boundaries represent rectangular zones, for which coordinates and the corresponding metal layer are given.

The implementation of this method has the form presented in Fig. 6.27 [92].

In such an approach, the metal layers have the structure presented in Fig. 6.28 [92].

For example, [94] applies the proposed method. The circuit on which the tests were performed has 100 available routing tracks for metals in one metal layer, but in this circuit, 172 pass through those tracks (Fig. 6.29a) [94]. With the application of the method, using 70% of the metal routing resources, metal connections were excluded in the given section of the circuit (Fig. 6.29b) [94].

However, the application of the above method first of all complicates the circuit design process, leads to repetitions of steps during the design, as well as complex management of the constraints intended for the circuit. As mentioned above, routing resources are used, which can cause additional connections in the metal layers,

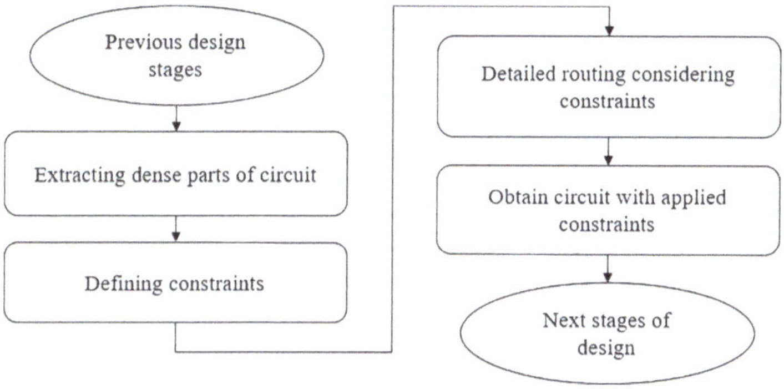

Fig. 6.27 Obtaining the circuit with the application of blockages

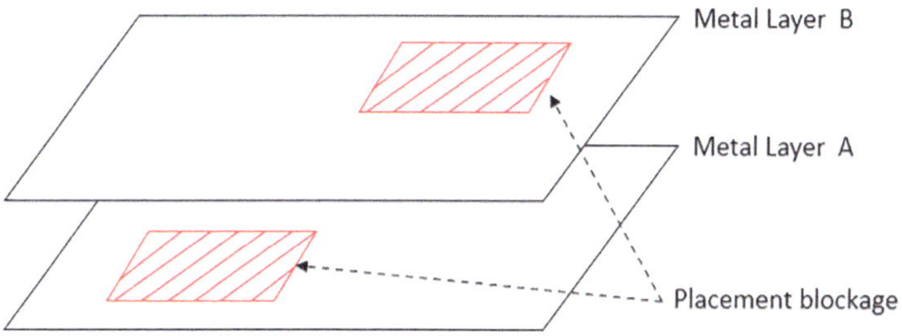

Fig. 6.28 View of metal layers with the application of blockages

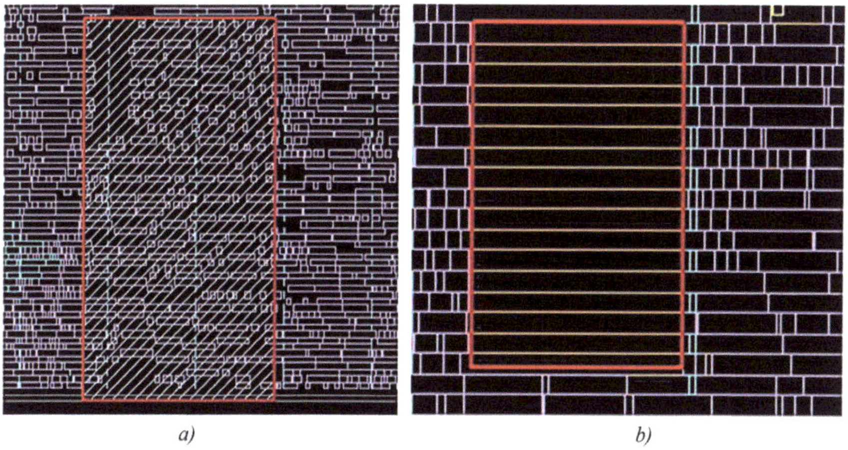

Fig. 6.29 Change of the circuit with the application of blocking

therefore leading to an increase in circuit area and power consumption, and also reducing the possibility of flexibility to make changes in the circuit later.

Thus, it is proposed to perform routing blockages for wires in the circuit, due to which those wires that are routed close to each other and can have interaction leading to failures will be placed far from each other. However, the application of such a method is complex, prolongs the IC design process, and also occupies the circuit's routing resources.

Existing AI-based Method for Detecting IBE Consequences

For detecting IBE in the initial stages of IC design, a method based on AI algorithms is proposed. The method involves identifying, at the element placement stage, points in the circuit that may have large mutual capacitance, effects on each other, and deterioration of timing parameters due to interactions.

In this flow, input features are extracted based on placement and post-layout data databases, which are then used as an ML database. By training a candidate ML model (e.g., XGBoost model) based on the database, and evaluating, the most effective feature sets and the best models are selected. These models perform the above three points (Fig. 6.30) [95].

To train such a model, the circuit's DEF format file, standard cell libraries, and static timing analysis results are needed. Features are extracted from all this data and used in the ML database. Based on that database, the most effective one is selected from the existing candidate ML models for the given project, and model training and evaluation are performed. As a result, a model is obtained that allows implementing the above requirements after feature extraction for a new project [95].

However, to use this method, besides the circuit's numerous data, such as its placement and post-layout data database, DEF format file, the standard logic cell libraries used in it, static timing analysis results, etc., an existing ML model is also needed that will allow implementing the above-described requirements.

Thus, for detecting IBE in circuits, a method based on ML algorithms is proposed, which does not depend on the arrangement of elements in the circuit. However, this method requires large databases of the circuit's data, libraries, analysis results, as well as existing models for ML.

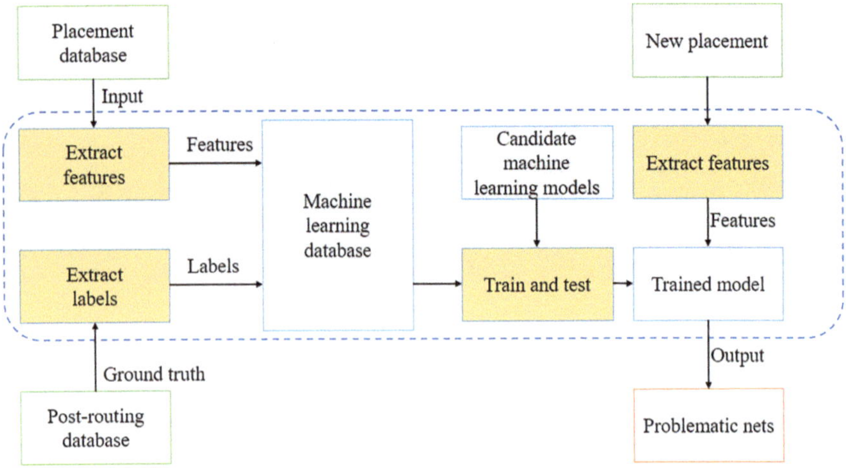

Fig. 6.30 IBE modeling flow

6.1.3 Proposed Principles of Means for Detecting and Mitigating the Crosstalk in Integrated Circuits Using Machine Learning

1. Prediction of IBE using ML algorithms by extracting information about the physical parameters of the circuit's interconnects and the signals passing through those nodes (clock signal or Signal Wire (SW)) (Fig. 6.31).

To implement the proposed method, from the files obtained as a result of the circuit's physical design (with nlib and GDSII extensions), information about the belonging of the circuit's wires to the CT or SW is extracted, based on which the signal frequency or switching activity is determined. The physical parameters of the circuit's interconnects are calculated: distance between wire pairs, overlap area, etc. Then, all the obtained information is consolidated and normalized.

An ML model is created, the training of which is performed on the already normalized information. After training, the model allows predicting IBE in the circuit without performing IBE analyses through other software tools. Due to the extraction of information necessary for IBE detection, the analysis related to it is performed only once; then, in case of any change in the circuit, such analysis is no longer needed.

2. Assessment of circuit's CT architecture and changing its wiring based on the information obtained as a result of IBE prediction (Fig. 6.32).

It is proposed to perform analyses of CT architecture in that IC based on the information obtained from the circuit's IBE prediction model and then make changes to its wire routing. The proposed changes depend on the mutual arrangement conditions of the clock-carrying and adjacent wires, the physical dimensions of the wires, and certain auxiliary information. Since the circuit's CT construction is performed before wire routing, the proposed method can be applied after that.

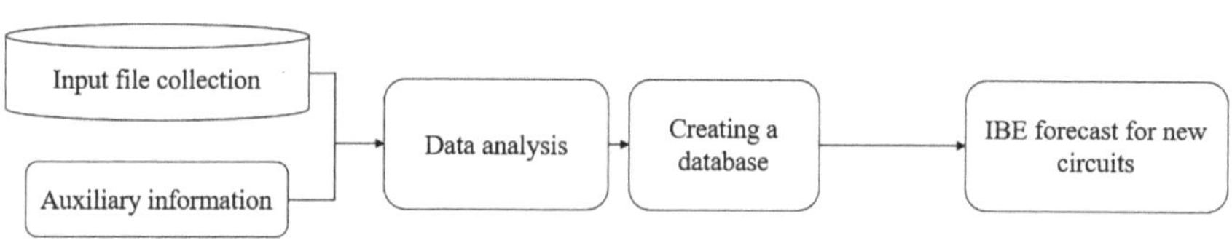

Fig. 6.31 General view of the principle of IBE prediction using ML algorithms

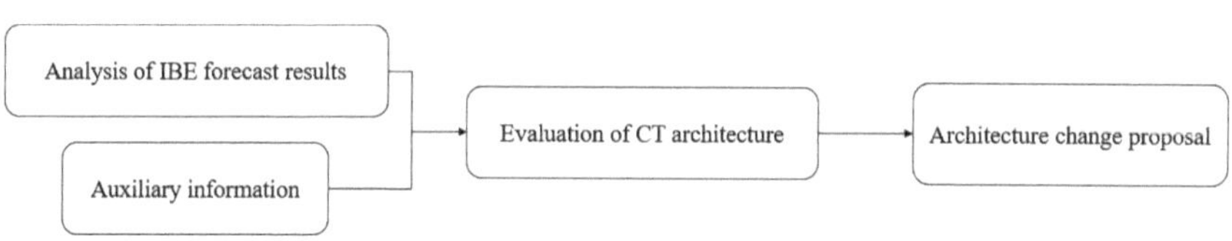

Fig. 6.32 General view of the method to assess CT architecture and change its wires

The assessment of the CT architecture and, if necessary, changing its wire routing will allow reducing IBE in the circuit, as well as improving the circuit's ST and HT ((6.6) and (6.7)).

3. Assessment of circuit's SW and their change based on the information obtained as a result of IBE prediction (Fig. 6.33).

It is proposed to perform assessment and change of interconnects in the circuit after IBE analysis, CT architecture assessment, and wire changes. The proposed changes depend on the physical parameters of the SW, their mutual arrangement conditions with other wires, and additional auxiliary information. Since the circuit's wire routing is performed after CT construction, in this case, the proposed method can be applied. The assessment of the wire mutual arrangement and, if necessary, change will allow reducing IBE in the circuit, improving the circuit's ST and HT ((6.6) and (6.7)), as well as NEXT and FEXT ((6.10) and (6.11)).

In general, the combination of the methods presents a sequence of steps that allows performing analyses from the perspective of the circuit's IBE using ML algorithms based on input information, and based on the obtained data, makes recommendations to mitigate the consequences of IBE (Fig. 6.34).

Conclusions

1. Detecting and preventing interactions between elements in integrated circuits has been and remains an important fundamental issue. There is a need to develop new methods that can detect them, allowing designers to consider interactions

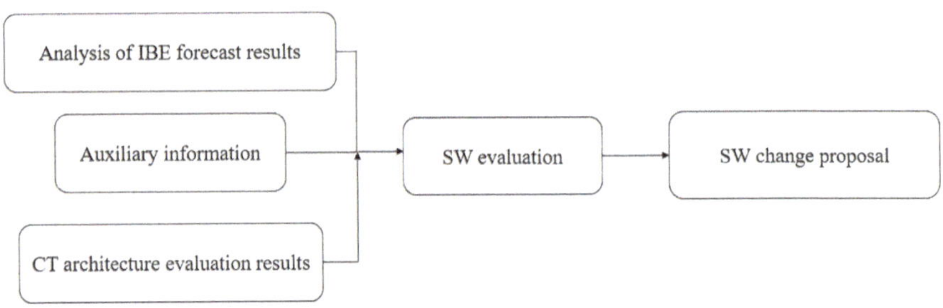

Fig. 6.33 General view of the principle of SW assessment and change

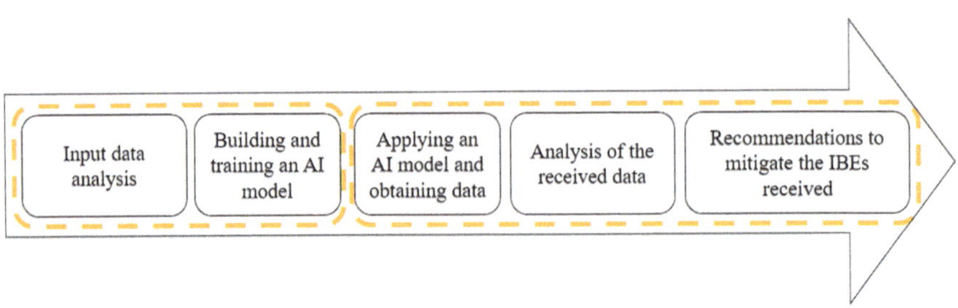

Fig. 6.34 Description of the combination of the proposed principles

during the design process and build reliable circuits. To speed up the design and testing process and exclude errors due to the human factor, it is preferable to use machine learning methods.

2. The existing solutions for detecting and mitigating the consequences of interactions between elements allow for mitigating them, but each of the solutions has its drawbacks. After changing the circuit, there is a need to detect and consider interactions, which does not satisfy the time requirements for circuit design, verification, and production.
3. Methods for detecting, analyzing, and mitigating interactions between elements have been proposed, which, with non-essential deterioration of their characteristic parameters, ensure time savings during detection and, with non-essential increases in area and power consumption, ensure mitigation of the consequences of interactions and overall timing parameters.

6.2 Proposed Solutions for Detecting and Mitigating the Crosstalk in Integrated Circuits Using Machine Learning

In Sect. 6.1, it was mentioned that detecting and preventing IBE in ICs plays a very important role in designing reliable circuits. The existing methods for IBE reduction and their drawbacks were also considered. Therefore, it is necessary to develop new methods that will allow predicting IBEs of the IC during the design process and deriving the possible causes of their occurrence, proposing possible solutions.

6.2.1 Method for Predicting Crosstalk in Integrated Circuits Using ML Algorithms

In case of applying the proposed method for detecting IBE in ICs, the files obtained as a result of the circuit's physical design (GDSII and nlib extension files) are provided, as well as the Element Encoding Compliance (EEC) text file corresponding to the given technological process. To save the resources used for method implementation, certain additional information is also applied. To implement the proposed approach, data collection is performed from the above files and auxiliary information, then based on that data, an ML model is created for predicting IBE in the IC. The novelty is the form of data collection, obtaining information from files obtained during the design process and their combination, obtaining the corresponding ML model, and the possibility of applying that model in the case of different circuits.

Decoding Input Information for IBE Prediction in Circuits

To implement the proposed method of IBE prediction in ICs, first, the decoding of input data and the collection of necessary data are performed (Fig. 6.35) [96].

First, to be able to perform operations with the GDSII extension file, it is converted into a text extension file (Fig. 6.36b) [97]. To obtain such a file, the GDS2GDT software tool is used [98]. The GDS2GDT tool provides encoding information for each wire in an IC (Fig. 6.36a) [97].

Then, data extraction for metals corresponding to metal layers is performed. From the EEC file, all codes corresponding to the circuit's metal layers are extracted. For example, the number 19 corresponds to metals located in the first metal layer, number 21 to the second one, and so on. In the EEC file, metal encoding is performed by combining the characters "M," "ME" with the number corresponding to the given metal layer [98]. If there is more than one metal intended for production with the same "mask" in the same metal layer, then they are given by the combination of corresponding characters and "1_2," "1_3," "2_2," etc. [98]. From (6.4), it can be seen that to have minimal interactions between elements in the circuit, it is

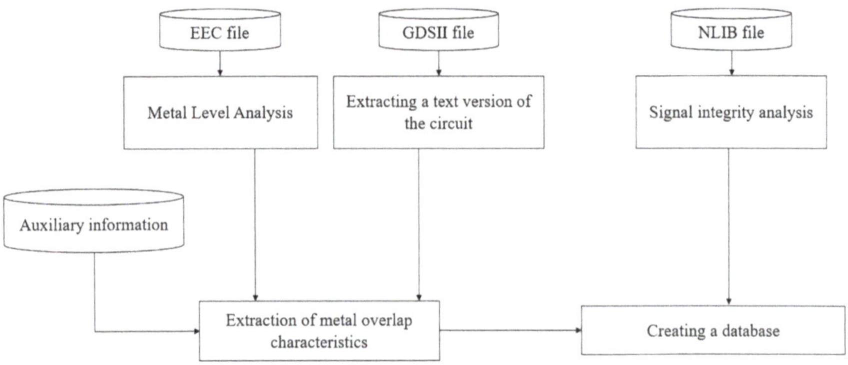

Fig. 6.35 Description of data collection to apply the proposed IBE prediction method in ICs

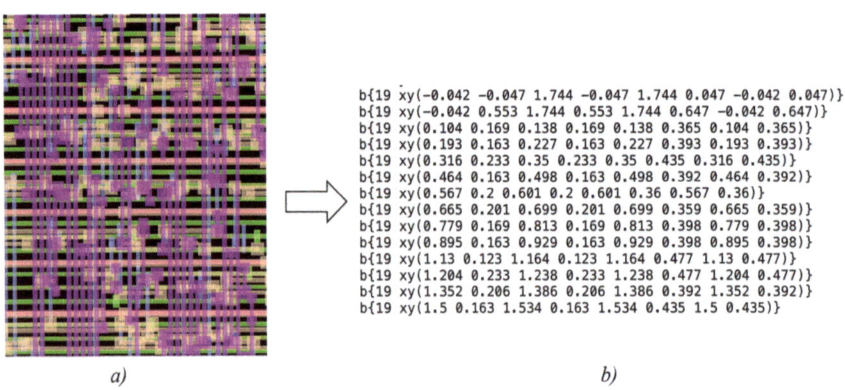

Fig. 6.36 Obtaining a GDT file from a GDSII extension file

preferable to construct wires in every next metal layer perpendicular to the wires of the previous level [99]. However, in the first metal layer of the circuit, wires can be implemented in either vertical or horizontal directions. Such a direction in the case of a given metal layer is called the preferred direction (Fig. 6.37) [99, 100].

Considering all this, auxiliary information is used in the proposed method: the preferred direction of wire construction for the first metal layer of the IC. The wires of every next metal layer are considered in a direction perpendicular to the previous one. However, based on the requirements placed on the circuit, there may be cases when a metal in a given metal layer has segments deviated from the preferred direction (Fig. 6.38) [99]. Considering this, there is also the possibility of detecting slanted metals if necessary. For this, another auxiliary information is used: an indicator of the presence of slanted metals in the circuit. If there is at least one metal in the circuit that is deviated from the preferred direction corresponding to its metal layer, then the corresponding indicator must be available. If the indicator of the presence of slanted metals provided at the input is available, then two more are added to the information decoded in the previous steps, corresponding to each element of the metal pair. Such an approach allows avoiding the decoding of parameters that will be constant for the entire circuit and, therefore, will not participate in the IBE prediction process. In the case of the slanted metals indicator being open, a check of the coordinate pairs of the wires is performed.

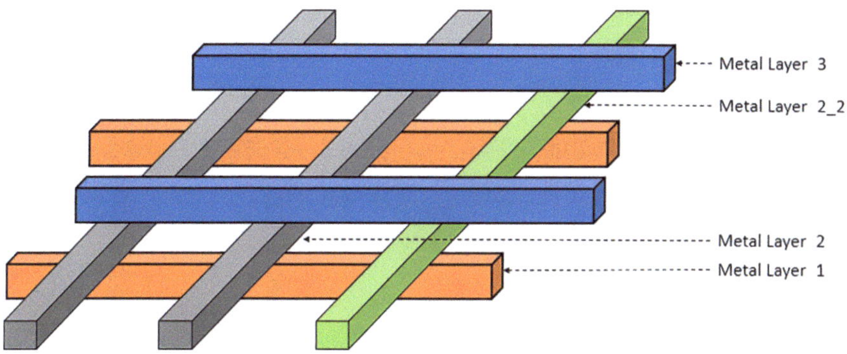

Fig. 6.37 Arrangement of metal layers and their preferred direction

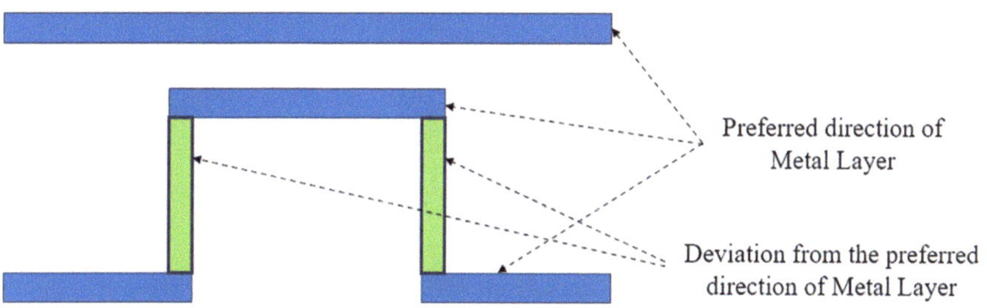

Fig. 6.38 Description of metal deviation from the preferred direction

The other auxiliary indicator needed for IBE prediction is obtained from the different forms of metal representation. Designers have the possibility to represent several metals located adjacent to each other as a combination of one metal and several cut layers (Fig. 6.39a) [101, 102]. In this case, the metal presented in the GDT file is converted into a number of metals corresponding to the cut layers (Fig. 6.39b) [101, 102], and IBE prediction is performed for each pair of those metals.

As a result of the described steps, the coordinates of all metal pairs located in every metal layer of the circuit are extracted. From those coordinates, for each pair, overlap characteristics are extracted. The distance between metal pairs and their overlap area are calculated (Fig. 6.40) [103]. Based on them, the mutual capacitance coefficient for the metal pair is derived.

In summary, from the input information, parameters corresponding to the circuit's physical design are extracted: the coordinates of metal pairs present in every metal layer, the distance between them, the overlap area, and the mutual capacitance coefficient between them. For the possibility of proposing mitigation of IBE consequences for the clock signal or SW as a result of model application, the minimum acceptable distance between wires established for the given metal layer [104] is also extracted, which is added to the database being created for IBE prediction.

Then, extraction of information corresponding to the circuit's logical design is performed. To perform such an operation, the Fusion Compiler (FC) [105] software tool developed by Synopsys is used. First, the nlib extension file mentioned above

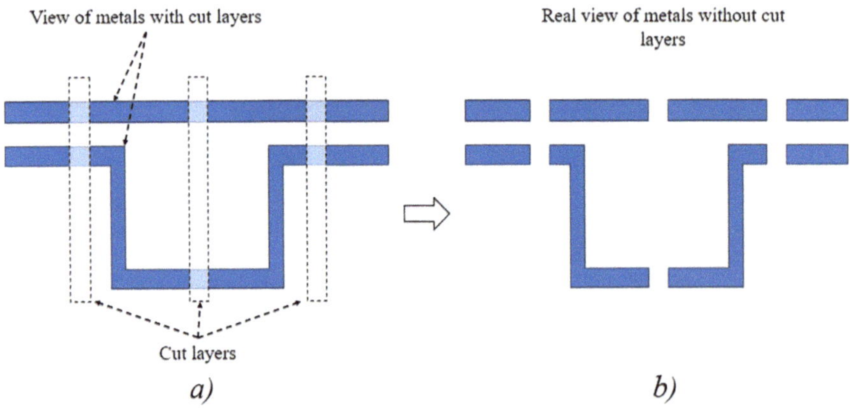

Fig. 6.39 View of metals with and without cut layers

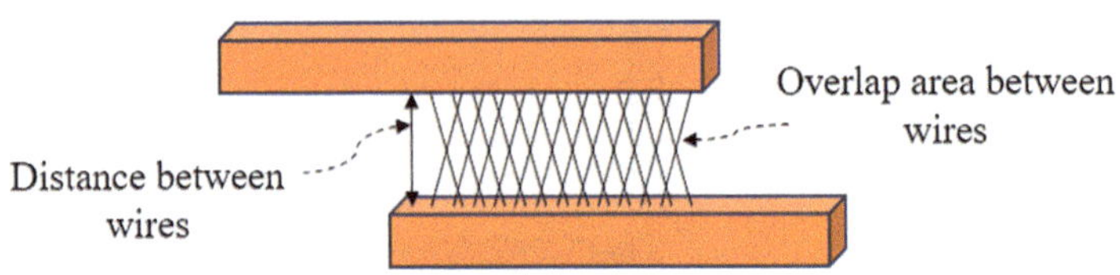

Fig. 6.40 Description of the distance between wires and the overlap area

is given to FC. Such a file contains information about the circuit's clock signals, input and output nodes, wires intended for shielding, power and ground wires, and other parameters of the circuit. Using the FC software tool, information about the existing clock signals in the circuit is extracted (Fig. 6.41). These include both input clock signals and those wires that are created from the inputs within the circuit. This information includes the coordinates of the clock-carrying wires, the metal layer, and the frequency of that signal. Besides such wires, the coordinates of the existing SW in the circuit, the metal layer, and its switching activity are also extracted. In the

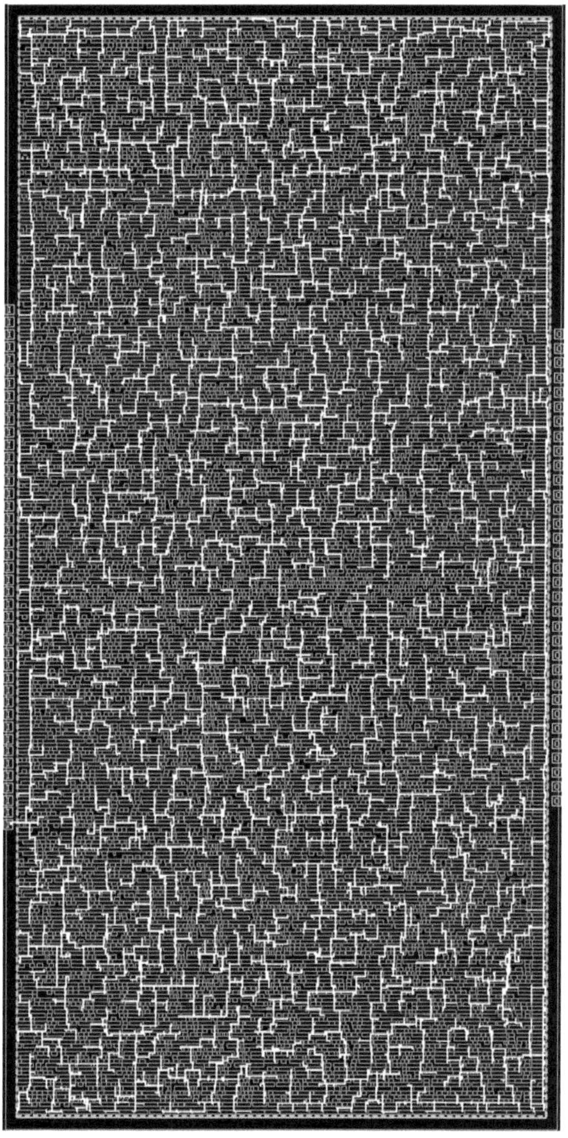

Fig. 6.41 Wires of circuit CT

case of this information, indicators for the clock signal and SW are respectively set for further work.

As a result of extracting the characteristics of the mutual arrangement of metals in the IC, the following features are obtained for each metal pair (Fig. 6.42):

- Coordinates of the first and second metals
- Their distance, overlap area, mutual capacitance coefficient, metal layer
- The minimum acceptable distance between wires for the metal layer
- In case of the availability of the indicator of the presence of slanted metals in the circuit, the indicators of the first and second metals are slanted
- Indicators of the first and second metals being part of the ct or being an information-carrying wire, as well as, in the case of being part of the clock tree, the clock frequency, and in the case of sw, the signal change activity.

Then, Signal Integrity Analysis (SIA) of the circuit is performed. This is also implemented through FC. SIA is performed between the circuit's input and output nodes. As a result of this analysis, the tool is configured so that it provides the interaction coefficient for the selected pair of metals, which shows the delay of one signal due to the change of another signal. Then, for those metals, their coordinates and information corresponding to the metal layer are extracted. This information is then matched to the same metal pairs obtained in the data collection and decoding phase. Therefore, the results obtained from SIA are also added to the above parameters. All the obtained data are placed in one database, based on which the formation and training of the ML model must be carried out.

Thus, it is proposed to perform the above-described collection of different types of data and SIA. Then, perform matching and grouping of that data in one place for the purpose of later using it as a database for the ML model for IBE prediction.

IsM1Clock	IsM2Clock	M1SwitActivity	M2SwitActivity	A_x0	A_y0	A_x1	A_y1	A_x2	A_y2	A_x3	A_y3	B_x0
0	1	0,56	2	0,72	4,72	0,72	4,8	0,74	4,8	0,74	4,72	1,02
0	0	0,56	0,71	0,72	0,72	0,72	4,8	0,74	4,8	0,74	4,72	1,02
0	1	0,56	2	0,72	4,72	0,72	4,8	0,74	4,8	0,74	4,72	1,08
0	0	0,56	0,65	0,72	4,72	0,72	4,8	0,74	4,8	0,74	4,72	1,08
0	0	0,56	0,92	0,72	4,72	0,72	4,8	0,74	4,8	0,74	4,72	1,17
0	1	0,56	2	0,72	4,72	0,72	4,8	0,74	4,8	0,74	4,72	1,2
0	0	0,56	0,7	0,72	4,72	0,72	4,8	0,74	4,8	0,74	4,72	1,26
0	1	0,56	2	0,72	4,72	0,72	4,8	0,74	4,8	0,74	4,72	1,32
0	0	0,56	0,42	0,72	4,72	0,72	4,8	0,74	4,8	0,74	4,72	1,38
0	1	0,56	2	0,72	4,72	0,72	4,8	0,74	4,8	0,74	4,72	1,38
0	0	0,56	0,35	0,72	4,72	0,72	4,8	0,74	4,8	0,74	4,72	1,44
0	1	0,56	2	0,72	4,72	0,72	4,8	0,74	4,8	0,74	4,72	1,44
0	0	0,56	0,66	0,72	4,72	0,72	4,8	0,74	4,8	0,74	4,72	1,5
0	1	0,56	2	0,72	4,72	0,72	4,8	0,74	4,8	0,74	4,72	1,5

Fig. 6.42 Features of the circuit's metal pairs

Building and Training an ML Model for Predicting IBE in ICs

When all the above information is already available and everything is collected in one place, operations intended for the corresponding ML model are performed (Fig. 6.43) [96].

First, preparation of all the obtained information is performed. The information obtained in text format for metal layers is encoded [106] depending on the metal layer or metals intended for production with different masks in one metal layer. All repeated data are removed from the data set, as well as those columns whose value is unchanged. In the case of providing only one input GDSII file, there cannot be repeated rows, but in the case of several input data, rows with the same parameters may occur.

Then data normalization is performed. For this operation, the linear normalization technique was used [107]. For its application, first the maximum and minimum values corresponding to each data are extracted; then, to obtain the updated data, the following operation is performed [107]:

$$x_{\text{updated}} = \frac{x - x_{\text{max}}}{x_{\text{max}} - x_{\text{min}}}, \tag{6.13}$$

where x_{updated} is the updated data after applying the technique, x_{min} and x_{max} are the minimum and maximum values of the data series, respectively. From (6.13) it can be seen that when the data has the largest value of the series, then as a result of normalization, x_{updated} will become 1, and in the case of having the minimum value, it will become 0. Therefore, through such a technique, all available data are converted into numbers in the 0..0.1 range. Through this, the model will work with numbers of the same range and will spend minimal time working with new data [107].

After complete data collection, checking, and normalization, the most important part of the method is implemented: the creation of the ML model. For model construction, the Branched Artificial Neural Network (BANN) architecture was chosen

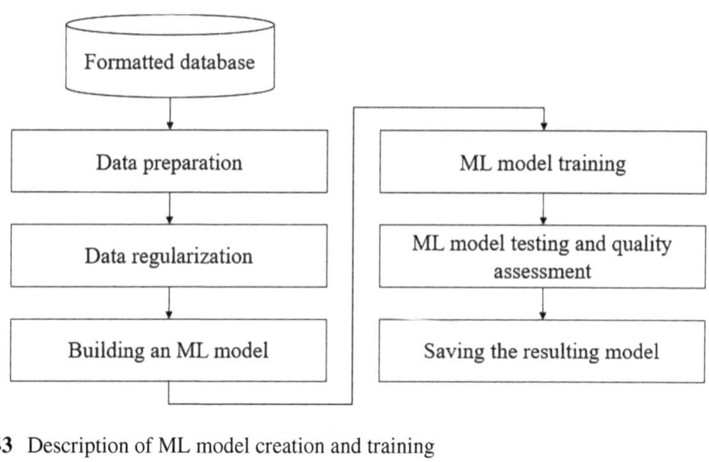

Fig. 6.43 Description of ML model creation and training

(Fig. 6.44) [96]. A fully connected version of the model was applied, where any neuron of any layer of the model is connected to all neurons of the layers preceding and/or following it.

There are several reasons for choosing the specific BANN architecture, among which are:

- BANN is flexible when working with different types of data. Therefore, different types of data can be managed without introducing many changes in the BANN [108].
- Each branch of the architecture can work with specific features, accelerating the model training process [109].
- The model can detect patterns in different branches based on the training data and, based on those patterns, make correct predictions [110].

Since all data are already normalized in the 0…1 range, for BANN neuron activation, non-linear continuous activation (Fig. 6.45a) [111] and sigmoid (Fig. 6.45b) [112] functions were chosen.

The non-linear continuous activation function [111] is used because, as already mentioned, after data normalization, they are in the 0…1 range. That function has the following form [111]:

$$x_{\text{new}} = \begin{cases} 0, \text{when } x <= 0, \\ x, \text{when } x > 0, \end{cases} \tag{6.14}$$

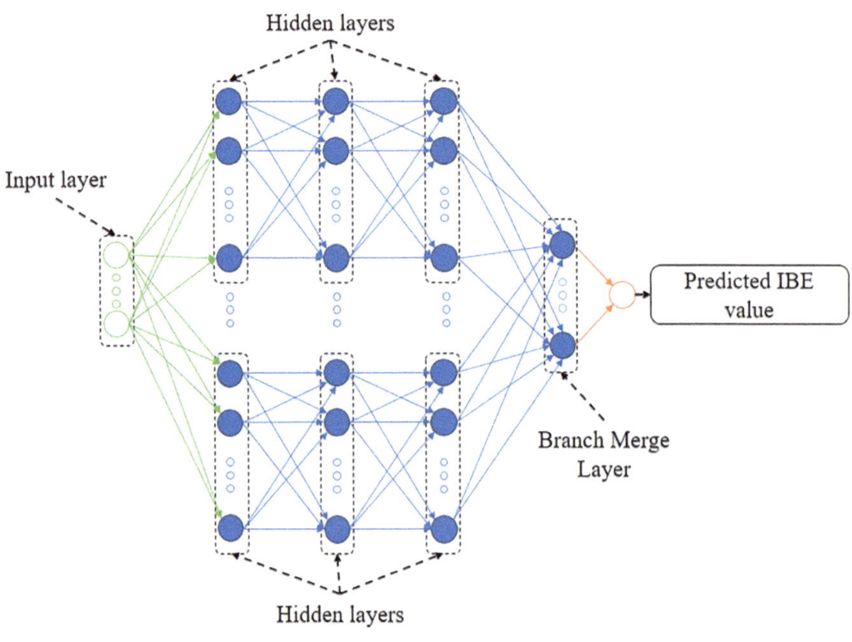

Fig. 6.44 Description of a fully connected BANN

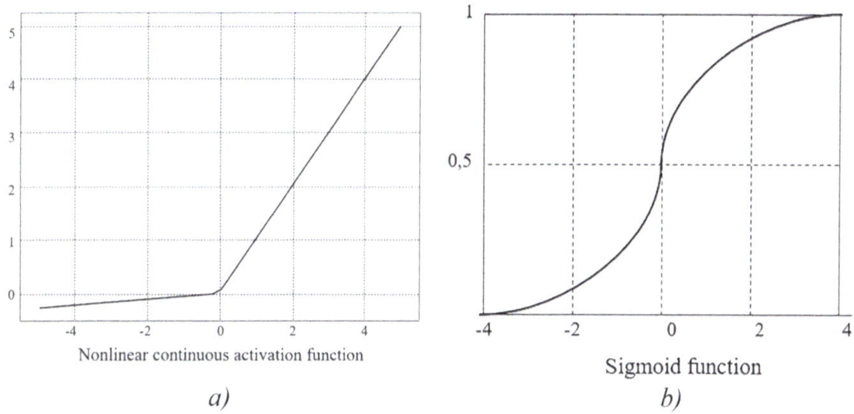

Fig. 6.45 View of non-linear continuous activation and sigmoid functions

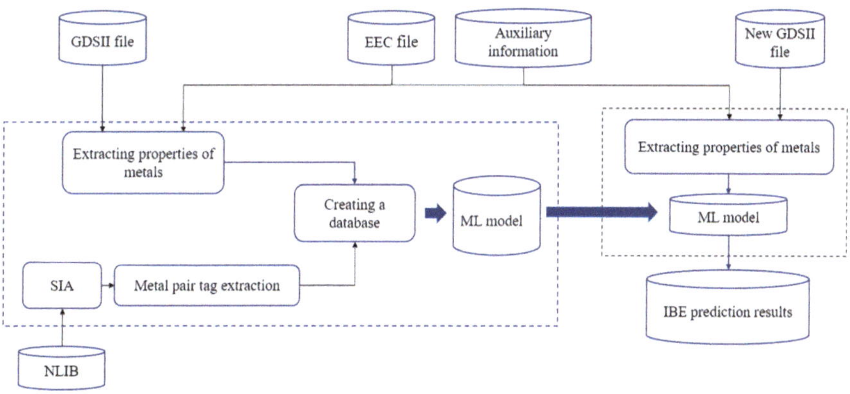

Fig. 6.46 General view of the proposed method

where x is the value of the given row of the feature, x_{new} is the output of the non-linear continuous activation function. The sigmoid function [112] is applied for the IBE prediction output. That function has the following form [112]:

$$x_{\text{new}} = \frac{1}{1+e^{-x}}, \tag{6.15}$$

where x is the value of the given row of the feature, x_{new} is the output of the sigmoid activation function.

After checking all the obtained information, normalization, and selection of neuron activation functions, a new database is obtained, which is already used for the creation and training of the ML model. ML model training is performed using the Keras sub-library [113] of the Tensorflow library [114, 115]. The overall structure of data collection, ML model creation, training, and application has the form presented in Fig. 6.46 [96].

Thus, it is proposed to perform data preparation and normalization after data collection, selecting corresponding functions and algorithms. Then use the obtained results as a database for training the ML model intended for obtaining IBE. And use the ML model in new circuits to predict IBE, based on the files obtained as a result of the physical design of those circuits.

Application and Evaluation of the Proposed Method for Predicting IBE in ICs Using ML Algorithms
To apply the ML model, first, an EEC extension file, auxiliary information about the metals in the circuit, and a GDSII extension file obtained from the new circuit for which IBE prediction should be performed are provided. Data collection is performed with the steps mentioned above; then, using the created ML model, IBE is predicted for the provided circuit. For training and testing the ML model obtained as a result of the proposed method for IBE prediction in ICs, circuits composed of cells from the SAED14 educational library were used, which do not contain intellectual property and in which the transistor gate width is 14 nm [116, 117]. After qualitative testing experiments of the model, analyzing the results of successful and unsuccessful experiments, the most acceptable ML model architecture was confirmed, based on which experimental research was performed to check the accuracy of IBE prediction in already used projects. Considering the amount of input data, in the case of the simplest memory with 4 rows and 4 columns, which is tested with low power processors [118–120], the number of metals is 2149, of which 1205 are in metal layer 1, and 944 in metal layer 2, ML model training was performed with the training technique presented in [121].

For ML model training, cross-validation, and loss function evaluation, and training information, a circuit designed with cells from the SAED14 educational library [116, 117] was used, in which the number of standard cells is 22,826, the area is 9140.6141 μm^2, the total length of wires constituting the CT is 7781.291 μm, the density of logic cells is 60.34%. The time spent by the circuit for SIA through FC was 8227.7 s. The circuit was designed with an input 2 GHz clock signal. To evaluate the model's effectiveness, the Mean Squared Error (MSE) of the IBE prediction results was calculated as follows [122]:

$$\mathrm{MSE} = \frac{1}{N} * \left(\sum_{k=1}^{N} \left(x_i - x_i^{'} \right)^2 \right), \tag{6.16}$$

where $x_i^{'}$ is the expected value of the given output, x_i is the predicted value, N is the number of all data to be predicted. The MSE itself was chosen as the loss function, which expresses the rate of decrease of the model's error. The latter reaches its minimum value within 48...53 cycles (Fig. 6.47) [96].

During the research, it was found that when training is performed with 48...53 cycles, the accuracy of the model's IBE prediction results is approximately 91% (Fig. 6.48) [96].

And to reduce the error between the model's expected and predicted values, the optimization algorithm described in [123] was applied. The model, to predict the

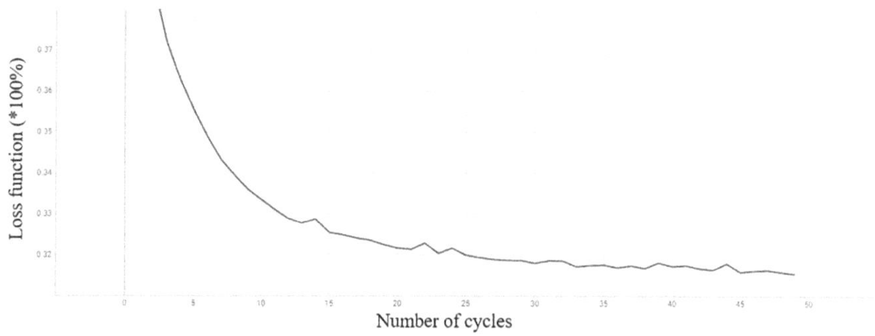

Fig. 6.47 ML model loss function depending on the number of cycles

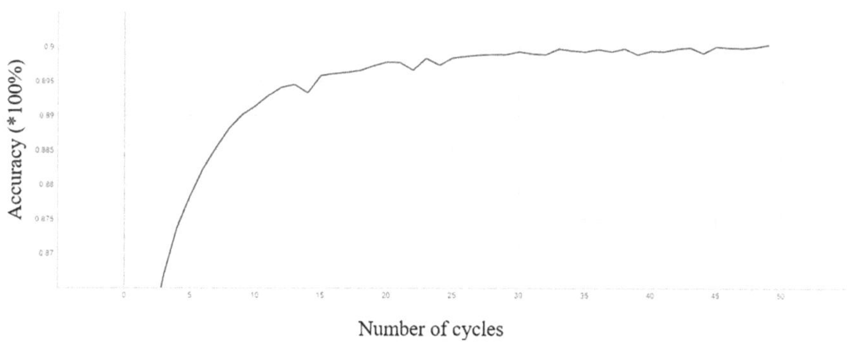

Fig. 6.48 Accuracy of the ML model depending on the number of cycles

circuit's IBEs during the application of the proposal, performed the above operations in 5216.4 s, which is 36.6% less than the time to obtain the same operation with FC (Fig. 6.49).

Thus, a method based on ML algorithms has been developed for predicting IBE in ICs, which, through as little data collection and circuit analysis as possible, ensures approximately 36.6% savings in time spent on SIA, at the cost of approximately 10% loss of predicted data. As a result of such an approach, an ML model was obtained that, having corresponding data about the circuit, provides the interaction coefficient between wire pairs located in every metal layer of the circuit.

6.2.2 Method to Assess the Architecture of the Clock Tree in Integrated Circuits and Change its Wiring

As mentioned above, IBE prediction in the circuit was implemented in the proposed method. It was shown that with the development of ICs, the frequency of their clock signals has already reached tens of GHz. From (6.3), it becomes clear that the mutual capacitance between elements in the IC decreases in the case of such

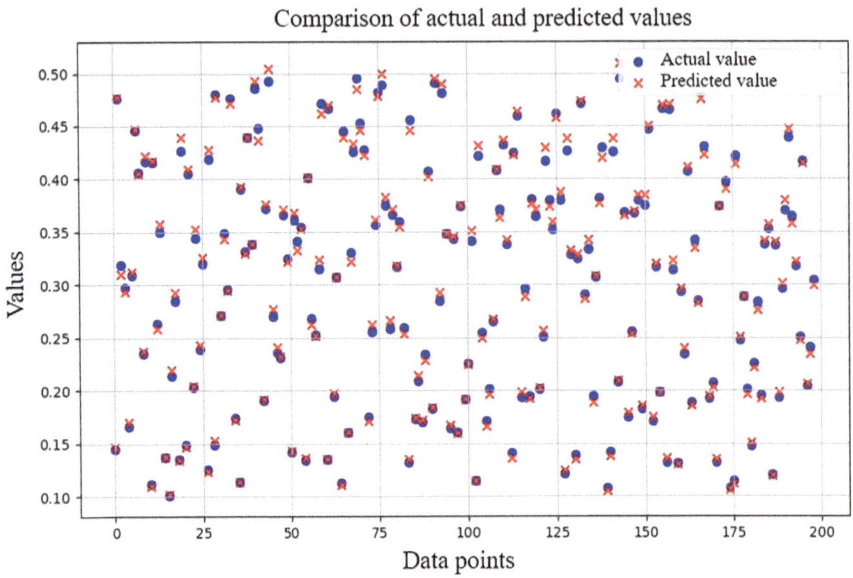

Fig. 6.49 Comparison between actual and predicted IBE values

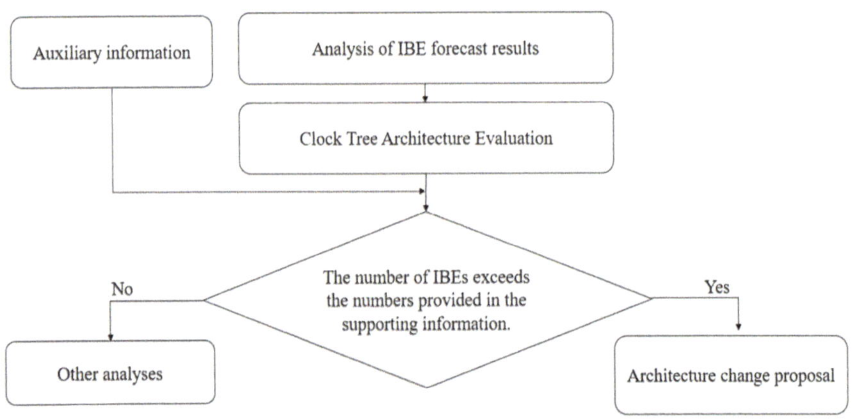

Fig. 6.50 Description of the method for assessing CT architecture and changing it

frequencies. Therefore, IBEs causing failures in the circuit will be present between metal pairs of which at least one belongs to the CT. Therefore, it is necessary first and foremost to consider such pairs.

Based on the obtained ML model, instead of a complete change of the CT architecture, it was proposed to perform transformations on its individual sections using improved combinations of existing IBE reduction methods (Fig. 6.50) [96].

The wire encoding presented above is performed during logical design, the presented methods can lead to deviations in timing parameters between several clock signals of the circuit, as well as changes in the arrival times of one clock signal at the inputs of different sequential elements. For that very reason, a combination of two methods was chosen for clock tree changes: the application of wire routing constraints and the screening method.

The essence of the method is as follows: for selected value ranges of the overlap area between elements, the distance between them is assessed (Fig. 6.51), and as a result of that assessment, corresponding changes to the CT wires are performed.

If the distance between the wires is smaller than three times the minimum acceptable distance between wires for the given metal layer [104], then for such wires, it

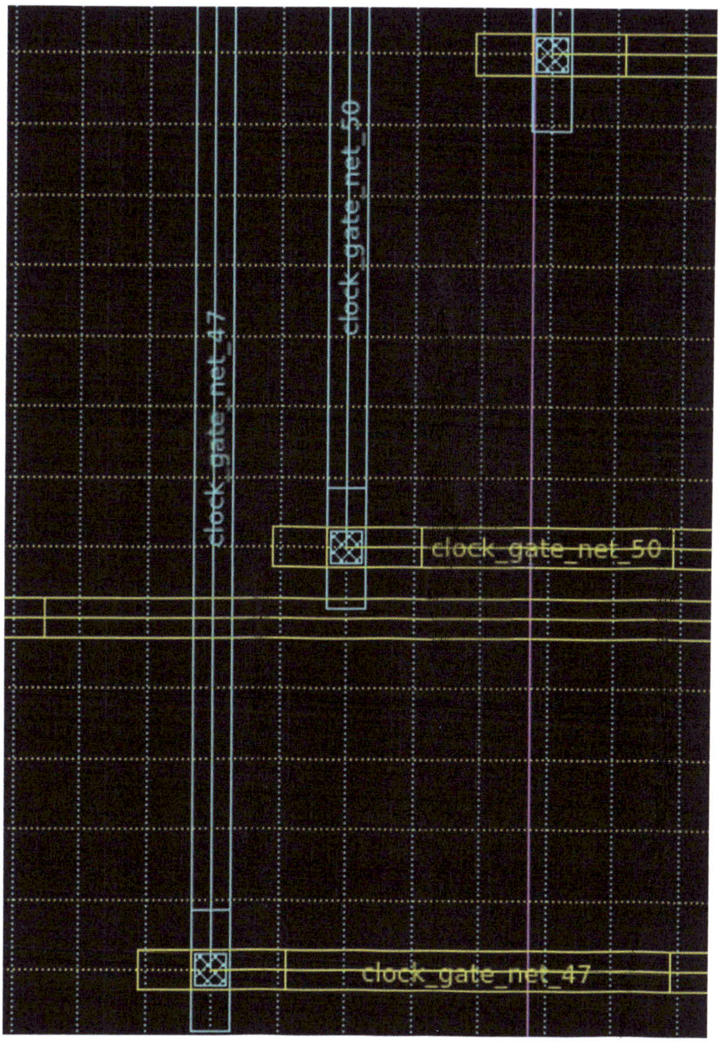

Fig. 6.51 Mutual arrangement of wires before method application

is necessary to apply the definition of blockages and perform the routing, taking that constraint into account [92].

In this way, the distance between them will increase; therefore, also the mutual capacitance (Fig. 6.52). Otherwise, when the distance between the wires exceeds three times the minimum acceptable distance between wires for the given metal layer [104], then for the given wires, it is necessary to apply the screening method.

Screening will allow converting the mutual capacitance formed between two wires into separate capacitances relative to the circuit's "0" point (Fig. 6.53). After performing the changes, due to the reduction of IBE, power consumption will also decrease [16]; therefore, self-heating [24], and the IC will age later [28].

For implementing the changes, from the obtained IBE prediction results, all those metal wire pairs of which at least one wire belongs to the CT are separated. Since the obtained database contains an indicator of the circuit wire belonging to the CT, which was obtained through FC, the operation of extracting CT wires is no longer performed (Fig. 6.54). Unlike the combination of methods known from the

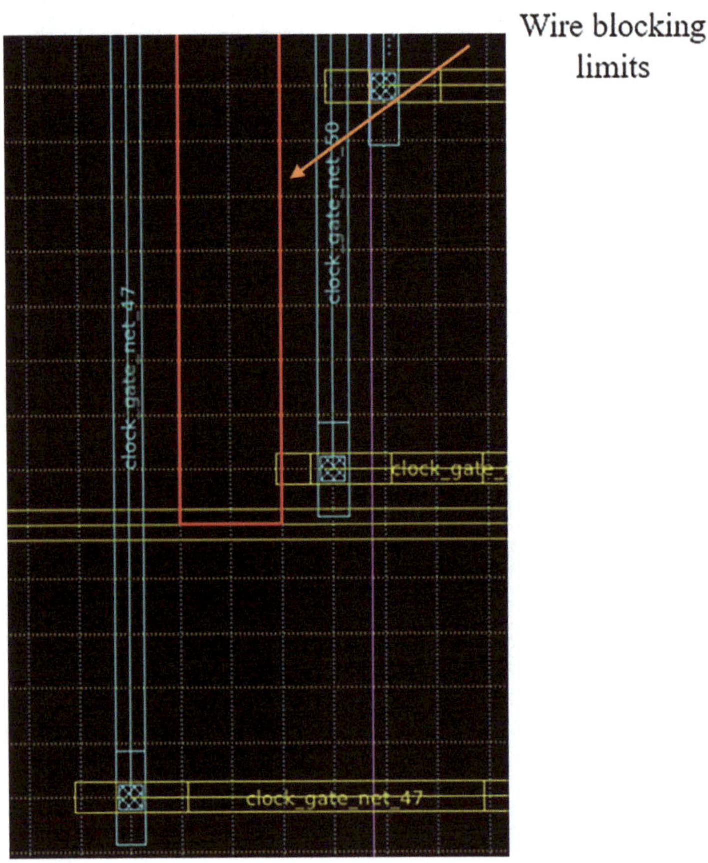

Fig. 6.52 Change in mutual arrangement of wires after applying blocking

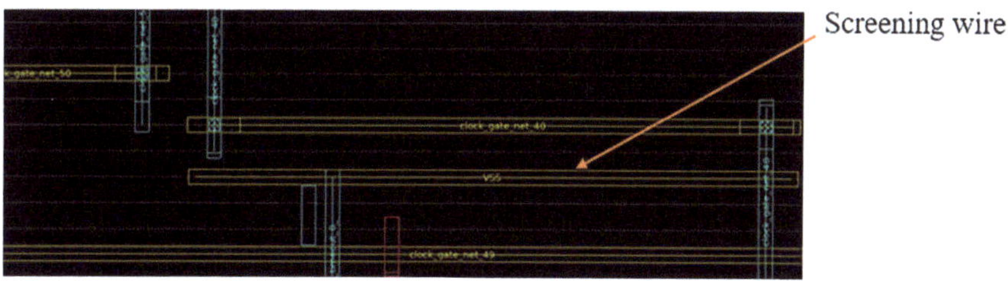

Fig. 6.53 Mutual arrangement of wires after screening

Clock presence indicator — Clock frequency or switching activity

IsM1Clock	IsM2Clock	M1SwitActivity	M2SwitActivity	A_x0	A_y0	A_x1	A_y1	A_x2	A_y2	A_x3	A_y3	B_x0
0	1	0,56	2	0,72	4,72	0,72	4,8	0,74	4,8	0,74	4,72	1,02
0	0	0,56	0,71	0,72	0,72	0,72	4,8	0,74	4,8	0,74	4,72	1,02
0	1	0,56	2	0,72	4,72	0,72	4,8	0,74	4,8	0,74	4,72	1,08
0	0	0,56	0,65	0,72	4,72	0,72	4,8	0,74	4,8	0,74	4,72	1,08
0	0	0,56	0,92	0,72	4,72	0,72	4,8	0,74	4,8	0,74	4,72	1,17
0	1	0,56	2	0,72	4,72	0,72	4,8	0,74	4,8	0,74	4,72	1,2
0	0	0,56	0,7	0,72	4,72	0,72	4,8	0,74	4,8	0,74	4,72	1,26
0	1	0,56	2	0,72	4,72	0,72	4,8	0,74	4,8	0,74	4,72	1,32
0	0	0,56	0,42	0,72	4,72	0,72	4,8	0,74	4,8	0,74	4,72	1,38
0	1	0,56	2	0,72	4,72	0,72	4,8	0,74	4,8	0,74	4,72	1,38
0	0	0,56	0,35	0,72	4,72	0,72	4,8	0,74	4,8	0,74	4,72	1,44
0	1	0,56	2	0,72	4,72	0,72	4,8	0,74	4,8	0,74	4,72	1,44
0	0	0,56	0,66	0,72	4,72	0,72	4,8	0,74	4,8	0,74	4,72	1,5
0	1	0,56	2	0,72	4,72	0,72	4,8	0,74	4,8	0,74	4,72	1,5

Fig. 6.54 Extraction of CT wires

literature, this approach is obtained based on the results from the ML model, and information for every metal layer is used.

To avoid numerous changes to the circuit as a result of the method's application, auxiliary information is used. First, it is possible to set the distance between elements, above which values wire changes are not performed. In a similar way, thanks to auxiliary information, a certain IBE coefficient is also set, below which values are not considered failure-causing coefficients for the given circuit. Therefore, the method's application will not extend to such a wire pair whose overlap and interaction corresponding numbers fall outside the boundaries of the numbers obtained through auxiliary information. The extraction of data from the circuit, the change proposal, its application, and obtaining the new circuit have the form presented in Fig. 6.55.

Thus, as a result of the improved combination of the results obtained from the IBE prediction ML model and the above methods, the possibility is created to mitigate the consequences of IBE, taking into account the features of CT construction.

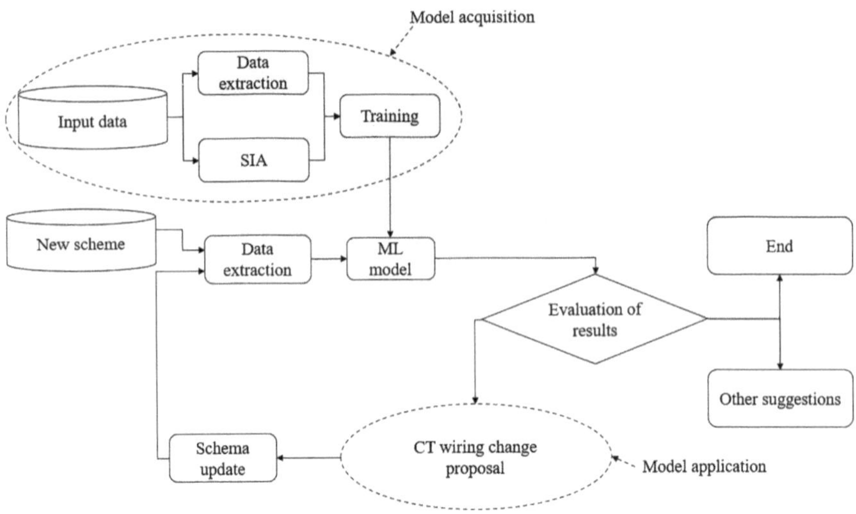

Fig. 6.55 Application of changes made for CT wires

Evaluation of the Effectiveness of the Proposed Method for Evaluating the Architecture of the CT in the IC and its Wiring Changes

To check the effectiveness of the mitigation proposal based on the information obtained as a result of IBE prediction in ICs, circuits composed of cells from the SAED14 educational library [116, 117] were used. First, the methods known from the literature were implemented separately for IBE mitigation, which were then combined with the results obtained by the ML model. Screening was applied between the circuit's metals for IBE reduction in the selected circuit, after which the circuit's timing parameters (ST, HT, due to IBE) were extracted, as well as the number of those sequential elements that have deviations from them. Then, IBE mitigation blockages were used, after which the same parameters were extracted. Finally, an improved combination of the two methods was performed, corresponding to the form presented in the proposed method (Table 6.4).

A comparison of the parameters of the circuits obtained as a result of the above changes was performed (Fig. 6.56).

After method implementation, a comparison of the change in the number of sequential elements with timing parameter deviations as a result of circuit change was performed (Fig. 6.57), as well as extraction of other circuit parameters (Table 6.5).

Table 6.4 Comparison of timing parameters before and after implementation of the proposed method

Design	Parameters					
	Number of elements with ST deviation	Total ST deviation (ps)	Number of elements with HT deviation	Total HT deviation (ps)	Number of elements with IBE	IBE deviation (ps)
Initial IC	937	−1121,6	600	−411,6	476	−10,9075
With screening	879	−1024,0208	568	−368,382	367	−10,0894
With blocking	898	−1049,8176	575	−384,0228	391	−99,258
Proposed form	856	−931,7131	541	−332,9432	310	−89,529

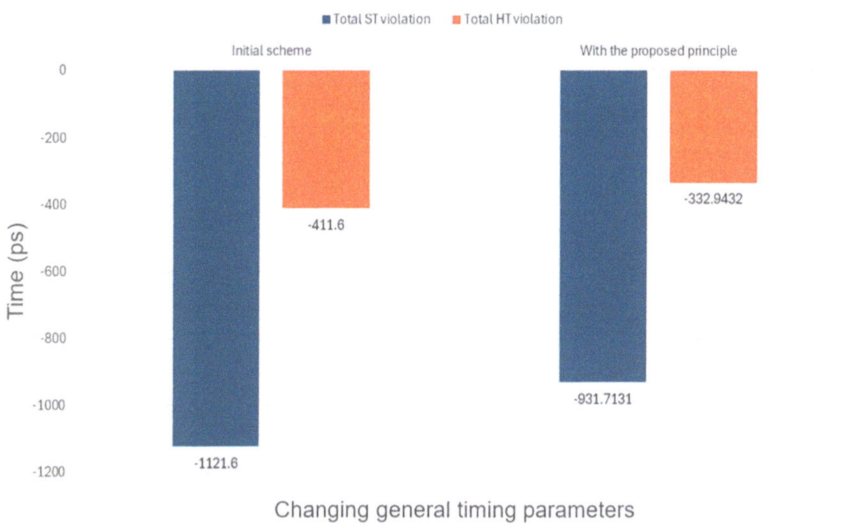

Fig. 6.56 Comparison of obtained timing parameters after method application

From the results, it can be seen that after the implementation of the proposed approach for CT wire change, the total area of the circuit increased by approximately 3.1%, the total power consumption increased by 5.66%. The total length of CT wires in the circuit increased by 17%, and the total wire length increased by 0.84% (Fig. 6.58).

As a result, IBE decreased by 17.92%, the circuit's total ST by 16.9%, and HT by 19.11% (Fig. 6.59).

The number of sequential elements in the circuit that had deviations from the above timing parameters also decreased by 34.87%, 8.64%, and 9.83%, respectively (Fig. 6.59).

Thus, a method has been proposed for assessing the CT architecture and changing the corresponding wires considering the features of the circuit design, which,

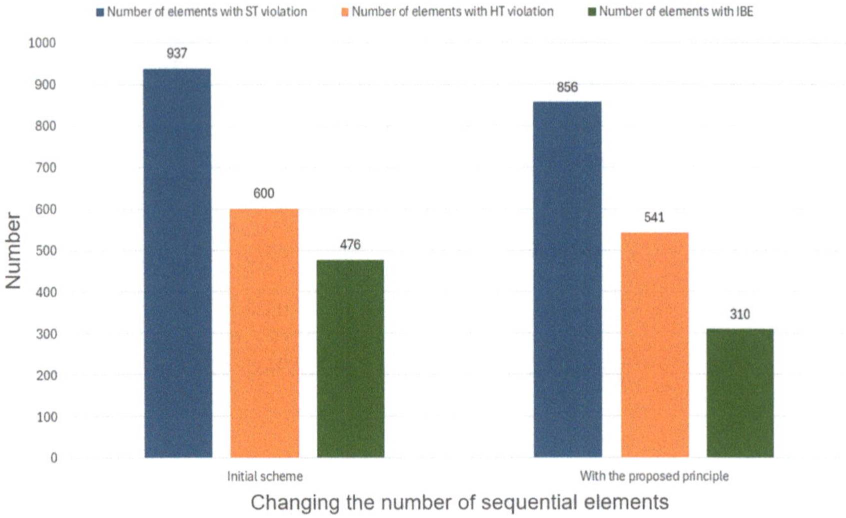

Fig. 6.57 Comparison of the number of sequential elements with timing parameter deviations after method application

Table 6.5 Comparison of general parameters before and after implementation of the proposed method

	Parameters				
Design	Area (μm²)	CT wire length (μm)	Total wire length (μm)	Total power consumption (mW)	Leakage power consumption (mW)
Initial IC	9140,6141	7781,291	86458,7889	1740,56	92,2
With screening	9140,6141	7781,291	86824,5096	1740,56	92,2
With blocking	9506,2387	9882,2396	87625,9826	1886,62	93,3064
Proposed form	9423,9731	9104,1105	87182,449	1839,07	92,8454

through an improved combination of screening and blockages by relocating CT wires, ensures approximately 17.9% mitigation of IBE consequences at the cost of approximately 5.66% loss in power consumption and 3.1% in area. As a result, ST and HT also decrease by 16.9% and 19.1%, respectively, and the number of sequential elements in which there were the specified deviations from the timing parameters decreases by 8.64% and 9.83%; however, the total length of CT wires increases by approximately 17%, which, however, constitutes about 0.84% of the increase in the total length of the circuit's wires.

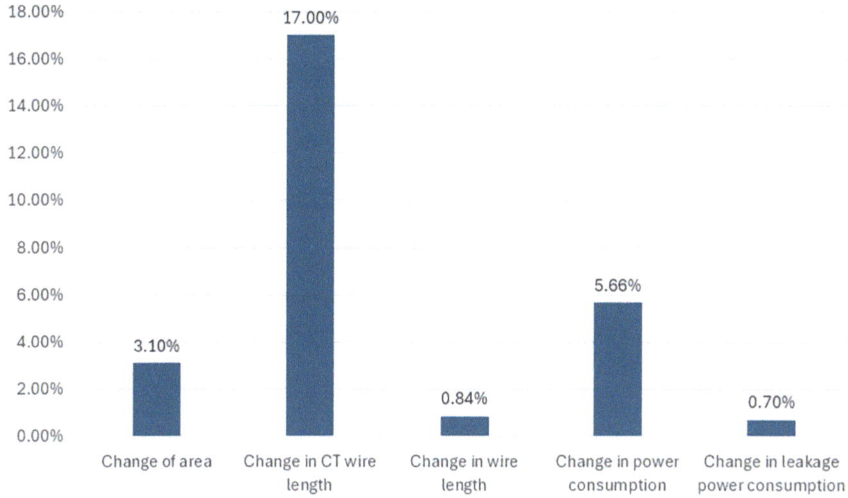

Fig. 6.58 Assessment of deterioration of circuit parameters after method application

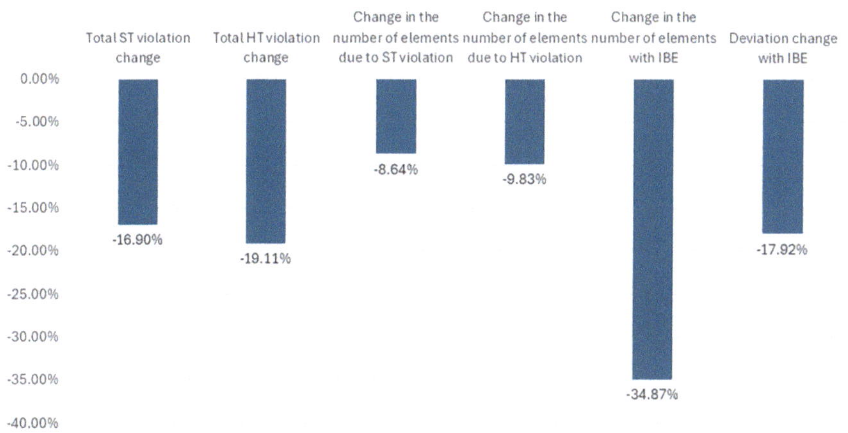

Fig. 6.59 Comparison of IBE change after method application

6.2.3 *Method to Assess and Change Data Wire in Integrated Circuits*

It was shown above that the development of ICs has led to the presence of now more than 100 billion transistors in the IC. Therefore, in the case of such a number of transistors in the circuit, more than 100 billion interconnects are obtained, in each of which there can be influence from other wires. Such metal pairs were considered of which at least one is a clock-carrying wire, and from (6.3) and (6.4) it follows that if the CT architecture assessment has already been considered, and its change predicted, if necessary, then it is necessary to consider the mutual arrangement of information-carrying metals.

Based on the obtained ML model it is proposed to perform additional transformations for the SW using improved combinations of existing IBE reduction methods (Fig. 6.60) [96] after making changes in individual sections of the CT.

Since metal routing in the circuit is performed after CT construction, it would not be preferable to perform routing constraints for other wires as well, as was done in the case of CT wires. Thus, for method implementation, repeater insertion and screening were chosen.

It is proposed to assess the distance between them for selected value ranges of the overlap area between elements (Fig. 6.61) and as a result of the assessment, perform corresponding changes to the SW.

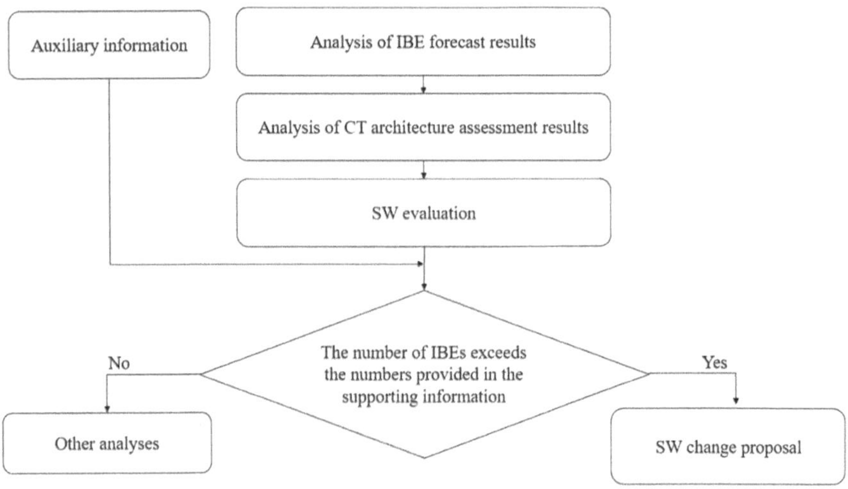

Fig. 6.60 Description of SW assessment and change

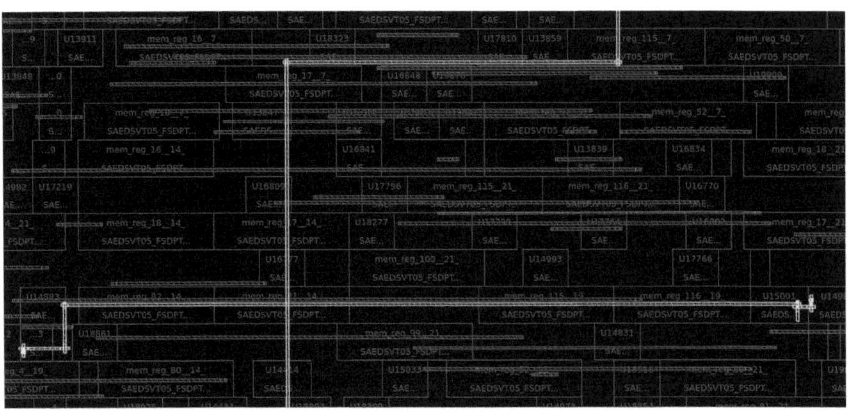

Fig. 6.61 Mutual arrangement of wires before method application

If the distance between the wires is smaller than three times the minimum acceptable distance between wires for the given metal layer [104], then it is necessary to install a repeater on the longer of these wires (Fig. 6.62).

Otherwise, when it is possible to place additional wires between the wires, wire screening is performed (Fig. 6.63).

Adding repeaters in the circuit implies adding additional transistors. In that case, after increasing the number of transistors, any arbitrary wire must be connected

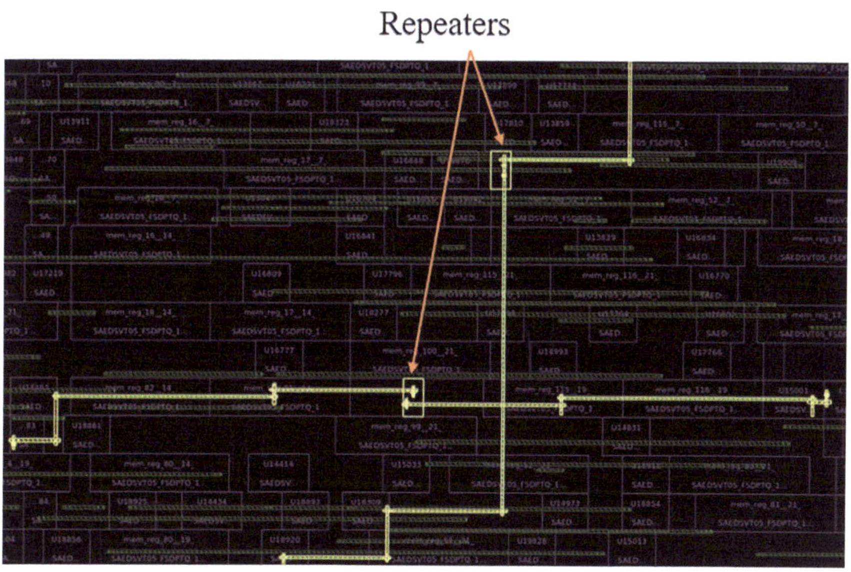

Fig. 6.62 Change in wire mutual arrangement after repeater installation

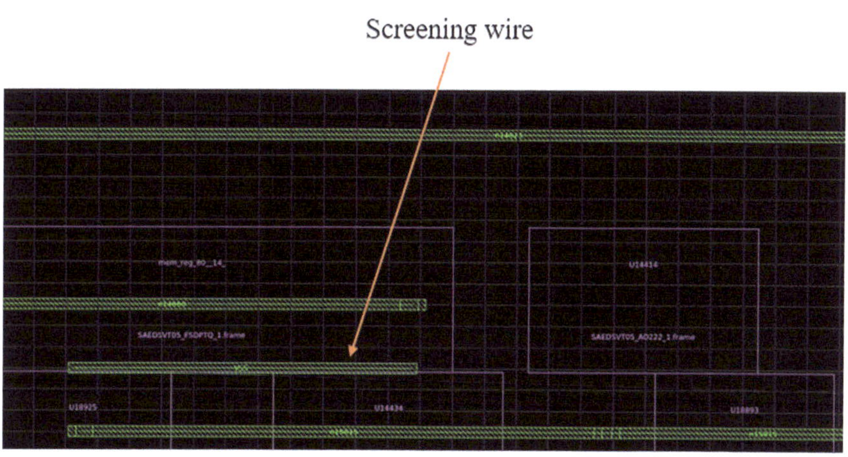

Fig. 6.63 Mutual arrangement of wires after screening

down to metal layer 1. Therefore, to improve the effectiveness of the proposed method's application, only repeater installation was performed in metal layer 1. Then, for the remaining levels, the above transformations were performed. After performing them, improvements in other factors affecting the circuit's reliability will be observed [16, 24, 28].

For the method's application, from the obtained IBE prediction results, all those metal wire pairs of which neither belongs to the CT are separated. Since the obtained database contains the indicator of the circuit wire being an SW, which was obtained through FC, therefore, the corresponding extraction operation is no longer performed (Fig. 6.64).

In this case as well, to avoid numerous changes to the IC as a result of the proposal's application, auxiliary information is used regarding the maximum acceptable value of the distance between elements and the minimum acceptable value of the IBE coefficient. The method is not applied to such a wire pair whose distance and interaction corresponding numbers fall outside the boundaries of the numbers given through auxiliary information. The extraction of data from the circuit, the change proposal, its application, and obtaining the new circuit have the form presented in Fig. 6.65.

Thus, as a result of the improved combination of the results obtained from the IBE prediction ML model and the above methods, the possibility is created to mitigate the consequences of IBE, taking into account the features of SW routing.

Signal Wire indicator

IsM1Clock	IsM2Clock	M1SwitActivity	M2SwitActivity	A_x0	A_y0	A_x1	A_y1	A_x2	A_y2	A_x3
0	0	0,56	0,34	0,72	4,72	0,72	4,8	0,74	4,8	0,74
0	0	0,56	0,71	0,72	4,72	0,72	4,8	0,74	4,8	0,74
0	0	0,56	0,28	0,72	4,72	0,72	4,8	0,74	4,8	0,74
0	0	0,56	0,65	0,72	4,72	0,72	4,8	0,74	4,8	0,74
0	0	0,56	0,92	0,72	4,72	0,72	4,8	0,74	4,8	0,74
0	0	0,56	0,62	0,72	4,72	0,72	4,8	0,74	4,8	0,74
0	0	0,56	0,7	0,72	4,72	0,72	4,8	0,74	4,8	0,74
0	0	0,56	0,34	0,72	4,72	0,72	4,8	0,74	4,8	0,74
0	0	0,56	0,42	0,72	4,72	0,72	4,8	0,74	4,8	0,74
0	0	0,56	0,44	0,72	4,72	0,72	4,8	0,74	4,8	0,74
0	0	0,56	0,35	0,72	4,72	0,72	4,8	0,74	4,8	0,74
0	0	0,56	0,69	0,72	4,72	0,72	4,8	0,74	4,8	0,74
0	0	0,56	0,66	0,72	4,72	0,72	4,8	0,74	4,8	0,74
0	0	0,56	0,49	0,72	4,72	0,72	4,8	0,74	4,8	0,74
0	0	0,56	0,74	0,72	4,72	0,72	4,8	0,74	4,8	0,74

Fig. 6.64 Extraction of SW

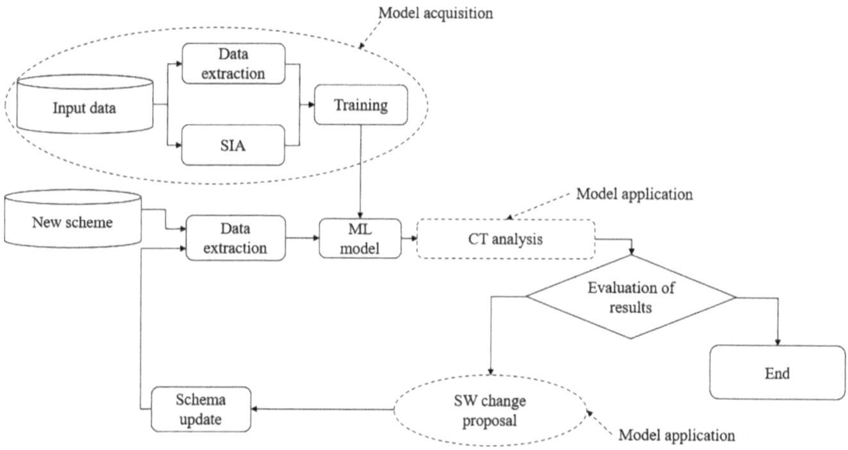

Fig. 6.65 Application of changes made for SW

Table 6.6 Comparison of timing parameters before and after implementation of the proposed method

	Parameters					
Design	Number of elements with ST deviation	Total ST deviation (ps)	Number of elements with HT deviation	Total HT deviation (ps)	Number of elements with IBE	IBE deviation (ps)
Initial IC	937	−1121,6	600	−411,6	476	−10,9075
With screening	914	−1036,3584	581	−373,3212	424	−10,024
With repeaters	870	−1022,8992	548	−367,1472	329	−98,386
Proposed form	885	−938,779	564	−339,16	362	−9053

Evaluation of the Effectiveness of the Proposed Method for Evaluating and Modifying SW in IC

To check the effectiveness of the mitigation proposal based on the information obtained as a result of IBE prediction in ICs, circuits composed with cells from the SAED14 educational library [116, 117] were used. First, the methods known from the literature were implemented separately for IBE mitigation, which were then combined with the data obtained as a result of the ML model. Screening was applied between the circuit's metals for IBE reduction in the selected circuit, after which the circuit's timing parameters (ST, HT, due to IBE) were extracted, as well as the number of those sequential elements that have deviations from them. Then repeater insertion was performed, after which the same parameters were extracted. Then an improved combination of the two methods was performed, corresponding to the form presented in the proposed method (Table 6.6).

A comparison of circuit parameters obtained as a result of the above changes was performed (Fig. 6.66).

After method implementation, a comparison of the change in the number of sequential elements with timing parameter deviations as a result of circuit change was performed (Fig. 6.67) as well as extraction of other circuit parameters (Table 6.7).

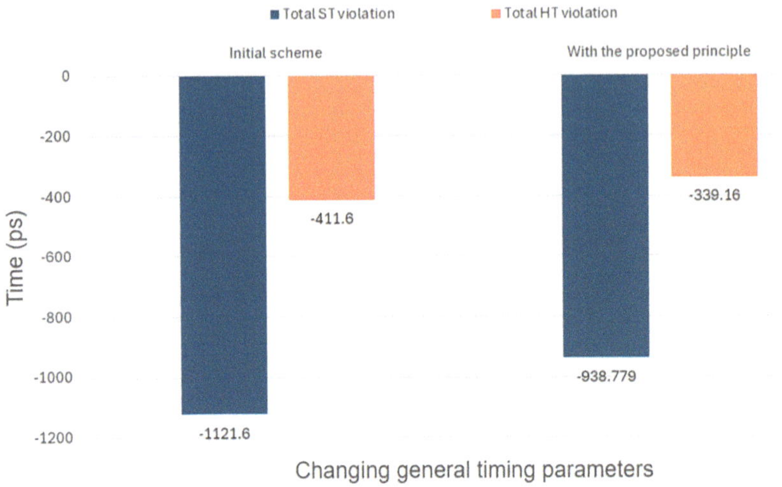

Fig. 6.66 Comparison of obtained timing parameters after method application

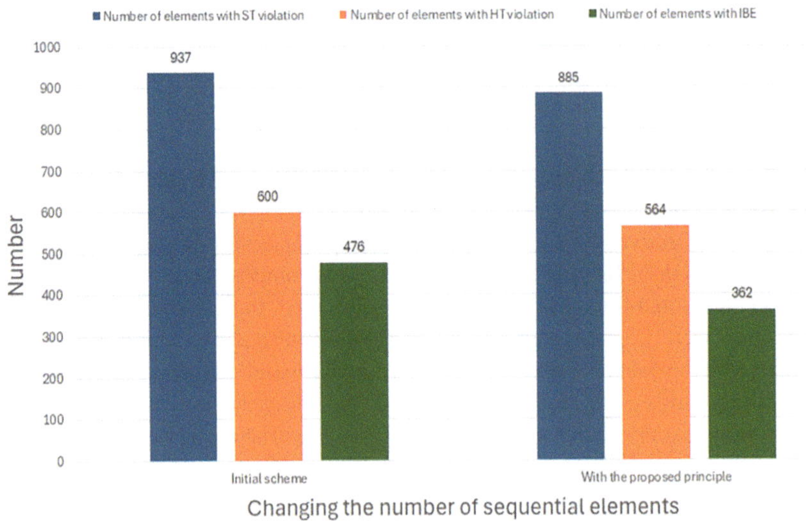

Fig. 6.67 Comparison of the number of sequential elements with timing parameter deviations obtained after method application

Table 6.7 Comparison of general parameters before and after implementation of the proposed method

	Parameters				
Design	Area (μm^2)	Cell density (%)	Total wire length (μm)	Total power consumption (mW)	Leakage power consumption (mW)
Initial IC	9140,6141	60,34	86458,7889	1740,56	92,2
With screening	9140,6141	60,34	91127,5635	1740,56	92,2
With blocking	9689,0509	62,1502	92856,7393	1843,41	93,3064
Proposed form	9407,52	61,5468	91646,3162	1791,03	93,122

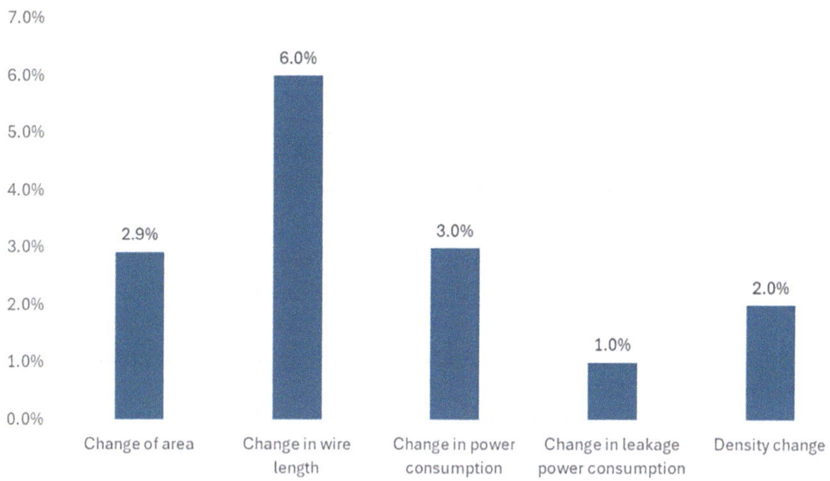

Fig. 6.68 Assessment of deterioration of circuit parameters after method application

From the results, it can be seen that after the implementation of the proposed method for CT wire change, the total area of the circuit increased by approximately 2.9%, the total power consumption by 3%. The total wire length in the circuit increased by 6%, and the cell density by 2% (Fig. 6.68).

The number of sequential elements in the circuit that had deviations from the above timing parameters also decreased, by 23.9%, 5.5%, and 6%, respectively (Fig. 6.69).

Thus, a method has been created for assessing the SW and changing the corresponding wires considering the features of the circuit design, which, through an improved combination of screening and repeaters by wire routing, ensures approximately 17% mitigation of IBE consequences at the cost of approximately 3% loss in power consumption, 2.9% in area, and 2% in cell density, as well as a 6% increase in total wire length. As a result, ST and HT also decrease by 16.3% and 17.6%

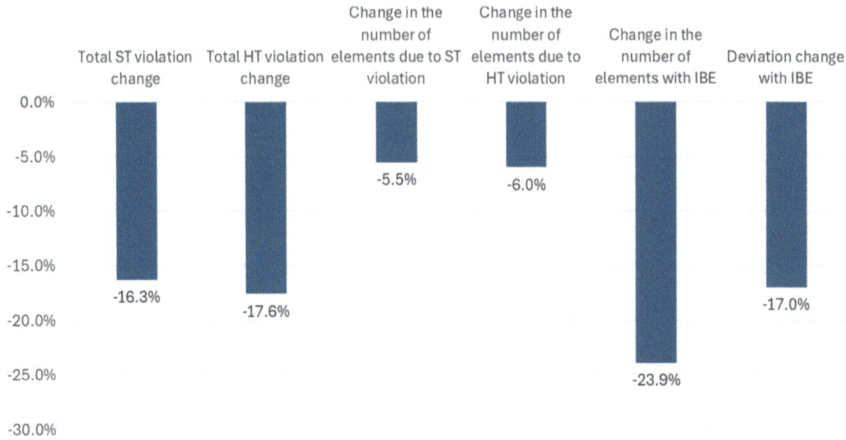

Fig. 6.69 Comparison of IBE change after method application

respectively, and the number of sequential elements in which there were deviations from the finally specified timing parameters, decrease by 5.5% and 6%.

Conclusions

1. A method based on machine learning algorithms has been developed for predicting interactions between elements in integrated circuits, which, through as little data collection and circuit analysis as possible, ensures approximately 36.6% savings in time spent on signal integrity analysis at the cost of approximately 10% loss of predicted data.
2. A method has been proposed for assessing the clock tree architecture and changing the corresponding wires considering the features of the circuit design, which, through an improved combination of screening and blockages by relocating the clock tree wires, ensures approximately 17.9% mitigation of the consequences of interactions, and a 16.9% and 19.1% mitigation of the total setup and hold times of the sequential element input, respectively, at the cost of approximately 5.66% loss in power consumption and 3.1% in area.
3. A method has been created for assessing the information-carrying wires and changing the corresponding wires considering the features of the circuit design, which, through an improved combination of screening and repeaters by wire routing, ensures approximately 17% mitigation of the consequences of interactions, and a 16.3% and 17.6% mitigation of the total setup and hold times of the sequential element input, respectively, at the cost of approximately 3% loss in power consumption and 2.9% in area.

References

1. A. Mocuta, P. Weckx, P. Demuynck, et al., Enabling CMOS scaling towards 3 nm and beyond, in *2018 IEEE Symposium on VLSI Technology. Honolulu, HI, USA*, (IEEE, 2018), pp. 147–148. https://doi.org/10.1109/VLSIT.2018.8510683
2. S.C. Song, Design/system technology co-optimization for 3 nm node and beyond, in *2021 International Symposium on VLSI Technology, Systems and Applications (VLSI-TSA). Hsinchu, Taiwan*, (2021), p. 1. https://doi.org/10.1109/VLSI-TSA51926.2021.9440089
3. E.P. DeBenedictis, It's time to redefine Moore's law again. Computer **50**(2), 72–75 (2017). https://doi.org/10.1109/MC.2017.34
4. D. Arbet, L. Nagy, V. Stopjakova, Ultra-low-voltage IC design methods, in *Integrated Circuits/Microchips*, (IntechOpen, 2020), 204p. https://doi.org/10.5772/intechopen.91958
5. G. Aishwarya, B. Patil, P.V. Joshi, et al., A survey on use of FPGA in automotive system, in *2022 International Conference on Distributed Computing, VLSI, Electrical Circuits and Robotics (DISCOVER). Shivamogga, India*, (2022), pp. 51–56. https://doi.org/10.1109/DISCOVER55800.2022.9974941
6. P. Salvatore, The integrated circuit industry at a crossroads: threats and opportunities. Chips **1**(3), 150–171 (2022). https://doi.org/10.3390/chips1030010
7. S. Nagavarapu, S.S. Rout, A review on VLSI implementation in biomedical application. Innov. Bio-Inspir. Comput. Appl. **1180**, 130–138 (2021). https://doi.org/10.1007/978-3-030-49339-4_14
8. P. Singh, L.K. Singh, Reliability and safety engineering for safety critical systems: an interview study with industry practitioners. IEEE Trans. Reliab. **70**(2), 643–653 (2021). https://doi.org/10.1109/TR.2021.3051635
9. G.A. Abgaryan, A.A. Galstyan, G.A. Harutyunyan, Reduction consequences of the wire interconnections in integrated circuits by optimizing of the clock tree. Proc. NAS RA NPUA Ser. Tech. Sci. **75**(4), 424–430 (2022). https://doi.org/10.53297/0002306X-2023.v76.1-93
10. T. Jerse, Introduction to crosstalk, in *2017 IEEE International Symposium on Electromagnetic Compatibility & Signal/Power Integrity (EMCSI). Washington, DC, USA*, (2017), pp. 1–45. https://doi.org/10.1109/ISEMC.2017.8078014
11. S. Hall, G. Hall, *High-Speed Digital System Design: A Handbook of Interconnect Theory and Design Practices* (Wiley, 2000), 327p
12. Electronics Tutorials, *AC Inductors, Inductive Reactance*, (2021), https://www.electronics-tutorials.ws/inductor/ac-inductors.html
13. Electronics Tutorials, *Capacitive Reactance*, (2021), https://www.electronics-tutorials.ws/filter/filter_1.html
14. S.S. Sapatnekar, Capturing the effect of crosstalk on delay, in *Wireless and digital imaging in the millennium: proceedings of 13th international conference on VLSI design. Calcutta, India*, (2000), pp. 364–369. https://doi.org/10.1109/ICVD.2000.812634
15. H. Li, Y. Zhang, X. Li, Delay test pattern generation considering crosstalk-induced effects, in *2003 Test Symposium. Xi'an, China*, (2003), pp. 178–183. https://doi.org/10.1109/ATS.2003.1250806
16. V. Melikyan, P. Petrosyan, N. Avagyan, G. Abgaryan, Self-heating analysis method of integrated circuits, in *International Conference on Microwave & THz Technologies, Wireless Communications and OptoElectronics (IRPhE 2022). Yerevan, Armenia*, vol. 41, (2022), pp. 63–65. https://doi.org/10.1049/icp.2022.2799
17. V.A. Chhabria, S.S. Sapatnekar, Impact of self-heating on performance and reliability in FinFET and GAAFET designs, in *20th International Symposium on Quality Electronic Design (ISQED). Santa Clara, CA, USA*, (2019), pp. 235–240. https://doi.org/10.1109/ISQED.2019.8697786
18. L. Yin, G. Du, X. Liu, Impact of ambient temperature on the self-heating effects in FinFETs. J. Semicond. **39**(9), 094011 (2018)

19. C.-W. Lin, T.-H. Hsu, X.-W. Shih, Y.-W. Chang, Buffered clock tree synthesis considering self-heating effects, in *2014 IEEE/ACM International Symposium on Low Power Electronics and Design (ISLPED). La Jolla, CA, USA*, (2014), pp. 111–116. https://doi.org/10.1145/2627369.2627640
20. D. Jang, E. Bury, R. Ritzenthaler, et al., Self-heating on bulk FinFET from 14nm down to 7nm node, in *2015 IEEE International Electron Devices Meeting (IEDM). Washington, DC, USA*, (2015), pp. 11.6.1–11.6.4. https://doi.org/10.1109/IEDM.2015.7409678
21. V. Melikyan, G. Abgaryan, P. Petrosyan, N. Avakyan, Low power analogue-digital converter model. J. Contemp. Phys. (Armen. Acad. Sci.) **56**(3), 275–279 (2021). https://doi.org/10.3103/s1068337221030178
22. J. Zhang, L. Lu, M. Lee, et al., Source of IC power consumption and low power design. Theor. Nat. Sci. **28**(1), 116–122 (2023). https://doi.org/10.54254/2753-8818/28/20230411
23. L.L. Ng, M. Brown, K. White, et al., *Power Consumption in CMOS Circuits* (IntechOpen EBooks, 2023), p. 190. https://doi.org/10.5772/intechopen.105717
24. V. Melikyan, P. Petrosyan, N. Avagyan, G. Abgaryan, Smart adjustment of transistor parameters to reduce temperature rise due to self-heating effect, in *2023 IEEE East-West Design & Test Symposium (EWDTS). Batumi, Georgia*, (2023), pp. 1–5. https://doi.org/10.1109/EWDTS59469.2023.10297080
25. J.H. Lau, T.G. Yue, Thermal management of 3D IC integration with TSV (through silicon via), in *2009 59th Electronic Components and Technology Conference. San Diego, CA, USA*, (2009), pp. 635–640. https://doi.org/10.1109/ECTC.2009.5074080
26. A. Gupta, F.J. Kurdahi, N.D. Dutt, et al., Thermal aware global routing of VLSI chips for enhanced reliability, in *9th International Symposium on Quality Electronic Design (ISQED 2008). San Jose, CA, USA*, (2008), pp. 470–475. https://doi.org/10.1109/ISQED.2008.4479779
27. J. Shanthi, D.G.N. Rani, S. Rajaram, Thermal aware floorplanner for multi-layer ICs with fixed-outline constraints, in *2021 2nd International Conference on Communication, Computing and Industry 4.0 (C2I4). Bangalore, India*, (2021), pp. 1–6. https://doi.org/10.1109/C2I454156.2021.9689353
28. Մխիթարյան Ա.Խ., Աբգարյան Գ.Ա., Պետրոսյան Պ.Ս., Ավագյան Ն.Ա.14 նմ տեխնոլոգիական գործընթացի համար ծերացման հատկությունների հետազոտումը արագագործ համեմատիչի օրինակով. Վանաձորի պետական համալսարանի գիտական տեղեկագիր. Երևան, 2021. էջ. 21–29
29. S. Zhan, C. Chen, Circuit reliability analysis with considerations of aging effect, in *2022 35th SBC/SBMicro/IEEE/ACM Symposium on Integrated Circuits and Systems Design (SBCCI). Porto Alegre, Brazil*, (2022), pp. 1–5
30. D. Kraak, J. Perez, Y.S. Obeng, R. Smith, et al., Device aging: a reliability and security concern, in *2018 IEEE 23rd European Test Symposium (ETS). Bremen, Germany*, (2018), pp. 1–10
31. A. Santana-Andreo, P. Martin-Lloret, E. Roca, et al., Characterizing aging degradation of integrated circuits with a versatile custom array of reliability test structures, in *2022 IEEE 34th International Conference on Microelectronic Test Structures (ICMTS). Cleveland, OH, USA*, (2022), pp. 1–6
32. Design & Reuse, *Optimization of Crosstalk Delta Delay on Clock Nets*, (2020), https://www.design-reuse.com/articles/48514/optimization-of-crosstalk-delta-delay-on-clock-nets.html
33. P. Patil, S. Pable, Crosstalk delay analysis in very deep sub micron VLSI circuits. Int. Res. J. Eng. Technol. **03**(5), 1695–1698 (2016)
34. Physical Design 4U, *Clock Tree Synthesis - Part 2. Crosstalk*, (2020), https://www.physicaldesign4u.com/2020/03/clock-tree-synthesis-part-2-crosstalk.html
35. VLSI Universe, *Setup and Hold: Basics of Timing Analysis*, (June 2013), https://vlsiuniverse.blogspot.com/2013/06/setup-and-hold-basics-of-timing-analysis.html
36. Cadence System Analysis Resources, *Crosstalk Analysis and Simulation in High-Speed PCB Design*, (2021), https://resources.system-analysis.cadence.com/blog/msa2021-crosstalk-analysis-and-simulation-in-high-speed-pcb-design

37. R. Mudavath, B.R. Naik, Estimation of far end crosstalk and near end crosstalk noise with mutually coupled RLC interconnect models, in *2018 International Conference on Communication and Signal Processing (ICCSP). Chennai, India*, (2018), pp. 182–185. https://doi.org/10.1109/ICCSP.2018.8524191
38. Polar Instruments. *AP8164: Signal Integrity Design Guidelines*, https://www.polarinstruments.com/support/si/AP8164.html
39. A.G. Bouazza, B. Bouazza, Crosstalk noise and signal propagation delay analysis in submicron CMOS integrated circuits, in *2012 6th International Conference on Sciences of Electronics, Technologies of Information and Telecommunications (SETIT). Sousse, Tunisia*, (2012), pp. 155–160. https://doi.org/10.1109/SETIT.2012.6481905
40. G. Shan, G. Li, Y. Wang, et al., Application and prospect of artificial intelligence methods in signal integrity prediction and optimization of microsystems. Micromachines (Basel) **14**(2), 344 (2023). https://doi.org/10.3390/mi14020344
41. Y. Huang, B. Benware, R. Klingenberg, et al., Scan chain diagnosis based on unsupervised machine learning, in *2017 IEEE 26th Asian Test Symposium (ATS). Taipei, Taiwan*, (2017), pp. 225–230. https://doi.org/10.1109/ATS.2017.50
42. Z. Zhao, Z. Cai, A review of intelligent design for test based on machine learning, in *2023 International Symposium of Electronics Design Automation (ISEDA). Nanjing, China*, (2023), pp. 116–120. https://doi.org/10.1109/ISEDA59274.2023.10218713
43. E.C. Barboza, N. Shukla, Y. Chen, J. Hu, Machine learning-based pre-routing timing prediction with reduced pessimism, in *2019 56th ACM/IEEE Design Automation Conference (DAC). Las Vegas, NV, USA. Article 106*, (2019), pp. 1–6. https://doi.org/10.1145/3316781.3317857
44. D. Yuan, Y. Zhang, X. Wang, An improved analytical model for crosstalk of SiC MOSFET in a bridge-arm configuration. Energies **14**(3), 1–30 (2021). https://doi.org/10.3390/en14030683
45. D. Amuru, H.V. Vudumula, P.K. Cherupally, et al., AI/ML algorithms and applications in VLSI design and technology. Integration **93** (2023). https://doi.org/10.1016/j.vlsi.2023.06.002
46. A. Sathish, L.-M. Madhavi, K.K. Lal, Data encoding technique for crosstalk delay reduction on fault tolerant data-bus in DSM technology. Procedia Eng. **38**, 2967–2972 (2012). https://doi.org/10.1016/j.proeng.2012.06.346
47. K.S. Sainarayanan, J.V.R. Ravindra, K.T. Nath, M.B. Srinivas, Coding for minimizing energy in VLSI interconnects, in *2006 International Conference on Microelectronics. Dhahran, Saudi Arabia*, (2006), pp. 166–169. https://doi.org/10.1109/ICM.2006.373293
48. S.K. Verma, K.K. Brajesh, Bus encoder design for crosstalk and power reduction in RLC modelled VLSI interconnects. J. Eng., Des. Technol. **13**(3), 486–498 (2015). https://doi.org/10.1108/JEDT-05-2013-0040
49. K.S. Sainarayanan, J.V.R. Ravindra, M.B. Srinivas, A novel, coupling driven, low power bus coding technique for minimizing capacitive crosstalk in VLSI interconnects, in *2006 IEEE International Symposium on Circuits and Systems (ISCAS). Kos, Greece*, (2006), pp. 4155–4158. https://doi.org/10.1109/ISCAS.2006.1693544
50. B. Victor, K. Keutzer, Bus encoding to prevent crosstalk delay, in *IEEE/ACM International Conference on Computer Aided Design (ICCAD 2001): IEEE/ACM Digest of Technical Papers (Cat. No.01CH37281). San Jose, CA, USA*, (2001), pp. 57–63. https://doi.org/10.1109/ICCAD.2001.968598
51. M. Madhu, Selective shielding technique to eliminate crosstalk transitions. ACM Trans. Des. Autom. Electron. Syst. **14**(3), 43 (2009). https://doi.org/10.1145/1529255.1529265
52. M. Madhu, Selective shielding: a crosstalk-free bus encoding technique, in *2007 IEEE/ACM International Conference on Computer-Aided Design. New York, NY, USA*, (2007), pp. 618–621. https://doi.org/10.1109/ICCAD.2007.4397333
53. S. Hong, U. Narayanan, K.-S. Chung, T. Kim, Bus-invert coding for low-power I/O - a decomposition approach, in *Proceedings of the 43rd IEEE Midwest Symposium on Circuits*

and Systems (Cat.No.CH37144). Lansing, MI, vol. 2, (2000), pp. 750–753. https://doi.org/10.1109/MWSCAS.2000.952865

54. Y. Zhang, J. Lach, K. Skadron, M.R. Stan, Odd/Even bus invert with two-phase transfer for buses with coupling, in *Proceedings of the International Symposium on Low Power Electronics and Design, Monterey, CA, USA*, (2002), pp. 80–83. https://doi.org/10.1109/LPE.2002.146715
55. U. Narayanan, K.-S. Chung, T. Kim, Enhanced bus invert encodings for low-power, in *2002 IEEE International Symposium on Circuits and Systems (ISCAS). Phoenix-Scottsdale, AZ, USA*, (2002), pp. 25–28. https://doi.org/10.1109/ISCAS.2002.1010631
56. Y. Shin, S.-I. Chae, K. Choi, Partial bus-invert coding for power optimization of application-specific systems. IEEE Trans. VLSI Syst. **9**(2), 377–383 (2001). https://doi.org/10.1109/92.924059
57. I.I. Basith, T.M. Supon, E. Abdel-Raheem, R. Rashidzadeh, Comparative study on bus-coding schemes and improvement on SINV coding, in *2016 28th International Conference on Microelectronics (ICM). Giza, Egypt*, (2016), pp. 17–20. https://doi.org/10.1109/ICM.2016.7847896
58. M. Alamgir, I.I. Basith, T. Supon, R. Rashidzadeh, Improved bus-shift coding for low-power I/O, in *2015 IEEE International Symposium on Circuits and Systems (ISCAS). Lisbon, Portugal*, (2015), pp. 2940–2943. https://doi.org/10.1109/ISCAS.2015.7169303
59. J. Natesan, D. Radhakrishnan, Shift invert coding (SINV) for low power VLSI, in *Euromicro Symposium on Digital System Design. Rennes, France*, (2004), pp. 190–194. https://doi.org/10.1109/DSD.2004.1333276
60. R.R. Rao, H.S. Deogun, D. Blaauw, D. Sylvester, Bus encoding for total power reduction using a leakage-aware buffer configuration. IEEE Trans. VLSI Syst. **13**(12), 1376–1383 (2005). https://doi.org/10.1109/TVLSI.2005.862718
61. Cadence System Analysis Resources, *The Advantages of Differential Signaling*, (2021), https://resources.system-analysis.cadence.com/blog/msa2021-the-advantages-of-differential-signaling
62. All About Circuits, *The Why and How of Differential Signaling*, (2016), https://www.allaboutcircuits.com/technical-articles/the-why-and-how-of-differential-signaling/
63. T.S. Balan, B. Devadas, G. Kumar Srivastava, Suppression of common and differential component of crosstalk coupling onto the differential line and field analysis of the crosstalk, in *2018 IEEE Symposium on Electromagnetic Compatibility, Signal Integrity and Power Integrity (EMC, SI & PI). Long Beach, CA, USA*, (2018), pp. 260–264. https://doi.org/10.1109/EMCSI.2018.8495377
64. M. Kim, Periodically corrugated reference planes for common-mode noise suppression in high-speed differential signals. IEEE Trans. Electromagn. Compat. **58**(2), 619–622 (2016). https://doi.org/10.1109/TEMC.2015.2513392
65. D. Seo, H. Lee, M. Park, W. Nah, Enhancement of differential signal integrity by employing a novel face via structure. IEEE Trans. Electromagn. Compat. **60**(1), 26–33 (2018). https://doi.org/10.1109/TEMC.2017.2725943
66. J. Wang, C. Xu, S. Zhong, et al., Differential via designs for crosstalk reduction in high-speed PCBs, in *2020 IEEE International Symposium on Electromagnetic Compatibility & Signal/Power Integrity (EMCSI). Reno, NV, USA*, (2020), pp. 145–149. https://doi.org/10.1109/EMCSI38923.2020.9191558
67. M.R. Rifhan Amira, M.M. Siti Marwangi, H.H. Azri, M. Zuhanis, Mitigation techniques for crosstalk in ICs. IOP Conf. Ser. Mater. Sci. Eng. **701**(1) (2020). https://doi.org/10.1088/1757-899X/701/1/012037
68. S. Majumdar, J. Sahota, Impact of buffer driver strength on crosstalk noise propagation in global interconnects. Int. J. Adv. Eng. Technol. **10**(4), 490–496 (2017)
69. S. Saini, A.M. Kumar, S. Veeramachaneni, M.B. Srinivas, An alternative approach to buffer insertion for delay and power reduction in VLSI interconnects, in *2010 23rd International*

Conference on VLSI Design. Bangalore, India, (2010), pp. 411–416. https://doi.org/10.1109/VLSI.Design.2010.53
70. Y.-H. Chen, P.-C. Huang, F.-W. Chen, et al., Crosstalk-aware TSV-buffer insertion in 3D IC, in *2019 32nd IEEE International System-on-Chip Conference (SOCC). Singapore*, (2019), pp. 400–405. https://doi.org/10.1109/SOCC46988.2019.1570539111
71. A. Khursheed, K. Khare, Optimized buffer insertion for efficient interconnects designs. Int. J. Numer. Model. Electron. Netw. Devices Fields **33**(4) (2020). https://doi.org/10.1002/jnm.2748
72. T. Pathade, U. Shah, Y. Agrawal, R. Parekh, Preeminent buffer insertion technique for long advanced on-chip graphene interconnects, in *2018 IEEE Electrical Design of Advanced Packaging and Systems Symposium (EDAPS). Chandigarh, India*, (2018), pp. 1–3. https://doi.org/10.1109/EDAPS.2018.8680882
73. Physical Design 4U, *CTS Part III: Clock Buffer and Minimum Period*, (2020), https://www.physicaldesign4u.com/2020/03/cts-part-iii-clock-buffer-and-minimum.html
74. W. Shi, Z. Li, A fast algorithm for optimal buffer insertion. IEEE Trans. Computer-Aided Des. Integr. Circuits Syst. **24**(6), 879–891 (2005). https://doi.org/10.1109/TCAD.2005.847942
75. K. Damanpreet, V. Sulochana, Crosstalk minimization in VLSI interconnects. Int. J. Reconfig. Embed. Syst. **2**(2), 89–98 (2013). https://doi.org/10.11591/ijres.v2.i2
76. R. Balakrishnan, S.A. Thomas, S. Sharan, Crosstalk and EMI reduction using enhanced guard trace technique, in *2018 IEEE Electrical Design of Advanced Packaging and Systems Symposium (EDAPS). Chandigarh, India*, (2018), pp. 1–3. https://doi.org/10.1109/EDAPS.2018.8680903
77. B. Semerdjiev, D. Velenis, Optimal crosstalk shielding insertion along on-chip interconnect trees, in *20th International Conference on VLSI Design Held Jointly with 6th International Conference on Embedded Systems (VLSID'07). Bangalore, India*, (2007), pp. 289–294. https://doi.org/10.1109/VLSID.2007.122
78. J. Zhang, E.G. Friedman, Crosstalk modeling for coupled RLC interconnects with application to shield insertion. IEEE Trans. VLSI Syst. **14**(6), 641–646 (2006). https://doi.org/10.1109/TVLSI.2006.878223
79. D. Mishra, S. Mishra, P. Agnihotry, B.K. Kaushik, Effect of distributed shield insertion on crosstalk in inductively coupled VLSI interconnects. J. Comput. Sci. Eng. **1**(1), 77–81 (2010). https://doi.org/10.48550/arXiv.1006.2820
80. M. Kachout, J. Bel Hadj Tahar, F. Choubani, Crosstalk reduction using nonuniform transmission lines, in *2015 23rd International Conference on Software, Telecommunications and Computer Networks (SoftCOM). Split, Croatia*, (2015), pp. 60–63. https://doi.org/10.1109/SOFTCOM.2015.7314081
81. M. Akhsham, A. Seyedolhosseini, Z. Navabi, Test adapted shielding by a multipurpose crosstalk avoidance scheme, in *2019 IEEE European Test Symposium (ETS). Baden-Baden, Germany*, (2019), pp. 1–2. https://doi.org/10.1109/ETS.2019.8791507
82. S. Kose, E. Salman, E.G. Friedman, Shielding methodologies in the presence of power/ground noise. IEEE Trans. VLSI Syst. **19**(8), 1458–1468 (2011). https://doi.org/10.1109/TVLSI.2010.2054119
83. H. Sankaran, S. Katkoori, Simultaneous scheduling, allocation, binding, re-ordering, and encoding for crosstalk pattern minimization during high-level synthesis. IEEE Trans. VLSI Syst. **19**(2), 217–226 (2011). https://doi.org/10.1109/TVLSI.2009.2031864
84. T.S. Chindhu, N. Shanmugasundaram, Clock gating techniques: an overview, in *2018 Conference on Emerging Devices and Smart Systems (ICEDSS). Tiruchengode, India*, (2018), pp. 217–221. https://doi.org/10.1109/ICEDSS.2018.8544281
85. AnySilicon, *Clock Tree Synthesis*, https://anysilicon.com/clock-tree-synthesis/
86. P. Huang, Y. Wang, Z. Zhao, D. Yue, CAUTS: clock tree optimization via skewed cells with complementary asymmetrical uniform transistor sizing. IEEE Trans. VLSI Syst. **32**(1), 137–149 (2024). https://doi.org/10.1109/TVLSI.2023.3328592

87. R. Ewetz, C.-K. Koh, Fast clock scheduling and an application to clock tree synthesis. Integr. VLSI J. **56**(3), 115–127 (2017). https://doi.org/10.1016/j.vlsi.2016.10.012
88. L. Yi-Chen, C. Wei-Ting, G. Deyuan, et al., RL-CCD: concurrent clock and data optimization using attention-based self-supervised reinforcement learning, in *2023 60th ACM/IEEE Design Automation Conference (DAC)*, (2023), pp. 1–6. https://doi.org/10.1109/DAC56929.2023.10248008
89. R.F.W. Ewetz, Synthesis of clock trees with useful skew based on sparse-graph algorithms. Doctoral dissertation, Purdue University, 2016, https://docs.lib.purdue.edu/cgi/viewcontent.cgi?article=2600&context=open_access_dissertations
90. R. Ewetz, S. Janarthanan, C.-K. Koh, Fast clock skew scheduling based on sparse-graph algorithms, in *The 20th Asia and South Pacific Design Automation Conference. Chiba, Japan*, (2015), pp. 472–477. https://doi.org/10.1109/ASPDAC.2015.7059051
91. C.-C. Hsu, Y.-T. Chang, L. Mark, Crosstalk-aware power optimization with multi-bit flip-flops, in *17th Asia and South Pacific Design Automation Conference*, (2012), pp. 431–436. https://doi.org/10.1109/ASPDAC.2012.6164987
92. J.A. Charles, Z. Li, G.-J. Nam, United States Patent US8930873B1. Creating regional routing blockages in integrated circuit design, 2015, https://patents.google.com/patent/US8930873B1/en
93. J.A. Charles, Z. Li, N.S. Chin, United States Patent US20130086544A1. Consideration of local routing and pin access during VLSI global routing, 2013, https://patents.google.com/patent/US20130086544A1/en
94. Design & Reuse, *Tackling Congestion and Shorts with Practical Approaches and PnR Tool ICC2*, https://www.design-reuse.com/articles/54881/tackling-congestion-and-shorts-with-practical-approaches-and-pnr-tool-icc2.html
95. L. Rongjian, X. Zhiyao, J. Jinwook, et al., Routing-free crosstalk prediction, in *Proceedings of the 39th International Conference on Computer-Aided Design (ICCAD '20). Article 163*, (Association for Computing Machinery, 2020). https://doi.org/10.1145/3400302.3415712
96. G. Abgaryan, Crosstalk prediction in integrated circuits based on machine learning techniques, in *2023 IEEE East-West Design & Test Symposium (EWDTS). Batumi, Georgia*, (2023), pp. 1–5. https://doi.org/10.1109/EWDTS59469.2023.10297079
97. Artwork, *Examples of GDSII Files with Different Extents*, (2011), https://www.artwork.com/gdsii/gds_extents/examples.htm
98. SourceForge, *GDS2*, (2015). http://sourceforge.net/projects/gds2/
99. VLSI Expert, *Metal Wire Orientation*, (2017), https://www.vlsi-expert.com/2017/11/metal-wire-orientation.html
100. Z. Yi, H. Cong, Y. Takeshi, TSV assignment of thermal and wirelength optimization for 3D-IC routing, in *2018 28th International Symposium on Power and Timing Modeling, Optimization and Simulation (PATMOS)*, (2018), pp. 155–162. https://doi.org/10.1109/PATMOS.2018.8464161
101. C.-H. Hsu, H.-Y. Chen, Y.-T. Hou, United States Patent US8875067B2. Reusable cut mask for multiple layers, 2014, https://patents.google.com/patent/US8875067B2/en
102. Y.-S. Wang, M.-Y. Lin, C.-H. Lu, United States Patent US20150286765A1. Cut mask design layers to provide compact cell height, 2015, https://patents.google.com/patent/US20150286765A1/en
103. J.R. Baker, *CMOS Circuit Design, Layout, and Simulation*, 3rd edn. (Wiley, 2010), 1173p
104. Signoff Semiconductors, *Standard Cell Library*, (2017), https://signoffsemiconductors.com/standard-cell-library/
105. Synopsys, *Fusion Compiler™ User Guide*, (2022), https://www.synopsys.com
106. GeeksforGeeks, *Label Encoding of Datasets in Python*, (April 2023), https://www.geeksforgeeks.org/ml-label-encoding-of-datasets-in-python/?ref=lbp
107. Analytics Vidhya, *Feature Scaling in Machine Learning: Normalization vs. Standardization*, (April 2020), https://www.analyticsvidhya.com/blog/2020/04/feature-scaling-machine-learning-normalization-standardization/

108. Z. Yu, Y. Qiang, An overview of multi-task learning. Natl. Sci. Rev. **5**(1), 30–43 (2017). https://doi.org/10.1093/nsr/nwx105
109. R. Cipolla, Y. Gal, A. Kendall, Multi-task learning using uncertainty to weigh losses for scene geometry and semantics, in *2018 IEEE/CVF Conference on Computer Vision and Pattern Recognition. Salt Lake City, UT, USA*, (2018), pp. 7482–7491. https://doi.org/10.1109/CVPR.2018.00781
110. L. Fontenot, J.R. Puri, A. Sgambellone, Hierarchical neural networks (HNN): using TensorFlow to build HNN. SMU Data Sci. Rev. **6**(2), Article 4 (2022)
111. J. Si, S.L. Harris, E. Yfantis, A dynamic ReLU on neural network, in *2018 IEEE 13th Dallas Circuits and Systems Conference (DCAS). Dallas, TX, USA*, (2018), pp. 1–6. https://doi.org/10.1109/DCAS.2018.8620116
112. DigitalOcean Community Tutorials, *Sigmoid Activation Function in Python*, (2022), https://www.digitalocean.com/community/tutorials/sigmoid-activation-function-python#relu-activation-function
113. Google, *TensorFlow Keras: build, train, and deploy machine learning models with ease*, (2022), https://www.tensorflow.org/guide/keras
114. R. Das, TensorFlow: revolutionizing large-scale machine learning. Complex Semicond. Des. Int. J. Comput. Eng. **5**(3), 1–9 (2024). https://doi.org/10.47941/ijce.1812
115. B. Pang, E. Nijkamp, Y. Wu, Deep learning with TensorFlow: a review. J. Educ. Behav. Stat. **45**(2) (2019). https://doi.org/10.3102/1076998619872761
116. V. Melikyan, M. Martirosyan, A. Melikyan, G. Piliposyan, 14nm Educational design kit: capabilities deployment and future. Small Syst. Simul. Symp. **9**, 1–5 (2018)
117. V. Melikyan, G. Markosyan, R. Goldman, et al., Synopsys' open educational design kit: capabilities, deployment and future, in *2009 IEEE International Conference on Microelectronic Systems Education*, (2009), pp. 20–24. https://doi.org/10.1109/MSE.2009.5270840
118. G.A. Abgaryan, A built-in self-test system for external DRAM. Math. Probl. Comput. Sci. **54**, 80–87 (2020). https://doi.org/10.51408/1963-0061
119. T. Tewary, S. Dey, S. Roy, Realization of built-in self test (BIST) enabled memory (RAM) using VHDL and implementation in Spartan6 FPGA board, in *2020 IEEE VLSI device circuit and system (VLSI DCS). Kolkata, India*, (2020), pp. 322–326. https://doi.org/10.1109/VLSIDCS47293.2020.9179879
120. Գ.Ա. Աբգարյան, Ա.Վ. Բաբայան, Generic bist architecture for memories. ՀՀ ԳԱԱ և ՀԱՊՀ-ի Տեղեկագիր. Տեխն. Գիտ. Սերիա. **74**(2), 204–209 (2021)
121. J. Olamendy, *Mini-Batch Gradient Descent: Optimizing Machine Learning* (Medium, 2023) https://medium.com/@juanc.olamendy/mini-batch-gradient-descent-optimizing-machine-learning-98ef238c5225
122. DataCamp, *Loss Function in Machine Learning*, (2023), https://www.datacamp.com/tutorial/loss-function-in-machine-learning
123. Built In, *Adam Optimization*, (2023), https://builtin.com/machine-learning/adam-optimization

Index

V. Melikyan, K. Li, *Parametric Optimization Techniques of Integrated Circuits*, https://doi.org/10.1007/978-3-032-20400-4

MIX
Papier aus verantwortungsvollen Quellen
Paper from responsible sources
FSC® C105338

FSC
www.fsc.org

If you have any concerns about our products,
you can contact us on
ProductSafety@springernature.com

In case Publisher is established outside the EU,
the EU authorized representative is:
Springer Nature Customer Service Center GmbH
Europaplatz 3, 69115 Heidelberg, Germany

Printed by Libri Plureos GmbH
in Hamburg, Germany